The Sea and the Second World War

The Sea and the Second World War

Maritime Aspects of a Global Conflict

Edited by Marcus Faulkner and Alessio Patalano

andarta books
an imprint of the University Press of Kentucky

Brécourt Academic

Published by Andarta Books
An imprint of The University Press of Kentucky

Scholarly publisher for the Commonwealth,
serving Bellarmine University, Berea College, Centre
College of Kentucky, Eastern Kentucky University,
The Filson Historical Society, Georgetown College,
Kentucky Historical Society, Kentucky State University,
Morehead State University, Murray State University,
Northern Kentucky University, Transylvania University,
University of Kentucky, University of Louisville,
and Western Kentucky University.

Editorial and Sales Offices: The University Press of Kentucky
663 South Limestone Street, Lexington, Kentucky 40508-4008
www.kentuckypress.com

Cataloging-in-Publication data available from the Library of Congress

ISBN 978-1-94-966804-9 (hardcover : alk. paper)
ISBN 978-1-94-966806-3 (pdf)
ISBN 978-1-94-966807-0 (epub)

This book is printed on acid-free paper meeting
the requirements of the American National Standard
for Permanence in Paper for Printed Library Materials.

Manufactured in the United States of America.

ASSOCIATION of UNIVERSITY PRESSES
Member of the Association
of University Presses

Contents

Introduction

Marcus Faulkner and Alessio Patalano

The Second World War marked the apex of industrial war and was nothing short of the most costly and destructive conflict ever experienced. It was total in its conduct and global in its scale—a true World War. The scale of the conflict may be explained by virtue of the fact that it was the product of numerous regional conflicts and theaters of operation that increasingly became woven into a contiguous war. In Western Europe the conflict began as a rerun of the Great War. In Eastern Europe it evolved into an ideological war of extermination between the polar opposites of fascism and communism. Parts of Africa and the Middle East became battlegrounds where European colonial ambitions clashed while other parts provided men and material. Maintaining access to resources more generally was indispensable for all belligerents in order to sustain their war efforts, thus attempting to stem the flow of their opponents' resources was a central facet of most wartime strategies. Farther east, Japanese imperial ambitions clashed with the dynamics of a civil war in China in the attempt to create a new "Asian" international system free of American and European encroachments. In this respect, the war in the Asia-Pacific region that broke out in December 1941 should be separated from the one conducted in China up until that point. Japanese operations in China and on the Mongolian border before 1941 certainly had an impact on, but were different from, the vast new front that opened up in the Pacific following the Japanese attacks on Anglo-American positions from Pearl Harbor to Singapore. It was the events of December 1941 that brought these disparate strands formally together and linked them to events in Europe, Africa, and elsewhere. The result was a single war with very few areas formally out of bounds to armed conflict.

From the first moments of the German invasion of Poland on 1 September 1939 through to the Japanese surrender on 2 September 1945, the sea shaped the course and conduct of the war. The scope and scale of naval operations took on ever increasing dimensions and were conducted in all of the world's oceans and seas. The environments in which navies and merchant fleets operated varied from the frozen, largely dark conditions of the Arctic in the winter months through to the tropics of the southwest Pacific. The naval war was never a forgotten front with engagements, warships, and technology woven into the fabric of the war's narrative. The Battle of Midway or the *Bismarck* hunt have always been seen as key moments in the war and resulted in myriad accounts through to this day. Major operations or protracted campaigns like Leyte Gulf or the "Battle of the Atlantic," too, have been cornerstones of many broader histories.

Yet, the role of the sea as a whole has been increasingly marginalized or compartmentalized in the wartime historiography. While in recent years our understanding of the Second World War as a global conflict has been greatly enhanced by comparative thematic treatments or international histories of the wartime experience, little attention has been devoted to the sea as a perspective of analysis. This is at odds with the core notion that winning a global war was dependent on the ability of nations to operate globally, which in turn was dependent on attaining command of the sea, harnessing its power, and conversely denying its use to opponents. Ultimately, war is won on land, and only those who could destroy enemy armies and occupy their capitals could be assured victory in a total war. Nonetheless, the maritime domain provided the necessary foundation for this victory to be achieved.

The role of the sea was not limited to naval or military matters nor merely to the duration of the war itself. Throughout history the role of the sea is often seen as both a bridge and a barrier to human political, economic, and military activity. It is a shield that can protect nations, empires, and entire continents, while at the same time it is a sword with which to project power and sustain campaigns ashore. This duality lies at the heart of its role during the Second World War, and the sea was the glue that bound together the disparate conflicts. The role and influence of the sea may, though, be detected long before the war formally broke out in Europe. The flashpoints that occurred along the path to war in September 1939, such as the Abyssinia Crisis or Spanish Civil War, had distinct maritime elements. The prelude and opening phases of the Second Sino-Japanese War involved naval forces. More generally, maritime and naval concerns were a significant feature of the international system during the interwar years. Attempts to limit and control arms development in the aftermath of the Great War experience only proved successful in relation to naval armaments. The Washington Treaty system, further developed by means of an international conference process, remained a unique occurrence. As a result, it placed naval affairs and competition at the heart of international relations, and the breakdown of the system in the mid-1930s was a contributory factor to international tensions and ultimately war.

The role of the sea during the war was rather different when examined from Allied or Axis perspectives. Navies and merchant fleets provided the link between theaters of war and enabled Allied military manpower and resources to be transported to the frontline. The redistribution between theaters, often in relatively short timeframes, was equally important. Beyond the passage of military force, the need to move raw materials, fuel, and food through the maritime arteries made control of the sea imperative to Allied wartime strategy. In particular, for Britain, naval power and maritime concerns were non-negotiable cornerstones of the imperial construct. For the United States, too, the sea was significant, if not the key to its geostrategic position and political choices. The protection afforded by two oceans and a strong navy allowed the adoption of an isolationist position and a retreat from the international stage. The sea provided a formidable, yet by no means impassable, barrier to any power seeking to attack the United States. This was a crucial factor in shielding it from the initial impact of the war and then provided it with the defense it needed while it created the arsenal of democracy. It was the vast expansion of the United States Navy that established the United States as a first-rate power and one which could project its strength across the globe on an unprecedented degree after 1945.

For the Axis powers, the situation was more complex. Although Germany had rivaled Britain in its naval and merchant fleets prior to the Great War, much of its maritime power disappeared in the interwar years. Yet the experience of the sea's impact between 1914–18 was recognized by politicians and the military long before the establishment of the Third Reich. As one of the great industrial powers, and one that lacked domestic sources of certain key raw materials, Germany depended on the sea even though continental European concerns always dominated. The drive towards an autarchic economic system, greatly accelerated during the late 1930s, was a response to its unfavorable geographic and naval position. In the event, Germany amassed an unprecedented degree of control over the European continent between 1939 and 1941, making its war against the western allies a contest of land versus sea power. The desired war in the east against the Soviet Union prevented the Germans from ever fully committing to a maritime front and allowed the Allies to lay the foundation for the liberation of Europe.

Japan and, to a lesser extent, Italy were more attuned to the maritime domain as a core platform to access raw materials to support resource-poor countries and their limited industrial capacity. For Imperial Japan, the sea was the route to its expanding position on the Asian mainland, but it also represented the avenue by which American, and to a lesser extent, British power could cure Japanese ambition. For Italy, the situation was similar, albeit geographically on a more modest scale and against a different set of threats. The fascist vision of Italian power saw it dominating the Mediterranean Sea. However, both Japan and Italy focused their attention more on seeking to deal with opposing fleets rather than actively protecting their sea-lanes and maritime communications in general. The former did so by courting a decisive engagement to eliminate enemy fleets with little further thought devoted to the wider strategic concerns. The latter focused on avoiding battle as a means of preserving its fleet.

More crucially for the wartime experience, as long as the Axis powers could deny Allied maritime power access to a theater, they could secure their territorial acquisitions. Despite numerical superiority, Italy struggled to attain control of the central Mediterranean against the British and by the time it eventually did, to a degree—with the infusion of German airpower and U-boats, coupled with British overstretch—it was no longer in a position to capitalize from briefly pushing British naval influence from the central Mediterranean. Following Pearl Harbor and the thrust into south-east Asia, Japan was able to amass a vast new empire and held onto its acquisitions marginally longer. Denial is easier than control and no more so than in an era of rapid technological progress. The foundation of Allied victory depended on the ability to develop and execute the three principal naval wartime roles of sea control, the protection of maritime communications, and the projection of power ashore to subdue Axis gains.

The war at sea, and from the sea, contributed to the global and total nature of the conflict, but the sea's influence was felt in many other areas well beyond the naval sphere. The unprecedented scale of the conflict resulted in significant population movements and dislocation of global trading patterns which had implications well beyond actual theaters of war. Ports and their communities experienced considerable change as a result of military action and reconfigurations of the industrial base and economies more generally. War is a driver of innovation and the era was no different in this regard, with naval and civilian shipping construction practices and research and development undergoing considerable transformation. Logistics is often singled out as the cardinal, yet generally overlooked, area of military history. Considering the global scale, a multitude of theaters of operation, and vast amounts of material shipped over the world's oceans, this remains one of the least examined and understood elements of the conflict. All these areas, and many more, offer potential for further research in order to fully understand the role the sea played.

The purpose of this volume is to examine some of these aspects. It makes no claim to provide a comprehensive account of the role the sea and maritime activity played in the run-up and during the conflict. Rather, it seeks to highlight topics that have either received little interest or to reexamine well-trodden ground by means of a new perspective. Although the chapters cover a considerable range of topics, these are linked by certain common themes. A number address the role and perception of the sea in contemporary military and political thinking. The changing nature of warfare, and in particular its vastly increased scale and scope, posed new challenges in organizing and administering the use of the sea and maritime assets. For the Allies, and Britain in particular, the vast number of roles for its maritime assets coupled with the complexity of new capabilities and operational requirements led to a considerable expansion of the administration of its maritime power. Finally, the use of the sea as a means to project power is another theme addressed to varying degrees throughout most chapters. The course of the war in many theaters necessitated the projection of force from the sea on an historically unparalleled level. Such undertakings are inherently risky to successfully conduct and both administratively and logistically challenging to sustain. For the Allies, the development of a large-scale expeditionary capability was fundamental to winning the war and one of its greatest achievements.

The first three chapters deal with the strategic considerations surrounding the use of the sea in pre-war thinking with an emphasis on the Far East. James Goldrick examines the evolution of British strategy in the region by means of charting the role the submarine played in overall warplanning. As a new platform that had reached maturity during the Great War, the submarine offered a potential solution to Britain's strategic dilemma of having too few forces to cover the considerable naval commitments inherent to a global empire. A force deployed in the Far East possessed the ability to deter or delay Japanese advances while the main battle force was deployed from European waters. During the interwar years, the British capability was expanded into a potent force, albeit one that was then redeployed owing to the outbreak of the European war. Evan Mawdsley compares how the American, British, and Japanese navies conceived naval strategy and prepared for conflict during the interwar era. Although much has been written on the topic, this chapter seeks to examine the issue from an international perspective, contrasting the different approaches and examining how naval strategy fitted into the wider national strategies. Further, to what extent did geostrategic realities on the eve of the Pacific War reflect pre-war considerations and planning? Alan Zimm's chapter examines the well-known Pearl Harbor raid through a reassessment of the operational and technological factors that underlined Japanese choices and planning. Although the Japanese strategic reasoning as to the value of the raid is often questioned, the operational execution is often seen as a seminal moment in the relationship between naval and air power. This chapter provides a more critical analysis of the Japanese operation that helps to underline that while the war was marked by a considerable degree of innovation, this was far from a linear process.

Although recent British historiography has emphasized the imperial character of British power, the focus has largely been on land and air forces. Over the past decades, Canadian and Australian scholarship has contributed significantly to the wider understanding of the evolution of naval power and the war at sea. The focus of this work, principally dealing with the development of the respective navies, has tended to be on the shaping of national and naval identities rather than the wider imperial dimension of British maritime power. Iain Johnston's chapter seeks to outline and examine this latter aspect by means of a comparative breakdown of British imperial maritime power. Harry Bennett examines how the Germans intended to defend occupied Europe from Allied assault after a maritime strategy to defeat Britain in 1941 was abandoned in favor of a purely U-boat-based campaign in the Atlantic. Britain and Germany were only separated by the relatively narrow English Channel and lower North Sea and for most of the war fought an intense campaign to protect their coastal shipping and interdict each other. As the war progressed, the *Kriegsmarine* was transformed materially and doctrinally from an ocean-going navy focused on offensive action against enemy maritime communications, to one configured to fight defensively in the littoral on the edges of the Reich. The chapter outlines how, despite certain technical innovations and tactical successes, when put to the test in 1944 the system was not up to the task of defending Fortress Europe.

The development of organizations, techniques, and material to project power from the sea was a key component of the Allied war effort. By the end of the war, Britain and the United States had attained the ability to successfully conduct large-scale contested assaults on enemy shore lines and deploy and sustain modern mechanized forces over transoceanic distances. Combined or amphibious operations rank among the most complex and risky military endeavors, and no more so than in the age of industrial warfare. Developing this capability became a necessity as a result of vast territorial gains the Axis powers attained in the opening phases of the conflict. Although Britain possessed a long, albeit variable, history of conducting expeditionary operations, the need to land armies on a far larger scale in the wake of the defeat of France necessitated a fundamental overhaul of the existing structures. In view of the scale and complexity, such operations needed to be undertakings between land, air, and naval forces, but in an era where joint operations were rare, the ability to coordinate the activities of the services needed to be learned. C. I. Hamilton's chapter examines the development of the British Combined Operations Headquarters that was established in 1940 to develop and coordinate amphibious activities. Being outside the existing Admiralty structure, the intra- and inter-service politics surrounding the existence of such a joint body were considerable and it took some time before an effective organization emerged.

The Pacific Theater has, from the outset, readily been linked to the conduct of amphibious operations. The nature of the theater combined with the rapid Japanese expansion in 1941–42 meant that islands throughout the vast area needed to be captured on a substantial scale. The narratives of the operations on the two main American axes of advance in the central and southwestern Pacific are well known, yet aspects of the development of amphibious techniques remain understudied to this day. The Australian contribution to the Southwest Pacific theater is often sidelined even in accounts of the campaign in the region. The development and experience of the Australian amphibious campaign have been even further removed from the historiography. Peter Dean's chapter explores this very aspect and focuses on the development of an Australian approach to combined and littoral warfare. Keith Mitchener explores one of the technical aspects related to amphibious operations, the provision of naval gunfire support, which has to date received very little attention. Forces landed during amphibious operations are vulnerable to enemy action and thus the role of the fleet in subduing enemy defenses is crucial. Yet technical and operational factors meant that the provision of naval gunfire support was generally limited at the outset of the conflict. Using the seminal Iwo Jima campaign, this chapter examines how far the U.S. Navy had refined its procedures.

The impact of the sea could be felt far beyond the shoreline as the arms and armies carried across the oceans were ultimately destined to wage war ashore. Where one side lacked the transport capacity to move forces, or the naval forces to escort them, the ability to operate in a given theater was often curtailed. The two largest land theaters of the war—the Soviet Union and China—were also indirectly influenced by events at sea. Francis Grice examines a hitherto largely neglected topic: the impact of naval power on the war in China and how Mao Zedong and the Communist Party saw its effects. The war itself was a mixture of the Civil War waged between the nationalist and communist forces coupled to a direct Sino-Japanese war between 1937 and 1945. As the Japanese conducted operations from the sea and along China's large network of inland waterways while Chinese forces depended on supplies of heavy equipment from abroad, the sea played its part in shaping the conflict. How the Communists saw the role of naval power was dependent on the character of the war they fought and had consequences beyond the end of the Second World War and the Chinese Civil War.

The wartime experience influenced post-war thinking and use of the sea as a strategic space in numerous ways. One of the principal areas which had seen significant innovation and change was aviation. Between 1939 and 1945, the capabilities of both land- and sea-based airpower had expanded and radically transformed the ability to attack enemy targets of all types. Throughout the conflict, existing pre-war interservice rivalry surrounding the control of air assets continued in Britain, Germany, and the United States. Particularly where interests overlapped, such friction often had detrimental effects on the development of material and conduct of operations. In the final chapter, George Monahan examines the impact such a debate had between the U.S. Army and Navy as a result of experiences gained during the Atlantic campaign.

1

Buying Time

British Submarine Capability in the Far East, 1919–1940

James Goldrick

British strategic policy in the Far East and its naval dimensions between 1919 and 1939 have been the subject of increasingly sophisticated analysis in recent years. A much better understanding has been achieved of the challenges that the British faced in managing the emerging threat within the region, and their attempts to provide against it, that has gone well beyond the subject of the Singapore base itself. The analysis has included detailed assessment of the plans for the dispatch of the "Main Fleet" and its employment, as well as its intended tactics.[1] Much less attention, however, has been given to the composition of the forces kept in the Far East or the plans for their use.

In relation to surface units, this is understandable, since the cruiser squadron, single aircraft carrier, and flotilla of destroyers assigned to the China Station for the majority of the inter-war period did not possess the numbers or weight of fire-power to pose a serious threat to the Japanese battle fleet. However, from within a year of the Armistice until well after the start of the Second World War, the largest single group of operational British submarines was to be found in the Far East. Rarely falling below twelve boats, regarded as the Navy's premier submarine unit,[2]

[1]Amongst relevant works, see Christopher M. Bell, *The Royal Navy, Seapower and Strategy Between the Wars* (London: Macmillan, 2000); Joseph Moretz, *The Royal Navy and the Capital Ship in the Interwar Period* (London: Frank Cass, 2002); and Andrew Field, *Royal Navy Strategy in the Far East, 1919–1939: Planning for War against Japan* (London: Frank Cass, 2004). There are also important essays in Keith Neilson and Greg Kennedy, eds., *Far Flung Lines: Studies in Imperial Defence in Honour of Donald Mackenzie Schurman* (London: Frank Cass, 1997). Regarding tactics, see particularly Jon Tetsuro Sumida, "The Best Laid Plans: The Development of British Battle Fleet Tactics, 1919–1942," *The International History Review* 14:4 (November 1992), pp. 681–700. See also James P. Levy, "Royal Navy Fleet Tactics on the Eve of the Second World War," *War in History* 19:3 (July 2012), pp. 379–95. I am grateful to Dr. Michael W. D. White for his comments on the draft of this chapter, as well as to Dr. Christopher M. Bell with his guidance on archival sources for the period and his generous provision of some of his own notes. This chapter was originally published in *Global War Studies* 11:3 (2014).

[2]Rear Admiral G. W. G. Simpson, *Periscope View: A Professional Autobiography* (London: Macmillan, 1972), p. 37.

and possessing for much of the period by far the most capable submarines available to the Royal Navy (RN), the force was of a size sufficient to provide some real defensive capability against a would-be invader during what was termed the "Period Before Relief," as well as being an important addition to the offensive operations that would become possible after the arrival of the Main Fleet.

The British were also never alone in their use of submarines in the region. A parallel strategy for submarine warfare was adopted by the Netherlands for the Dutch East Indies, whose first boats had arrived in the Far East in 1916. Their local fleet in 1939 included twelve of their most modern submarines,[3] while the Americans in the Philippines also maintained a substantial flotilla. Until the early 1930s, most new RN submarine designs were intended for the Far East. It was only later in the decade and the return of a European threat that the priority moved away from the larger (and faster) boats which were required in the Asia-Pacific.[4] That threat, however, did not bring about a reduction in the submarine commitment to the Far East until the very end. In July 1939, in addition to the modern depot ship *Medway*, there were no less than fifteen large submarines on the China Station. By comparison, there were only seven operational submarines in the Mediterranean, three of which were clearly reinforcement units for the Far East, and five in the Home Fleet.[5] Although much of the surface force was withdrawn from the Far East at the outset of war with Germany, fear of Japan taking advantage of the situation was a key reason that submarines remained in theater until well into 1940.[6]

---

[3]Marc A. van Alphen and Anita M. C. van Dissel, eds., *Chronicle of the Royal Netherlands Navy: Five Hundred Years of Dutch Maritime History* (Amsterdam: De Bataafsche Leeuw, 2006), p. 392 and p. 451 particularly, although this book only lists events and has no analysis. An early exposition of the Dutch concept is available in English: Lieutenant First Class P. Post Uiterweer, "The Submarine and Aeroplane Defence of the East-Indies," *The Naval Review* 9:4 (November 1921), pp. 542–51. There is some guidance regarding additional Dutch-language material at the website www.dutchsubmarines.com.

[4]Between 1928 and 1938 a total of thirty-one boats (including the short-lived "X-1" cruiser submarine) were built for the RN and RAN for the Far East. This represented a substantial commitment of resources in a period of retrenchment and disarmament. See H. T. Lenton, *British Submarines* (London: MacDonald, 1972), p. 4 and pp. 39–53. See also David Henry, "British Submarine Policy, 1918–1939," in Bryan Ranft, ed., *Technical Change and British Naval Policy, 1860–1939* (London: Hodder & Stoughton, 1977), p. 92.

[5]See *The Navy List* issued between 1920 and 1939 (London: HMSO, 1920–1939). In 1939, there were twenty more submarines operating out of Portsmouth and Portland on training duties, plus eleven more boats in reserve. The division of submarine numbers between the three fleets (China, Mediterranean, and Home) and the training force had been roughly the same since the early 1920s.

[6]Admiralty Historical Section, *Naval Staff History of the Second World War: Submarines*, Vol. III, *Operations in Far Eastern Waters* (CB 3006(3)) (London: Admiralty, 1955), p. 2. An MLRS Books reprint from 2011 is available.

The Royal Navy has been criticized for its apparent focus on the blind alleys of "cruiser" submarines and "fleet" submarines.[7] Such criticism is to some extent justifiable, but the inter-war building program overall devoted much more resources to patrol units and minelayers than it did to the other types, suggesting where the long-term priorities really lay. Furthermore, the problems of submarine employment during the "Period Before Relief" did not involve close cooperation with the surface fleet. Many of the issues arising with submarines were either largely local in nature—and thus a matter for the China Station staff—or ones which could generally be managed between that staff and the leadership of the Fourth Submarine Flotilla and the largely independent organization of the Rear Admiral (Submarines). They thus did not create the complex tactical issues that appear in retrospect to have occupied Royal Navy planners with regard to such "fleet" operations, and did not require discussion in the Tactical School or test by the principal Mediterranean or Home Fleets. The archival record is thus somewhat thinner, not helped by the loss of China Station records arising from the fall of Singapore.[8] In this respect, it is particularly significant that many senior officers with submarine experience served in the Far East between 1919 and 1941. This included three Commanders-in-Chief China (Waistell 1928–31, Little 1936–38, and Layton 1940–41) and one on the East Indies Station (Dunbar-Nasmith 1932–34).[9] The Commander of the Australian Fleet in 1922–24, Addison, was also a submariner. In addition, several submariners served as Chiefs of Staff in both China and the East Indies (Laurence in the East Indies 1923–25. In China: Layton 1931–33 and G. P. Thomson 1933–34. Thomson's successor, Attwood 1934–36, was a torpedo specialist who had experience on the staff of Rear Admiral (Submarines)).[10]

---

[7]Henry, "British Submarine Policy, 1918–1939," pp. 85–86. For his more detailed analysis, see David Henry, "British Submarine Development and Policy, 1918–1939," Ph.D. thesis, King's College London, 1976, especially chapters 2–6.

[8]Or by the lack of consistency in the registry treatment of 4th Submarine Flotilla reports. While the majority of these are accessible in The National Archives (UK) in different series, the archives of the Submarine Museum at Gosport appear thin on the subject of the Far Eastern deployments. Perhaps the most useful sources outside The National Archives are the various personal recollections of the submariners themselves.

[9]Dunbar-Nasmith and Little also served as Rear Admiral (Submarines) as did Laurence after his promotion to flag rank.

[10]In addition to *The Navy List*, the website www.unithistories.com is particularly useful for its summaries of various officers' careers in the period. In fact, the only one of the first submariners to achieve senior flag rank who did not serve in the Far East seems to have been Max Horton. There were also a number of former submariners who commanded cruisers in the Far East between the wars and Cecil Talbot commanded the aircraft carrier *Hermes* during her first commission in China.

The Far Eastern Requirement

The requirements of the Far East were the subject of concern well before the end of the First World War.[11] The 1919 Naval Mission of Admiral Jellicoe, commissioned immediately after the Armistice, was the result of the earlier request by the Dominion Prime Ministers for advice on naval requirements in the Pacific. Even at this stage, submarines were part of the equation in developing a policy for the defense of the region. There were certain immediate practical attractions to their deployment. Although very expensive to build "ton for ton" by comparison with surface warships[12] and requiring depot ship support, they had relatively small crews. Furthermore, they used little fuel. The 1921 statement of fuel requirements highlighted this—a force of thirty submarines would have an annual consumption of only 36,000 tons of oil, compared with 960,000 for no more than eighty destroyers—some ten times as much per hull.[13] Given their offensive and defensive capabilities, these characteristics made them an economical option for such a distant station as China. The first Admiralty plans after hostilities provided for a six-boat submarine flotilla,[14] but this was soon increased to twelve L-class boats, supported by two submarine tenders, *Titania* and *Ambrose,* and a target ship and "maid of all work," the minesweeper *Marazion.*[15] The three old C-class boats which had been based at Hong Kong during the war were sold for scrap.[16]

Although it was to be rapidly disillusioned as to their material condition and capabilities, Australia had already accepted the gift of the six large and fast J-class submarines, which had left the UK in April 1919. While Jellicoe's much more ambitious proposals would be quietly ignored by the Admiralty, the latter's first formal post-war statement on "Imperial Naval Defence," submitted to the Committee of Imperial Defence at the end of October 1919, declared that "A strong submarine

---

[11]J. Kenneth McDonald, "British Naval Policy and the Pacific and Far East: From Paris to Washington, 1919–1922," DPhil thesis, University of Oxford, 1975, pp. 59–67.

[12]The best direct cost comparison available is that between the first two post-war boats ordered for the Royal Australian Navy and the first of the 10,000-ton cruisers. "All up," including ammunition and stores, *Oxley* and *Otway* were contracted to cost 439,942 Australian pounds each, the cruiser *Australia* 2,324,174 Australian pounds. This creates a ratio of approximately 1:5 for unit heavy cruiser/submarine costs—but does not include depot ships or other support elements. See Michael W. D. White, *Australian Submarines: A History* (Canberra: AGPS, 1992), p. 129.

[13]Nicholas Tracy, *The Collective Naval Defence of the Empire, 1900–1940* (Aldershot: Navy Records Society and Ashgate, 1997), p. 294. See also Field, *Royal Navy Strategy in the Far East, 1919–1939,* p. 55.

[14]Stephen Roskill, *Naval Policy Between the Wars,* Vol. 1, *The Period of Anglo-American Antagonism, 1919–1929* (London: Collins, 1968), p. 106.

[15]*The Navy List for January 1922* (London: HMSO, 1922), p. 714.

[16]F. J. Dittmar and J. J. Colledge, *British Warships, 1914–1919* (London: Ian Allan, 1972), p. 83.

flotilla based on the Island [Hong Kong] would, however, not only strengthen its power of resistance very considerably, but would act as a threat on the flank of any Japanese expeditions directed further South." The Pacific Dominions were also considered to be best protected by local submarine flotillas guarding against surface raids on their ports. In consequence, as part of any Dominion naval effort, "A start should be made in all cases with submarines."[17] The Dominion element of this design was to bear little fruit, but the concept of a submarine capability as the tripwire during what became known as the "Period Before Relief" and the arrival of the Main Fleet had already been formulated. It was to remain key to British naval policy in the Far East for more than twenty years.

Fleshing out the Submarine Strategy

The Admiralty expanded on its ideas for the employment of naval forces against Japan in its "War Memorandum" of January 1920. During the "Period Before Relief," the priority of the units in theater had to be upon defense of the key British bases.[18] While policy as to whether such essential—and practically defensible—bases should include Hong Kong would change at intervals in the years ahead,[19] the fundamental nature of the submarines' role "Before Relief" would not. The concept for their employment was to form a cordon around the defended base, at a distance and at intervals which would force an invading force to run the gauntlet. The issue would always remain, however, whether other missions could be taken on by the submarines in theater without excessively weakening the forces required to execute this task effectively.

This War Memorandum was examined in detail in 1921 by the first conference of the local Commanders-in-Chief, at which the China and East Indies Stations and Australia were represented. In their examination of the requirements to buy time for reinforcements to arrive, the admirals acknowledged the importance of the submarine capability, recommending that two mine-laying boats be added to the twelve L-class already on station, with eighty-three mines for ready use. The submarines would be divided between Singapore and Hong Kong, with the latter being allocated six boats for offensive patrols around Formosa and Japan. Two of the units allotted to Singapore would initially be dispatched to Labuan in Borneo. Their need for a

---

[17]The National Archives, London (hereinafter TNA), Admiralty Records (hereinafter ADM) 116/3104, Admiralty Memorandum "Imperial Naval Defence" of 31 October 1919. Accessed National Library of Australia (hereinafter NLA), Australian Joint Copying Project (hereinafter AJCP) microfilm.

[18]Bell, *The Royal Navy, Seapower and Strategy Between the Wars*, p. 69.

[19]Ibid., p. 79. See also McDonald, "British Naval Policy and the Pacific and Far East," pp. 203–04 for the use of Hong Kong as a submarine base and the initial plans of 3 August 1919 in the Memorandum "Imperial Naval Defence," ADM 167/59.

secure oil supply was identified as a "difficult and anxious" problem for the Japanese and the Borneo oil fields would be a primary target for capture. Other submarines would patrol the Sunda Strait to prevent Japanese surface forces outflanking the British and entering the Indian Ocean.[20] Plans for the Australian units were less ambitious at this stage, reflecting some realism as to the condition of the J-class boats, and they were allocated to local defense of the major Australian ports. In the event of the conflict involving a land war between Japan and China, the War Memorandum proposed that submarines be employed to interdict troop movements between the Japanese islands and the mainland of Asia, but the Conference delegates had a clearer idea of the difficulties of distance involved, as well as the limited number of boats available.[21] The Admiralty accepted the report but, now under increasing financial pressure, explained that it did not "propose to increase materially the naval forces in Eastern waters."[22] The twelve L-class would remain the local force in being.

The Concept Developed

The Washington Treaty negotiations created something of a hiatus in policy development, particularly as the United Kingdom made an attempt to have submarines banned outright, but the signing of the Treaty meant that planning could be resumed with a clear understanding of the strategic context. With an Imperial Conference proposed for 1923, the Admiralty produced a statement on naval cooperation within the British Empire which emphasized that "the Naval Staff are convinced . . . of the great capabilities of submarines for legitimate naval operations, and this opinion, based on war experience, has been confirmed in fleet exercises which have since been carried out." Submarines and fast light cruisers were identified as being the most valuable contribution that the Dominions could make for the Pacific, the former because "only by prolonged operations could our adversary hope to deal effectively with these vessels."[23]

Canada would not provide any material support, but Australia and New Zealand endorsed the concept. The latter had a submarine force only as a very long-term goal,[24] but, despite the failure of its second venture into the capability,

[20]ADM 116/3100, "Report of Conference held at Penang March 1921 between the Commanders-in-Chief of the China, East Indies and Australian Stations," ACJP microfilm. See also Field, *Royal Naval Strategy in the Far East, 1919–1939,* p. 69.

[21]ADM 116/3100, "Report of Conference held at Penang March 1921 between the Commanders-in-Chief of the China, East Indies and Australian Stations," ACJP microfilm.

[22]ADM 116/3100, Admiralty Letter to C-in-C China dated 17 August 1921, AJCP microfilm.

[23]ADM 116/3165, Admiralty Memorandum "The Washington Conference and its Effect upon Empire Naval Policy and Co-operation" of 28 July 1922, ACJP microfilm.

[24]See Tracy, *The Collective Naval Defence of the Empire,* pp. 322–23, and *New Zealand Parliamentary Debates,* 18 August 1922, Vol. 196, p. 482.

the Australian Government was prepared to renew its commitment and agreed to acquire up to six submarines. The difficulty was that the Australian flotilla would have to be reconstituted from scratch. The J-class had proved an expensive disappointment and were already being removed from service. There had been a number of valiant rearguard proposals to retain one or two of the boats for training, but it was eventually accepted that the J-class "were too obsolete and worn out ever to be brought forward for further service."[25]

In the wake of the Imperial Conference, the Admiralty continued to refine its policy. The British Government's drive for economy soon reduced the original plan for a force of eighty submarines, with new orders for eight replacement boats a year to six boats a year with a twelve year life and thus a total force of seventy-two submarines, of which sixty would be the overseas type and the remainder Fleet or minelaying units.[26] Design work for the first of the new overseas boats was already in hand, with the prototype planned for the 1923 Estimates.

The Evolution of the War Plans

The operational element was further considered by a second Far East naval commanders' conference in 1925. This confirmed the importance of the submarines in any plan for the "Period Before Relief" and not only recommended twenty boats in full peacetime commission, with eight ready to go to Hong Kong and twelve to Singapore, but suggested that a force of thirty was desirable. These numbers were in recognition not only of the potential of Hong Kong as a forward operating base—and the requirement to contribute to its defense—but also the need to close the archipelagic passages such as the Sunda Strait to Japanese surface raiders which might interfere in the Indian Ocean with the military reinforcement of Singapore.[27] It was in the latter role that an Australian flotilla of six boats based on Darwin could make a substantial contribution and this became the planned war employment for the new Royal Australian Navy (RAN) units.

The Admiralty's initial response to the conference report was to confirm that the intended force for China was some twenty-one RN submarines in full commission (two more were intended in reserve). However, this reinforcement depended upon the addition of a third depot ship and the date on which a new unit would be available was moving steadily into the future—first to 1927 and then 1928. In the event, construction of *Medway* was not authorized until the 1926 Estimates and

[25]First Naval Member Minute to Minister 16 November 1922. Cited White, *Australian Submarines: A History*, p. 117.

[26]Ship Department, Ministry of Defence BR 3043, *The Development of HM Submarines: From Holland N01 (1901) to Porpoise (1930)* (London: MOD, January 1979), p. 12.9.

[27]ADM 116/3121, "Interim War Plan Proceedings of the Singapore Conference 1925," AJCP microfilm.

she did not go out to China until 1930. The 1925 Admiralty plans proposed the increases to begin in 1929, with the addition of the first six of the new patrol submarines, followed by progressive replacement of the L-class, bringing the numbers up to twenty-one operational boats by 1931.[28]

Operational Concepts and Design Issues

There was a clear distinction in the expected roles of submarines during the "Period Before Relief" and the later stages of the naval campaign which would follow the arrival of the Main Fleet from Singapore. The characteristics of the boats themselves, particularly in terms of endurance and communications capabilities, had to meet the demands of the extended campaign and the long-range operations which would be involved in reconnaissance and in interdicting Japanese surface forces and merchant shipping. For the last role, the Admiralty was already experimenting with the concept of an anti-shipping "cruiser submarine" and the huge *X1* was laid down in 1921. Completed in 1924, but not fully operational until nearly two years later, *X1* was plagued by unreliable diesels and never fulfilled her potential.[29] When the Admiralty made its first experiment with a long-range single boat "cruiser submarine" deployment to the Far East, it was with the last, much-modified, steam-powered K-class boat. *K26* sailed for the Mediterranean in January 1924 and went on to Singapore, returning to home waters seven months later.[30]

X1 was to be the only one of her type and the backbone of the British effort would be the "Overseas Patrol Type" submarine design. Key elements were a very high surface endurance (10,000 nautical miles), habitability for the tropics, and a desirable wireless range of 1,000nm.[31] In later boats, the latter goal was increased to 1,500nm,[32] a more realistic estimate of the likely distances to key patrol areas in the region[33] and a target that was generally achieved in practice (and more, with

[28]ADM 1/8715/815, Director of Plans Memorandum "Future Disposition of S/M Flotillas" dated 8 December 1925.

[29]David K. Brown, *The Grand Fleet: Warship Design and Development, 1906–1922* (Barnsley: Seaforth, 2010), p. 168.

[30]C. N. Robinson, "Naval Forces of the British Empire," in Alexander Richardson and Archibald Hurd, eds., *The Naval and Shipping Annual 1925* (London: Clowes, 1925), p. 13. The captain was Commander G. P. Thomson, later Chief of Staff on the China Station. A new study of *X1* has recently appeared: Roger Branfill-Cooke, *X.1: The Royal Navy's Mystery Submarine* (Annapolis: Naval Institute Press, 2013), but this is largely a technical history of the ship and does not provide much analysis of the wider operational and strategic requirements of the Far East.

[31]BR 3043, *The Development of HM Submarines*, p. 14.1.

[32]Ibid., p. 29.2.

[33]The quality of the radio equipment was explicitly commented on by the Australian naval staff in 1929. See White, *Australian Submarines: A History*, p. 160.

the radio station at Rugby in the UK being in regular contact). Although they embodied all the experience of 1914–18, the prototype boats, *Oberon, Oxley,* and *Otway,* were not a complete success. The lead boat, *Oberon,* "soon acquired a reputation for mechanical unreliability."[34] The two Australian units were no better and suffered a lengthy and embarrassing delay on their voyage out to Australia in 1928. Both boats were forced to remain at Malta for nine months while their engine columns were rebuilt. (Similar work was required for *Oberon's* somewhat less powerful machinery.)[35] For the Australians, it was an ominous start for a small submarine force in a navy under considerable financial pressure.

The design characteristics of the Overseas type of submarine were refined after *Oberon* and the requirement for extended endurance confirmed, despite some disquiet. The first classes of Overseas submarines were to share a defect that proved very difficult to remedy. The external riveted oil tanks tended to leak, creating a trail on the sea surface.[36] There were repeated efforts to fix the problem, but it was never fully cured and may have contributed to the loss of several units in the Mediterranean in 1940–41. The Admiralty's view was that greater endurance provided greater operational flexibility, an important issue when transit distances, even from Hong Kong, were so great. The existing L-class were capable of getting from there to Japan's Inland Sea, but only for a seven-day patrol and with 15% fuel available as a margin. This was clearly insufficient and the 10,000 mile surface endurance requirement stood.[37]

The travails of the first trio meant a cautious approach to the dispatch of any new units to China and the changeover was programmed to align with the completion of *Medway,* which was well equipped to deal with any mechanical issues and could provide a much higher level of support than the elderly *Titania.* Many of the prototype problems were ironed out in the interim, although the O-class were to remain a concern.[38] Three out of four of the submarines experienced defects with their main motors during their passage out in 1930. Unlike the diesel breakdowns of the RAN O-class, however, they were rectified within a week at Malta.[39]

---

[34] J. J. Tall and Paul Kemp, *HM Submarines in Camera, 1901–1996* (Stroud: Sutton, 1996), p. 74. Alastair Mars went further when he labeled *Otway* an "electro-mechanical monstrosity." *British Submarines at War, 1939–1945* (London: William Kimber, 1971), p. 119.

[35] White, *Australian Submarines: A History,* pp. 145–50.

[36] David K. Brown, *Nelson to Vanguard: Warship Design and Development, 1923–45* (Annapolis: Naval Institute Press, 2000), p. 109.

[37] BR 3043, *The Development of HM Submarines,* pp. 12.6–12.7.

[38] Rear Admiral Ben Bryant, *Submarine Command* (London: William Kimber, 1958), p. 30.

[39] Lieutenant (later Rear Admiral) G. B. H. Fawkes, RN (writing as "G.B.H.F."), "Narrative in Log Form of Passage of Fourth Submarine Flotilla to Hong Kong, 1930," *The Naval Review* XIX:1 (February 1931), pp. 69–89.

The End of the Australian Contribution

The fate of the RAN effort was not so happy. The Australian naval staff confirmed the plan for a six-boat force with a new depot ship to the Australian Council of Defence as late as the end of 1928,[40] but there had already been internal efforts to change the focus to destroyers.[41] The Admiralty hastened to affirm its continuing support for the RAN capability,[42] but the whole program was well behind schedule, since a third boat had been intended for order in 1926 and three more in 1928. Under increasing budgetary pressure, the Australian Government was in no position to approve any new construction and the Australian squadron was cut to the bone. *Oxley* and *Otway* were placed in immediate reserve in May 1930. By the end of the year, Australia had offered the boats to the Admiralty, an offer that was quickly accepted. By late 1931, both were in the Mediterranean and the RAN submarine program was at an end.[43] What funds Australia had would be focused on cruisers.

Holding the Line

There was another aspect to the British acceptance of the situation. Under the terms of the 1930 London Naval Disarmament Treaty (as it had been for the earlier Washington Treaty), British Empire naval strength was treated as one. With submarine tonnage now limited to 52,700 in total, the RN could not afford to waste it on less-than-effective units. If the RAN could not operate the two O-class submarines properly, they were better off in British hands.[44]

Although the flotilla in China was to be maintained at twelve boats, the RN was also under financial pressure. In 1927 came a decision to reduce the British submarine force, slow the building rate, and delay indefinitely the reinforcement of the China Station.[45] The building rate dropped by half and, until the 1936

---

[40] B. N. Primrose, "Australian Naval Policy, 1919–1942: A Case Study in Empire Relations," Ph.D. thesis, The Australian National University, 1974, p. 123.

[41] Australian Archives (hereinafter AA), MP 1049/5, "RAN Submarines—General," Captain G. F. Hyde Minute "Submarine Building Programme—Possible Substitution of Destroyers" dated 1 April 1926. This was written shortly before Hyde was appointed Commodore First Class commanding the Australian Squadron. Hyde continued to campaign against submarines over the next two years.

[42] AA, MP 1049/5, Director of Plans (Captain W. A. Egerton) Letter to First Naval Member Australian Commonwealth Naval Board (Rear Admiral Napier) of 30 May 1927.

[43] White, *Australian Submarines: A History*, pp. 163–64. See also ADM 1/8743/109.

[44] The British had been loath to accept the boats when the possibility of their transfer had been raised earlier.

[45] ADM 1/8715/185, Secret Office Memorandum M0933/27 of 22 August 1927.

Estimates, only three new boats would be ordered each year.[46] The limitation imposed by the London Treaty three years later was more a recognition of the situation than a new constraint[47] because the building rate was insufficient even to maintain the treaty limit of tonnage in modern boats.[48]

Gathering War Clouds

The RN was to remain bound by economy for the next few years, but Japanese aggression in Manchuria in 1931–32 confirmed the increasing fragility of the strategic situation in the Far East. This was well understood by the new Commander-in-Chief China, Frederic Dreyer, who arrived in March 1933. He was soon bombarding the Admiralty with requests for reinforcements, including a minelaying submarine.[49] Largely at Dreyer's behest, the Far East naval commanders' conference was resumed in January 1934. In addition to Australian and New Zealand representatives, this gathering had significant submariner expertise in the C-in-C East Indies, Dunbar-Nasmith, and Dreyer's chief of staff, Thomson.

Not surprisingly, the conference recommended restoration of the goal of twenty-one submarines on station. Given that Singapore and protection of the Indian Ocean sea lanes had to be the priority, all that could be spared for Hong Kong without greater numbers were the two boats normally in refit there. The remaining ten would be divided between a patrol line more than 500 miles from Singapore, positions in the key straits of the archipelago, and four as far away as Japan. In assessing the report of the conference, the Admiralty staff were concerned that some of the dispositions were over-ambitious, since the submarines were so dispersed that, without external guidance, they could not be sure of intercepting hostile forces. This was a fair point and one that had already been identified by the conference, which was very critical of Royal Air Force (RAF) plans for air patrols that were too close to Singapore to be sure of locating an enemy carrier force before the latter was in a position to launch a strike.[50]

---

[46] Lenton, *British Submarines,* p. 14.

[47] The annual program for submarines was unchanged by the Treaty. See Stephen Roskill, *Naval Policy between the Wars,* Vol. II, *The Period of Reluctant Rearmament, 1930–39* (London: Collins, 1976), p. 73.

[48] Lenton, *British Submarines,* p. 18.

[49] ADM 116/3122, C-in-C C Signal 1945 27 July 1934, This repeats an earlier request, AJCP microfilm.

[50] ADM 116/3121, C-in-C China N053/01526 "Report of the Singapore Conference January 1934" and covering Naval Staff comments, AJCP microfilm. See also ADM 116/3862, "Naval Arrangements in the Event of War in the Far East," Director of Plans Minute dated 24 July 1936. The exchange between the conference delegates and the local RAF commander is remarkable for the accuracy of the 1934 naval assessment of the Japanese carrier capability—

The Admiralty found Dreyer's constant bombardment of signals and memoranda somewhat tedious,[51] but took the point and three new R-class boats arrived on station in July 1935. The addition was relatively short-lived, since four submarines were dispatched to the Red Sea and thence to the Mediterranean during the Abyssinian crisis. Nevertheless, they were amongst the first units to return to their parent station after the prospect of war diminished. Once the Admiralty had a chance to draw breath, the requirement for twenty-one boats was assessed in 1936 as "by no means an excessive demand."[52] However, although two minelaying boats were sent out in early 1937, they only replaced two of the oldest O-class and the overall number on station remained at fifteen.

Another word on submarines and the "Period Before Relief" came from the submariner C-in-C, Sir Charles Little, in July 1937. Little acknowledged that the priority for the boats on station had to be anti-invasion patrols in the South China Sea, in combination with RAF maritime assets, but pointed out that even a single submarine in Japanese waters would create the need for countermeasures and drain the enemy's resources. He also noted that it was ironic that submarines designed for long-range operations should be required to operate "in restricted waters."[53] Clearly, Little had the smaller S-class boats just entering service in mind as being more suitable for the defensive role in the southern South China Sea that the Admiralty demanded, with the older Overseas boats made available for the longer-range offensive patrols. For its part, the Admiralty could only plan on the dispatch of ten additional boats in the event of hostilities, so limited were the numbers available as the situation in Europe deteriorated.[54]

Little's influence was apparent in the version of "Emergency" orders issued by his successor, Noble, in April 1939. The situation envisaged was one in which Britain had become involved in a European war and in which the Japanese position was "uncertain" and the submarines represented a key element in the defensive plan. Four boats, together with any available of the three units expected to be in refit, were to operate out of Hong Kong—but with the *proviso* that their patrols had to be conducted with sufficient reserve of fuel to get back to Singapore. Although

---

suggesting that a mass strike of a hundred aircraft could be launched up to 200 miles from Singapore and that the Japanese would do so at dawn after a night approach. Hence the need for the very distant patrols to achieve detection and response before the launch. The delegates were citing the authority of the British naval attaché in Tokyo, who seems on this point to have been extremely well informed.

[51] Roskill, *Naval Policy between the Wars*, Vol. II, p. 186.

[52] ADM 116/3862, Director of Plans Minute dated 24 July 1936.

[53] ADM 116/3863, C-in-C China "Appreciation of the Situation of the Eastern Forces During the Period Before Relief" Letter 1077/01515 dated 10 July 1937.

[54] Henry, "British Submarine Development and Policy, 1918–1939," p. 390. See also ADM 116/3863.

forces were provisionally allocated to patrol the Sunda and Lombok Straits, the defense of Hong Kong and Singapore was the clear priority, even if the orders were realistic in their assessment that Hong Kong's situation was likely to be "untenable."[55] Even as late as June 1939, there was consideration of reinforcing the submarines in the Far East as the only practicable method of impeding a Japanese attack, but the boats available in home waters were considered too few to deal with Germany and the Admiralty refused to reduce their strength.[56]

Submarines on the China Station

The L-class began to arrive in the China Station in late 1919. There were early difficulties. *Titania* and six boats were initially retained in the Mediterranean because of the uncertain situation in Turkey and the Crimea. Although the flotilla was soon brought up to strength by the arrival of *Titania* and her brood, manning issues meant that first four and then six boats were put into reserve. The complexity of the submarines appears to have overwhelmed local deep maintenance capabilities, despite the experience of the Hong Kong dockyard with the older C-class. In 1923, the new Captain(S) was concerned to discover that only three out of twelve boats were operational.[57] Matters were not helped by the sinking of *L9* at Hong Kong during a typhoon in August 1923, although she was salved a month later and refitted. From this point, availability rose and over the following two years all the boats in reserve were progressively recommissioned.[58] Despite her repairs, *L9* never quite recovered from her sinking and was scrapped in 1927, her place in China being taken by *L27*. Force levels dropped only temporarily over 1929–31 with the changeover from the L-class to the new O- and P-class boats. After the loss of *Poseidon* in a collision with a merchant ship in June 1931, *Rainbow* was sent out as her replacement as soon as possible after completion.

The operational program was largely, but not wholly, routine. The submarines were based at Hong Kong in the colder months (October–April) and then went with the rest of the fleet to the northern Chinese anchorage of Wei-Hai-Wei, which had a very pleasant climate in summer. Local area operations, with an emphasis on practice attacks, were interspersed with cruises to other countries in the region. Economic restrictions tightened as the 1920s wore on, but earlier in the period there were clearly systematic efforts to develop a comprehensive understanding of the regional environment, with patrols off ports around the South and East China Seas, as well as communications exercises and experiments in long-range under-

[55] ADM 116/4635, "China Submarine Emergency Instructions" dated 24 April 1939.

[56] Admiralty, *Naval Staff History of the Second World War: Submarines*, Vol. III, p. 2.

[57] Rear Admiral C. G. Brodie (writing as "Sea Gee"), "Women Aboard," *The Naval Review* 49:4 (October 1961), p. 389.

[58] *The Navy List*, January 1922–January 1926 (monthly).

water signaling to support cooperative operations between submarines.[59] The first serious experiments with sustained submerged operations in tropical conditions were conducted as early as 1923,[60] although many patrol exercises were too short at only a few days and became over-constrained by peacetime safety.[61] Certainly, submerged time was limited and remained so for many years—in 1931 the distance transited submerged was an average of only 2.5% of the total distance run for fully operational boats.[62]

Money was even tighter at the end of the decade. A key effect of the financial restraints was to limit extended operations and regional cruises in favor of "day running" to save fuel. The result was to focus the submarines on basic tactics, particularly their torpedo attacks and "gun action surface" drills. The fact that the torpedoes could be recovered and reused encouraged practice firings. The flotilla launched some 469 in 1935.[63] With an average of nine boats operational, this meant that each deployed approximately fifty weapons, indicating a substantial number of practices. While more complex operational scenarios did not receive enough attention, the Fourth Submarine Flotilla was in this way a vital training ground for the future wartime submarine commanders.[64]

There was another problem for the submarines and their place in the China Fleet. The role of the Royal Navy in the region in the 1930s was one of presence and crisis management in an increasingly tense environment. The submarines were unsuited to this role, which occupied much of the surface force's time and resources.[65] They were employed on anti-piracy patrols around Hong Kong, another role for which the

---

[59]See ADM 1/8700/129, "4th Submarine Flotilla—Quarterly Reports March 1925–June 1926."

[60]Brodie, "Women Aboard," p. 389.

[61]Vice Admiral Sir Hugh Mackenzie, *The Sword of Damocles* (Penzance: Periscope Publishing, 2007), p. 36. HMS *Rainbow's* "war patrol" off Horsburgh Light in 1936 was only four days long, with the boat burning navigation lights on the surface at night. See also Bryant, *Submarine Command*, p. 39, which makes the comment that he did not dive a submarine at night between 1927 and 1938.

[62]ADM 116/2885, "4th Submarine Flotilla Quarterly Report of 1st October 1931." Figures for *Osiris, Olympus, Pandora,* and *Perseus.* While submerged speed and endurance were inherently limited by comparison with surface performance, this is a very low total and suggests less than forty-eight hours submerged over a 92-day period. The figures had not increased for normal running in 1935, see ADM 116/3564.

[63]Mackenzie gives 459 in *The Sword of Damocles*, p. 49, which suggests that he had kept copies of or consulted the Quarterly 4th Submarine Flotilla Reports. The figure from ADM 116/3564 is 469 for the year—a significant increase from 1934.

[64]It is difficult to find a successful RN wartime submarine commander who qualified in submarines before 1936 who did not serve in the Fourth Submarine Flotilla.

[65]Martin H. Brice, *The Royal Navy and the Sino-Japanese Incident, 1937–41* (London: Ian Allan, 1973), p. 28.

submarines were less than ideal,[66] but otherwise tended to operate by themselves. This accentuated the natural isolation of the submariners and also limited their exposure to high quality opposition. At the end of his command time in 1936, Dreyer saw the problem and pressed successfully in his final report for the dispatch of small destroyers to Hong Kong to take on the anti-piracy role which had consumed so much of the submarines' operational availability.[67] These began to arrive in 1937 and also took on as their war role some of the planned minelaying effort.

Nevertheless, after 1934 British operational concepts became more sophisticated, a process accelerated by the influence of Little as C-in-C from 1936. There was a renewed emphasis on extended patrols and sorties of up to a month in length were practiced.[68] *Oswald's* training war patrol in 1935 showed the new approach, with submerged distance run rising to 18.5% of the total over the quarter involved, indicating that the boat was underwater for the majority of her weeks simulating hostilities.[69] Although they themselves would have a surveillance and reconnaissance role in their own right, the submarines would not operate unsupported. In this period the RN and the local Royal Air Force began to conduct joint exercises in which flying boats were required to locate a naval force and home a submarine flotilla on to it. Four flying boats had been transferred from the United Kingdom to Singapore as far back as 1927 and formed into the new 205 Squadron,[70] but it was only with reinforcements from 1936 onwards, and more funds for training, that joint exercises stepped up. Such sea-air exercises, which also involved the aircraft embarked in the China Station's aircraft carrier and cruisers, soon demonstrated the vulnerability of the submarines in certain conditions to detection from the air at periscope depth and new, darker tone paint schemes were trialed.[71] Group or "wolf pack" coordinated submarine tactics were developed as a method of overwhelming the defenses of Japanese fleets and troop convoys.[72] Covert submarine

---

[66]Mackenzie, *The Sword of Damocles*, p. 38.

[67]ADM 116/3081, "China General Letter No. 35: Report of Proceedings of the Commander-in-Chief China Station 1st—11th January 1936."

[68]Admiral Sir Frederic Dreyer, *The Sea Heritage: A Study of Maritime Warfare* (London: Museum Press, 1935), pp. 318–19; and Mars, *British Submarines at War*, pp. 21–22.

[69]ADM 116/3564, "4th Submarine Flotilla Quarterly Report 1st April—30th June 1935."

[70]James J. Halley, *Famous Maritime Squadrons of the Royal Air Force* (Windsor: Hylton Lacey, 1973), p. 30.

[71]This included Home Fleet dark grey. ADM 116/3564, "4th Submarine Flotilla Quarterly Report 1st April—30th June 1936."

[72]Mars, *British Submarines at War* is particularly useful on British submarine training and operations in the Far East between 1937 and 1940. See pp. 21–23, 37–64, and 239–47. Much of this commentary is based not on Mars' memory, but the diary of his CO in HMS *Regulus*, Commander G. C. P. Menzies. See ADM 116/3564, "4th Submarine Flotilla Quarterly Report 1st April—30th June 1936" for an early six-boat effort.

reconnaissance techniques were also refined in order to monitor the progress of the increasingly secretive Japanese, particularly after 1936 when no more foreign warship visits were allowed to Japanese ports.[73]

War and the End

The outbreak of war with Germany saw the China Station rapidly denuded of surface ships. The Fourth Submarine Flotilla initially remained, however, not only to provide some defense against a Japanese intervention, but to intercept German merchant traffic. Four units were soon detached to Colombo to patrol for German surface raiders, while the remaining boats went as far north as Vladivostok to watch for any cooperation between the Soviets and the Germans. Not until March 1940, as the situation in Europe deteriorated, were *Medway* and her brood sent to the Mediterranean.[74] By the middle of 1940, there were no British submarines in the Far East and when Japan finally attacked in December 1941, there was only one—the refitting *Rover.* For the defensive operations that followed, the British would be hard pressed to muster even two T-class boats. Their losses had been too heavy and the demands of other theaters too great for anything more.

Conclusion

The disappointing performance of other Allied submarines in the Far East in 1941–42 must cast some doubt on the potential of the British submarine force to provide a meaningful defense against the Japanese, even had the Fourth Submarine Flotilla been available and at full strength when the time came. Nevertheless, there were key differences between the British units and their American and Dutch contemporaries. Even before 1939, there was much more combat experience available to them, while the drive to higher operational standards, particularly after 1937, would have made a difference. Finally, British torpedoes generally worked, and the evidence of the early months of the war against Germany was that the constraint on British success was more usually restriction in the rules of engagement than the submariners' ability to mount an attack. The submarines of the Fourth Flotilla might not have been able to stop a seaborne invasion by the Japanese, but they would not have allowed it to be a bloodless one.

There is another historiographical problem in assessing the significance of the British submarine capability at this time and its potential as a force element in the Far East. What is striking, apart from the "fleet submarine" issue, is the silence on

[73]Arthur J. Marder, *Old Friends, New Enemies: The Royal Navy and the Imperial Japanese Navy,* Vol. 1, *Strategic Illusions, 1936–1941* (Oxford: Oxford University Press, 1981), pp. 354–56.

[74]Admiralty, *Naval Staff History of the Second World War: Submarines,* Vol. III, pp. 2–5; and Mars, *British Submarines at War,* p. 64.

submarine employment within the wider naval operational and tactical debate over much of the period between the wars. That there was a much greater emphasis on anti-submarine warfare in the main fleets than was once believed has now been made clear by recent scholarship,[75] but the greater part of the planning effort for submarine warfare by the Royal Navy appears to have been left in the hands of relatively independent authorities such as the C-in-C China and Rear Admiral (Submarines). Perhaps this was inevitable, given the distribution of the main fighting strength of the submarine fleet in the 1920s and 1930s. Certainly the Admiralty itself, or at least the Plans Division, seems to have understood what was intended in the event of conflict.[76] In reality, and this may have been appreciated all along by the war-hardened leadership of the submarine arm,[77] submarines were always going to operate independently of the main surface fleets. The main problems of their employment were fundamentally either unique to their type or required cooperation with other independent arms, such as maritime patrol aircraft, not the surface ship navy.

Silence in the historical record may therefore be much more a reflection of an absence of the need to explain and debate externally to the submarine force than any other factor. Perhaps the most striking evidence that submarines did occupy and retain a key position within the Royal Navy's order of battle is that from the 1926 Estimates on, there were orders for new construction every year. Submarines were never cheap and investment in them not a trivial matter, particularly in a period of such financial constraint. Despite the efforts to ban the type in which the United Kingdom repeatedly engaged, the fact was that submarines were always recognized as a key component of the total force and as the most practicable and cost-effective solution to the problem of delaying a Japanese onslaught in the Far East long enough to allow the Main Fleet to arrive and begin operations. Had they been given the chance, British submarines may well have made a difference.

---

[75]See George Franklin, *Britain's Anti-Submarine Capability, 1919–1939* (London: Frank Cass, 2003).

[76]Submariners were well represented in the Admiralty between the wars. Apart from Little serving as Deputy Chief of Naval Staff (1933–35), he and Dunbar-Nasmith were Directors of the Trade Division (1920–22 and 1924–26), Brodie was Director of the Operations Division (1932), and Laurence was Director of the key Training and Staff Duties Division (1927–28, having been Assistant Director 1926–27). Horton served in the Mobilisation Division and Layton in Plans. Talbot served as Director of Naval Equipment (1932–34).

[77]Until Rear Admiral B. C. Watson in 1938, all the Rear Admirals (Submarines) from 1929 were experienced wartime submarine commanders. Little had been too senior to command a boat, but had commanded a flotilla from a seagoing ship. Watson was relieved by the submariner Horton in 1940.

2

Naval Strategies in Collision

Britain, the USA, and Japan at the Beginning of the Pacific War

Evan Mawdsley

Much has been written about Pacific strategy in the lead-up to the Japanese attacks on Malaya, Hawaii, and the Philippines in December 1941. Most studies have been concerned either with one of the three major participants or with bilateral rivalry or relationships—between Japan and the United States,[1] between Japan and the British Empire,[2] and even between the USA and Britain.[3] The objective here is to produce a more "international" perspective to what was essentially a three-sided relationship. I also want to examine the evolution of national strategy, and specifically *naval* strategy, over time. How far did the strategies of the 1920s and 1930s evolve, and to what extent were they still in place when the Pacific War actually broke out?

[1]Stephen E. Pelz, *Race to Pearl Harbor: The Failure of the Second London Naval Conference and the Onset of World War II* (Cambridge: Harvard University Press, 1974); Edward S. Miller, *War Plan Orange: The U.S. Strategy to Defeat Japan, 1897–1945* (Annapolis: Naval Institute Press, 1991); David C. Evans and Mark R. Peattie, *Kaigun: Strategy, Tactics, and Technology in the Imperial Japanese Navy, 1887–1941* (Annapolis: Naval Institute Press, 1997); Sadao Asada, *From Mahan to Pearl Harbor: The Imperial Japanese Navy and the United States* (Annapolis: Naval Institute Press, 2006). This chapter was originally published in *Global War Studies* 11:1 (2014).

[2]Stephen Roskill, *Naval Policy Between the Wars*, Vol. II, *The Period of Reluctant Rearmament, 1930–1939* (London: Collins, 1976); Arthur. J. Marder, *Old Friends, New Enemies: The Royal Navy and the Imperial Japanese Navy*, Vol. I (Oxford: Clarendon, 1981); Christopher Bell, *The Royal Navy, Seapower and Strategy Between the Wars* (hereinafter RNSS) (London: Macmillan, 2000).

[3]James R. Leutze, *Bargaining for Supremacy: Anglo-American Naval Collaboration, 1937–1941* (Chapel Hill: University of North Carolina Press, 1977); David Reynolds, *The Creation of the Anglo-American Alliance, 1937–1941: A Study in Competitive Cooperation* (London: Europa, 1981); Ian Cowman, *Dominion or Decline: Anglo-American Naval Relations in the Pacific, 1937–1941* (Oxford: Berg, 1996).

The "Classic" View of Pacific Strategy: the 1920s and 1930s

The Paris (1919) and Washington (1922) conferences established a strategic situation in the Pacific, which prevailed through the 1920s and 1930s. The strength of the three navies was set on roughly a 5:5:3 ratio; in the simplest terms, this meant that in the final version Japan had nine capital ships, and the Americans and British each had fifteen.[4] Britain and Japan were no longer allied, and Tokyo was awarded control of the German island colonies in the central Pacific as League of Nations mandates.

A broad-brush treatment of Pacific strategy in the 1920s and 1930s would lay out three elements. The American War Plan ORANGE envisaged a drive by the U.S. Navy west across the vast ocean. The Imperial Japanese Navy foresaw a strategy in which the American fleet would be worn down as it advanced, with this attrition followed by an ambush and a decisive battle. Finally, the "Singapore strategy" of the British Royal Navy anticipated the transfer of the "main fleet," in time of emergency, from European waters to a new naval base to be developed at the tip of the Malayan peninsula.

Although the first American plans for war with ORANGE (the codename for Japan) date back to 1906, planning became more formal only in 1924 with work approved by the (Army-Navy) Joint Board. Despite the existence of a "joint" national plan, the two American services had major disagreements. The U.S. Navy was more interested in planning for a war with Japan. The main force of the "United States Fleet" was based on the West Coast, where its only likely enemy could be found (only in February 1941 would the "U.S. Fleet" be renamed the "Pacific Fleet"). Variants of the war plan were developed in the 1920s, including a deep thrust to relieve the American colony in the Philippines should the Japanese capture it. From the mid-1930s, however, the plans assumed a more cautious advance and a readiness to accept the loss of the Philippines in the early stages of the war.[5]

The U.S. offensive, as laid out in the ORANGE war plans, was divided into three phases. In Phase I, the Japanese overran American possessions in the western Pacific—essentially the Philippines and Guam. In Phase II, the U.S. Navy advanced to the western Pacific and recaptured Manila or developed some other forward base. In Phase III, operating from this western position, American forces established further bases to the north, from which Japan could be blockaded or bombed, and forced to capitulate. There would be a decisive fleet battle at some point, probably in Phase II or Phase III.

---

[4]Capital ships were battleships and battle cruisers. Cruisers were a different and more complex story, which caused ill feeling on all sides; for these vessels the ratio determined by the 1930 London Naval Treaty was 10:10:7.

[5]Miller, *Orange*, passim. Some material from the Joint ORANGE plans for 1923, 1924, 1928, and 1938 is reproduced in Steven T. Ross, ed., *American War Plans, 1919–1941* (hereinafter AWP), Vol. 2 (New York: Garland, 1992), pp. 1–227.

The "Route of Advance to the Western Pacific" in American Phase II was oriented around the "Mandate Islands," the central Pacific island groups of the Marshalls and the Carolines. A more northerly route, via the Aleutians and the Kuriles, was ruled out by geography and bad weather; the southern route, the Gilberts and the Solomons, involved territory of the British Empire. In the mandates, the Marshall Islands, about 2,500 miles southwest of Pearl Harbor and the easternmost Japanese territory, were inevitably the first step; it was expected that an American forward base would be established there.[6]

War with Japan was also central to the British Admiralty's inter-war planners, at least until the rise of the German threat in the 1930s. The strategy, laid out in various versions of the "War Memorandum (Eastern)," was not unlike that of the ORANGE plans. The difference was mainly in the line of approach; for the British, Singapore occupied the place of Hawaii, and Hong Kong the place of Manila.[7] What might more accurately be called the "Singapore/Hong Kong strategy" (rather than the "Singapore strategy") in the 1931 version of the "War Memorandum (Eastern)" had three phases: Phase I was the "Period before Relief" (of Singapore and Hong Kong), or "the concentration of the British Main Fleet in the East." The difference from the American situation was that while the "U.S. Fleet" was already based on the American West Coast—or from 1940 in Hawaii—the ships of the Royal Navy had to be moved a very long distance from the bases of the Atlantic/Home Fleet or the Mediterranean Fleet. In a hypothetical crisis, in which British possessions in the Far East were under threat, it was important that the Japanese not capture Hong Kong or Singapore before the "main fleet" arrived.

Phase II was the "Period of Consolidation." Phase III was the "Period of Advance," and for this a range of options for "possible large scale operations against the Japanese" were discussed. As with the American Phase III, this would require the establishment of forward bases between Japan and the Asian mainland, which in the British case meant some location north of Hong Kong. The approach was a traditional one: "Whatever form the operations in Phase III may take they must be influenced by the primary object of destroying or disabling the Japanese Main Fleet."[8]

---

[6]For one inter-war view, see AWP, Vol. 2, pp. 122–23, "Joint Army and Navy Basic War Plan ORANGE [1928]," pp. 52–53 of original document. On the Marshall Islands and their place in pre-war U.S. planning, see Miller, *Orange*, pp. 108, 186–202, 228–38. The Marshalls were the target of an early U.S. carrier raid in February 1942 and a successful invasion two years later; the best known assault was at Kwajalein Atoll.

[7]A good survey of British planning is in Bell, RNSS, pp. 59–98. "War Memorandum (Eastern)" 1921 is in The National Archives, London (hereinafter TNA), ADM 116/3124, the 1923 version is in ADM 116/3124, that for 1924 in ADM 116/3125.

[8]ADM 116/3118, "War Memorandum (Eastern)," July 1931; ADM 116/3475, "War Memorandum (Eastern)," 1933.

In the final version of the "War Memorandum (Eastern)," the Royal Navy planners developed Phases I and II in considerable detail, envisaging a movement of the Atlantic and Mediterranean fleets to Singapore, in a deployment that would take a maximum of seventy days. This version of the "War Memorandum" was updated over time, but the earliest elements are dated December 1937, five months after the Sino-Japanese War began. The 1937 plan had no Phase III; Phase II was now simply "Naval Operations subsequent to the arrival of the Main Fleet at Singapore." Operations in this phase, including a possible fleet action, were taken to depend on whether Hong Kong was still available as a forward base. The measures listed to force Japan to capitulate were essentially economic, disrupting Japanese shipping on a global basis using the Royal Navy's large fleet of cruisers, and ultimately blocking trade with the Asian continent.[9]

Japanese naval planning in the 1920s and 1930s concentrated almost entirely on a war against the United States. Surprisingly, the Imperial Navy did not conduct planning for a war against Britain until 1938, and such a conflict was never a major strand of Japanese naval preparations.[10] Japanese thinking about war with America was heavily influenced by the full-scale 1904–05 naval war with Russia, in which two enemy fleets were destroyed, one based in the Far East at the outbreak of war and the other brought all the way from Europe. This experience developed into what might be called an "attrition/decisive battle" strategy (attrition of the advancing enemy fleet, followed by its destruction in a decisive battle). As with its operations against the Russians in 1904, the Imperial Navy expected to deny the Americans a forward base, in this case by capturing the Philippines and Guam early in the war.[11]

---

[9]Sources for British naval strategy in 1937–39 are more detailed than for earlier years. The December 1937 "War Memorandum (Eastern)" was updated through 1939 (ADM 116/4393, ADM 1/9767). In addition, at the time of the Tientsin Incident, the Chiefs of Staff were presented with an assessment of deployments and options: TNA, CAB 53/50, COS(39)931, "Situation in the Far East," 24 June 1939.

[10]Donald M. Goldstein and Katherine V. Dillon, eds., *The Pacific War Papers: Japanese Documents of World War II* (hereinafter PWP) (Washington, DC: Potomac Books, 2004), pp. 17–18 (Oi); Tsunoda Jun, "The Navy's Role in Southern Strategy," in James W. Morley, ed., *The Fateful Choice: Japan's Advance into Southeast Asia, 1939–1941* (New York: Columbia University Press, 1980), p. 262; Agawa Hiroyuki, *The Reluctant Admiral: Yamamoto and the Imperial Navy* (Tokyo: Kodansha, 1979), p. 198. The first Japanese map exercises directed against the British Empire were held only in 1938. The lack of IJN planning for a war against Britain in the 1920s and most of the 1930s is surprising, given the extent of planning by the RN and the level of the general British imperial presence in the Far East. The causes may have included the long-term close relations between the two navies. In addition, the main strength of the RN remained in Europe (despite the slow construction of base facilities at Singapore), while the "U.S. Fleet" was already based in the Pacific.

[11]The fullest account of inter-war strategy is in Evans and Peattie, *Kaigun,* especially pp. 187–91, 286–91. See also Ozawa Jisaburō, "Development of the Japanese Navy's Operational Concept against America," in PWP, pp. 71–75. Evans and Peattie called this a

For the Japanese, as with the other powers, the balance was seen primarily in terms of capital ships. Heavy cruisers and destroyers, however, had an important support role, and in these categories the qualitative and quantitative comparison with potential enemies was more favorable to Japan than it was for battleships. In the 1920s, the Imperial Navy developed a preliminary "attrition strategy" (*zengen sakusen*), in which light forces would observe and attack the oncoming U.S. battle fleet.[12] The air element, so important in 1941–45, was still secondary even in the 1930s, in view of the short range and small size of aircraft. Submarines, however, became a major element in the mid-1920s, as the Imperial Navy began to receive a number of long-range cruiser submarines; it was expected that these high-speed boats would watch, pursue, and attack the American fleet after the departure from its base in Hawaii.[13]

Several features of these "classic" war plans of the three Pacific powers deserve comment. First of all, they were essentially naval war plans rather than national war plans. Also, war between these major powers, their home territories literally on different sides of the planet, was unlikely, and the plans were substantially hypothetical. They had much to do with bureaucratic in-fighting; a plausible enemy and war mission were required to justify naval appropriations in the national budget. There was an intellectual element, too, of a staff-training exercise. Furthermore, each plan was based on the assumption that the navy in question would fight essentially on its own, in two senses: with limited support by the army (or in the British case, the Royal Air Force) and without the support of the navy of an ally.[14] They were also developed at a time when there was no dominant European continental power. On the whole, the assumption was that only one enemy would be directly involved, although the British studies from 1937 took into account the need to leave some ships in home waters to deal with the small—but growing—fleet of the Third Reich.

Technically, all three plans were built around the dominant naval technology of the day, the capital ship and its super-heavy artillery. All of the three navies could be described as essentially "Mahanist," assuming that in the last analysis "command of the seas" would be achieved by a decisive battle involving big-gun ships. Their admirals thought of capital ships as the measure of relative power. Air power was

---

"wait-and-react strategy" (*Kaigun,* p. 464). Gordon Prange used the term "Great All-Out Battle" for the Japanese *kantai kessen;* see *At Dawn We Slept: The Untold Story of Pearl Harbor* (Harmondsworth: Penguin, 1991 [1982]), p. 13.

[12]Evans and Peattie, *Kaigun,* pp. 199–212, 220–37.

[13]Ibid., pp. 212–19.

[14]This lack of attention to allies may have been because discussion, on paper, of this subject would have gone beyond the remit of military staffs. It is hard to believe that the British and American planners in the 1920s and 1930s, given the experience of World War I, did not take the possibility of future mutual support into account.

recognized to be of growing importance, but it was seen as an auxiliary element; aircraft carriers were still at the experimental stage into the early 1930s. The Royal Navy, however, did assume a strong "economic warfare" element in its planning against Japan, especially in the later 1930s; for this cruiser warfare would be very important. Blockade was also a feature of American Phase III operations, although the U.S. Navy was generally more confident of its ability to fight a fleet action.

Changes in Pacific Strategy: 1937 to 1940

One unfortunate effect of the 1937–40 period was to make the wars of the future decidedly less hypothetical. It would be over-neat to suggest that 1 January 1937 was the critical moment, but that was the first day after the end of the era of the Washington and London naval treaties, and it was followed by a shipbuilding "race." The Japanese rejection of the treaty limits proved self-defeating, as the Americans and British, with their superior industrial capacity, began unlimited construction. A realization that the American naval program, in particular, would soon outstrip their own would turn out to be a factor spurring Tokyo into drastic military action in 1940–41.

The year 1937 was also one of important developments on the international scene, especially the beginning of the Sino-Japanese War. Often neglected, this full-scale conflict expanded the role of the military in the Japanese government. The inconclusive war in China was eventually one of the causes of the Japanese decision in 1940–41 to attack the European colonies in East Asia—to cut China completely off from outside support.[15] This attack would require a radically different use of the Imperial Navy than had been anticipated in the 1920s and 1930s. Japanese aggression worsened relations with the U.S., and in the end—in November 1941—a *modus vivendi* was made impossible by the continued presence of Japanese armies in China. Relations with Britain became more tense and, as already mentioned, the Japanese Naval General Staff (NGS) began to incorporate conflict with the Royal Navy into its "annual plan" and in map exercises. One specific and significant result of the Sino-Japanese war was the occupation of Guangzhou (Canton) in October 1938, which made much less likely the use of Hong Kong—seventy-five miles to the southwest at the mouth of the Pearl River estuary—as a Phase II advance base by the Royal Navy. Incidents in the China war also led to the tentative beginnings of strategic cooperation between Washington and London.

The post-treaty naval race of the later 1930s, and tension over China, sharpened existing tensions in the Pacific between the Western powers and Japan and, to some extent, was compatible with the strategic war plans of the past two decades. On the

---

[15]On the relationship between the war in China and the later war with the Western powers, see Akira Iriye, *The Origins of the Second World War in Asia and the Pacific* (Harlow: Longman, 1987), pp. 99–101, 170–72.

other hand, Europe (and not the Far East and the Pacific) was made the center of attention for both Britain and the U.S. as a result of the acute European tension of 1938–39, the war that began in September 1939, and the capitulation of France.

It is true that the German occupation of the metropolitan territory of two of the major colonial powers, France and the Netherlands, had a fatally destabilizing effect in Southeast Asia; so, too, did the commitment of any available British Empire forces to the eastern Mediterranean rather than the Far East. Even so, the danger of Hitler's consolidation of his hegemony in Europe was regarded as far more important. For the Americans, there was the nightmare of Nazi Germany dominating the European continent and eventually the Atlantic Ocean. For the British, there was the immediate question of national survival. The secret Plan DOG memorandum of November 1940, prepared by the American Chief of Naval Operations (CNO), Admiral Harold Stark, is rightly seen as a seminal document. Stark, in a paper aimed at the highest levels of American government, had outlined various strategic orientations and explicitly contrasted the "classic" Plan ORANGE—the one-on-one offensive against Japan—with what he called "Plan (D)," "an eventual strong offensive in the Atlantic as an ally of the British, and a defensive in the Pacific." He recommended adoption of Plan DOG ("D" was "Dog" in the Navy's phonetic alphabet), and what came to be known as "Germany first" became, in effect, the strategic policy of the United States.[16]

The impact on American and British Pacific strategy was highly significant, but this was still in the realm of theory rather than that of practice. Japan and America were, after all, neutral, and the British commitment of the "main fleet" to Singapore had always been hypothetical. The British Chiefs of Staff (COS) Appreciation of 31 July 1940 (famously captured by a German raider aboard SS *Automedon* off Penang in November) indicated that, for the moment at least, the possibility of sending major naval forces to reinforce the Far East had to be abandoned. "[U]ntil we have defeated Germany and Italy or have drastically reduced their naval strength, we are faced with the problem of attempting to defend our Far Eastern interests in the absence of an adequate fleet."[17] As for the Americans, they refrained from early action. In the crisis summer of 1940, concerned that Japan would take advantage of the situation in Europe, the President ordered that the Pacific Fleet be kept at Pearl Harbor, 2,000 miles out into the Pacific, after the annual maneuvers, rather than returning to the normal home bases on the West Coast (mainly San Pedro, near Los Angeles). In November 1940, Admiral Stark stated that Plan DOG "does not mean the immediate movement of the fleet into the Atlantic." For the

---

[16]AWP, Vol. 2, pp. 225–50, Stark to Knox, 12 November 1940. The well-known Plan DOG memorandum does not require extensive comment here; for a fuller discussion, see Maurice Matloff and Edwin M. Snell, *Strategic Planning for Coalition Warfare, 1941–1942* (Washington, DC: U.S. Army Center of Military History, 1990 [1980]), pp. 25–28.

[17]CAB 66/10, COS(40)592, 31 July 1940, "The Situation in the Far East in the Event of Japanese Intervention against us."

moment, he was content to await developments: "I would make no further moves until war should become imminent, and then I would recommend redistribution of our naval forces as the situation then demanded. I fully recognise the value of retaining strong forces in the Pacific as long as they can profitably be kept there."[18]

Most important, the Japanese armed forces also re-examined their strategic plans after the defeat of France and the Netherlands, and the weakening of Britain. The Japanese Army, which hitherto had concentrated on a future war with the USSR, became more interested in mid-1940 in a campaign directed towards Southeast Asia. In late July 1940, civilian and military (Army and Navy) leaders agreed on a policy of extending Japanese influence into Southeast Asia including the establishment of bases in Vichy French Indochina, and meanwhile developing firmer relations with Germany and Italy. For the first time, the Japanese government began to look seriously at an offensive "Southern Operation."[19]

The Japanese Naval General Staff developed, with the Army, a complex series of operations directed at the occupation of Southeast Asia, while maintaining the core strength of the Combined Fleet to deal with an advance by the U.S. Navy.[20] The Navy did assume for planning purposes that occupation of the Netherlands East Indies (NEI) or operations against British possessions in Asia (Hong Kong, Borneo, Malaya) would probably lead to war with the United States. (The Japanese Army, in contrast, thought it might be possible that Britain and the United States could be kept detached from one another.) A "preparatory fleet mobilisation" (*suishi jumbi*) conducted between November 1940 and March 1941 brought personnel to full strength, allowed completion of ship refits and conversions, and built up supplies of fuel; as much as half the Japanese merchant marine was incorporated into naval service.

The NGS strategy did involve stripping out, at least for a short time, a considerable element of the "attrition" forces for service in Southeast Asia, mainly the 2nd Fleet heavy cruisers and the Navy's land-based bombers. The assumption was that other forces left in home waters—battleships, carriers, destroyers, submarines—could deal with any rash American movement directly across the Pacific; they had been training for this for the last two decades. Nevertheless, it was more likely that the Americans would proceed cautiously. In the spring of 1941, the "Plan of Operations for the Current Year" being developed by the Operations Section of the NGS had as its prime mission the seizure of Southeast Asia. "The attrition/decisive battle" strategy was still in place, with a special role to be played by the submarine force in wearing down the American Pacific Fleet from the time it left Hawaii.[21]

---

[18]AWP, Vol. 3, p. 249, Stark to Knox, 12 November 1940.

[19]Tsunoda, "Navy's Role," pp. 241–95.

[20]Evans and Peattie, *Kaigun*, 447–71. The Combined Fleet was the major sea-going command of the Imperial Navy and incorporated all naval forces in the Pacific. By early 1941, these forces were organized in six numbered warship fleets and two "air fleets."

[21]Prange, *Dawn*, p. 194.

The Naval General Staff was headed from April 1941 by Admiral Nagano Osami, an officer who—perhaps unfairly—has not been treated sympathetically by historians.[22] Nagano's (correct) view was that the advance across the Pacific planned by the Americans would be slow, beginning with an attempt to establish bases in the Marshalls; it would not involve a threat to vital Japanese areas. There would be time, after the Southern Operation, to concentrate the Combined Fleet for the long-planned decisive battle in the central or western Pacific.[23] As the Japanese knew, in the spring of 1941 the U.S. Navy had transferred a quarter of the Pacific Fleet—three battleships, a carrier, and five cruisers—to the Atlantic.[24] As a result, when Admiral Nagano later outlined the overall situation of the Navy at the Imperial Conference of 5 November 1941 he could point out that forty percent of the U.S. Navy was currently in the Atlantic. It would take the Americans considerable time to transfer their forces back: "[A] part of [the American] naval force might be able to obstruct our operations in the South, although the strength of that naval force would probably be insufficient to engage in a decisive battle." Britain would be unable to send a large fleet to reinforce the ships already based in the Far East ("a battleship, ten or more cruisers, and some aircraft"). Nagano thought the

---

[22]Nagano had replaced the long-serving Admiral Prince Fushimi as Chief of the NGS on 9 April 1941. He was Navy Minister in the Hirota government in 1936, and then commanded the Combined Fleet until December 1937. After this he had served in a number of very senior naval administrative posts in Tokyo. Nagano is assessed very critically in Asada, *Mahan,* pp. 246–47. Gordon Prange's account, based on post-war interviews, referred to his "slow moving mind" and described him as a person who, at age 62 (in 1941), had "lost much of the strength and drive of youth" and was a "tired old man" who fell asleep in the middle of reports (Prange, *Dawn,* pp. 102, 299). Admiral Yamamoto is said to have remarked that Nagano "was the kind of man who was convinced he was a genius, even though he wasn't" (Agawa, *Reluctant Admiral,* pp. 117–18, 213). Evans and Peattie are also generally dismissive of Nagano's abilities (*Kaigun,* pp. 528–29).

[23]Fukudome Shigeru, "Hawaii Operation," in Raymond O'Connor, ed., *The Japanese Navy in World War II: In the Words of Former Japanese Naval Officers* (Annapolis: Naval Institute Press, 1969), p. 7. This is the recollection of Admiral Fukudome (chief of the 1st Bureau of the NGS). He was discussing objections made by the NGS to Admiral Yamamoto's Pearl Harbor plan in October 1941; these related to NGS strategy as developed in earlier months. "It seemed unlikely that the enemy would come straight to our homeland seeking a decisive battle; rather, it [*sic*] would first establish advance bases in the Marshall Islands and then attempt the island-hopping strategy by gradually advancing its bases. In that case, we would be given ample time to concentrate all our available strength in a decisive contact engagement [*sic*] for which our Navy had long studied and trained. It would be wiser to seek a decisive battle in familiar waters. . . ." For background to Fukudome's discussion in October 1941 with the Combined Fleet's Captain Kuroshima (an advocate of the Hawaiian operation), see Prange, *Dawn,* pp. 297–98.

[24]The transfer of U.S. ships from the Pacific to the Atlantic is discussed more fully in the next section.

enemy surface fleets would probably retreat, and the main danger to Japanese operations would be from submarines.[25]

The clearest available presentation of the NGS perception of Japanese strategic intentions in 1940–41, and how the Imperial Navy fit into it, was Nagano's speech to the Imperial Conference on 6 September 1941. The probability of a prolonged war was very high, the admiral argued. The best prospect for Japan—but one that could not readily be expected—was a rash American or British attempt to attain a quick victory, with an attack early in the war into waters controlled by Japan.[26] But Nagano maintained that even a "decisive [Japanese] victory here would not bring the war to a conclusion. The position of the Americans was impregnable." As he admitted, "[O]ur Empire does not have the means to take the offensive, overcome the enemy, and make them give up the will to fight."[27]

In Nagano's view, the most important thing was, at the outset of the war, to take enemy bases (Singapore, Manila) and resource areas (Malaya, NEI) "making our operational position tenable and at the same time obtaining vital raw materials. . . ." "[O]ur Empire will have secured strategic areas in the Southwest Pacific [by which he presumably meant the Philippines, the NEI, and Malaya], established an impregnable position, and laid the basis for a prolonged war. . . ."[28]

The final test of the Southern Operation came at a large map exercise (war game) held at the Naval Staff College in Tokyo on 11–20 September 1941. The exer-

[25]Ike Nobutaka, ed., *Japan's Decision for War: Records of the 1941 Policy Conferences* (hereinafter JDW) (Stanford: Stanford University Press, 1967), pp. 233–34 (5 November). Nagano did sketch out a "worst case" in which the Americans completed the transfer of the Atlantic Fleet and the British might send "a part of their fleet [i.e. in addition to the single battleship and ten cruisers already there] because they would not want to lose Singapore." "In this case," Nagano admitted, "it is possible that Great Britain and the United States might combine their fleets. Our navy has plans for this contingency. . . . The combined force of Great Britain and the United States has weak points. We are, therefore, confident of victory [in a decisive battle]."

[26]Nagano was evidently referring, on the one hand, to the classic encounter with the U.S. Navy in mid-Pacific and, on the other, to a naval campaign in Southeast Asia ("in the areas of ocean we have in mind") against the limited naval forces Britain could move to the Far East.

[27]Asada, *Mahan,* p. 265; JDW, pp. 138–39.

[28]JDW, p. 140 (6 September). The problem with the strategy of Admiral Nagano, and that of the Japanese military establishment as a whole, was a lack of control over events *after* the operational victory. "What happens thereafter," Nagano noted, "will depend to a great extent on overall national power—including various elements, tangible and intangible—and on developments in the world situation." Probably the expectation followed the lines of the "Draft Proposal for Hastening the End of the War against the United States, Great Britain and the Netherlands, and Chiang [*sic*]" of 15 November 1941. According to this, Germany, Italy, and Japan would defeat Britain by occupying or overthrowing its Empire and perhaps even carrying out an invasion of the British Isles; this would "destroy the will of the United States to fight" (JDW, pp. 247–48).

cise covered the attacks on the Philippines, Thailand, and northern Malaya. It did not take in the second-phase attack on the NEI nor, evidently, did it include an encounter with the U.S. Pacific Fleet. It did envisage the early capture of Guam and Wake in the central Pacific, to establish favorable positions for the decisive battle to come.[29] In the final outline of the Southern Operation, laid out to the Imperial Conference on 5 November by General Sugiyama, the Chief of the Army General Staff, the main Japanese effort on the outbreak of war would be simultaneous operations in Malaya and the Philippines, followed by an advance to the NEI. He envisaged a five-month campaign, with operations in the Philippines lasting fifty days, in Malaya 100 days, and in the NEI fifty days. These times might be extended "in case the American fleet comes to attack us" or Russia supported the U.S., but a protracted war could be won as long as the bases and resources in Southeast Asia were held.[30]

On the Brink: Strategies in 1941

For both the Americans and the British, the secret military staff "conversations" (ABC-1) which were held in Washington, mainly in February and March 1941, marked the beginning of another strategic turn. The agreed general principles of strategy were summed up in the ABC-1 Report of 27 March 1941: "Since Germany is the predominant member of the Axis powers, the Atlantic and European area is considered to be the decisive theatre. The principle United States Military effort will be exerted in that theatre. . . ." American interests were stated to be primarily in the Western hemisphere. The stress was on fighting Germany and specifically the "security of sea communications." The ABC-1 "Basic War Plan" envisaged strengthening the U.S. Atlantic Fleet, when war with Germany began, with three battleships, a carrier, eight cruisers, and fourteen destroyers sent from the Pacific.

All the same, the American and British planners at ABC-1 devoted considerable attention to the Pacific and Far East. Because the overall resources of the partnership were greater than those just of Britain alone, it was in theory possible—even with a "Germany first" strategy—to reestablish deterrent or war-fighting forces facing Japan. The British stressed the need to defend the United Kingdom, but also the "ultimate security" of the whole Empire. "A cardinal feature of British strategic policy," the Report stated, "is the retention of a position in the Far East such as will ensure the cohesion and security of the British Commonwealth and the maintenance of its war effort."

The Report took in the problem of neutral Japan.

> Even if Japan were not initially to enter the war on the side of the Axis powers, it would still be necessary for the Associated Powers to deploy their

[29]Prange, *Dawn*, p. 224.

[30]JDW, pp. 232–33.

> forces in a manner to guard against eventual Japanese intervention. If Japan does enter the war, the Military strategy in the Far East will be defensive. The United States does not intend to add to its present Military strength in the Far East but will employ the United States Pacific Fleet offensively in the manner best calculated to weaken Japanese economic power, and to support the defense of the Malay barrier by diverting Japanese strength away from Malaysia. The United States intends so to augment its forces in the Atlantic and Mediterranean areas that the British Commonwealth will be in a position to release the necessary forces for the Far East.[31]

The "Joint Basic War Plan," an annex to the ABC-1 Report, stipulated that in the Far East the task of naval forces would be to raid enemy sea communications. As for land and air forces, an important distinction was made between what they would "defend" and what they would "hold." In the former category were the Philippines, Hong Kong, and the Netherlands East Indies (as a whole); in the latter were Malaya, Singapore, and Java (the center of the NEI). In the Pacific, U.S. naval forces would "[s]upport the forces of the Associated Powers in the Far East Area by diverting enemy strength away from the Malay barrier through the denial and capture of positions in the Marshalls, and through raids on enemy sea communications and positions."[32] ABC-1 was formally embodied into U.S. war plans, issued in May. RAINBOW No. 5 was the joint Army-Navy Plan, and WPL-46, the U.S. Navy element of this.[33]

Something very important had been put in place. In effect, the two western powers now anticipated watered-down versions of their individual pre-war (or pre-June 1940) strategies: a combination of the ORANGE war plan and the Singapore strategy would be enough to deter or contain the Japanese. A clear statement of their expectations was put in the reply by the British Chiefs of Staff to the Australian Prime Minister, Robert Menzies, shortly after the completion of the ABC talks. Menzies had complained about the lack of naval strength in the Far East. The COS stressed the potential American contribution: "Though nominally neutral, [the United States] are now so closely identified with our cause that the potential threat of the United States Fleet at Hawaii must alone impose a most powerful restraining influence on Japanese freedom of action to move southwards. It would, we feel, certainly prevent the use of the whole Japanese fleet in support of an expedition to the South Seas."[34]

---

[31] *Hearings before the Joint Committee on the Investigation of the Pearl Harbor Attack* (hereinafter PHA) (Washington, DC: GPO, 1946), part 15, pp. 1491–92, "United States—British Staff Conversations: Report," 27 March 1941.

[32] PHA, part 15, pp. 1511, 1518–19.

[33] U.S. National Archives and Records Administration, College Park, MD (hereinafter NARA), RG 38, Box 147F.

[34] CAB 80/27, COS(41)230, "Visit of the Australian Prime Minister," 11 April, pp. 3–4.

Admiral Pound, the First Sea Lord, made the same general point in October 1941: "What would deter [the Japanese] . . . would be the presence at Singapore of a force . . . of such strength that to overcome it they would have to detach the greater part of their Fleet and thus uncover Japan"—by which Pound presumably meant uncovering Japan to attacks by the U.S. Navy.[35] The same general view was being taken in London as late as 28 November 1941, when the Joint Intelligence Committee (JIC) presented a final pre-war assessment to the COS. Singapore was 2,400 miles from Japan ("where are the only large Japanese repair docks"). The main American fleet was based in Honolulu; there were American light naval forces in the Philippines, the Dutch fleet was in the East Indies, and a British fleet was growing in "Eastern and Australian" waters ("being strengthened, particularly as regards capital ships"): "In view of the above and their characteristic caution it is questionable whether the Japanese would be prepared to split their main fleet by sending to southern waters a force, including capital ships, of sufficient strength to protect seaborne expeditions against Malaya or possibly the Netherlands East Indies."[36] Rear Admiral V. H. Danckwerts, one of the senior Royal Navy representatives in Washington, had put the concept neatly in talks with senior American civilian and military officials on 29 April 1941: "[G]enerally speaking to check Japan it was necessary to have in the whole of the Pacific area either a fleet of the strength of 9 capital ships concentrated at Singapore or Hawaii or a fleet at each base of not less than 6 ships."[37]

No sooner had the Washington ABC staff talks finished at the end of March 1941, than President Roosevelt and Admiral Stark became convinced that the establishment of a significant naval presence in the Atlantic was an immediate task; this was true even when the United States was not at war, as the country committed itself to exercising control at least over the western Atlantic. Washington decided to transfer powerful forces from the Pacific. The first elements of this were three battleships (*Idaho, Mississippi,* and *New Mexico*), an aircraft carrier (*Yorktown*), five modern light cruisers, and a number of destroyers. The United States Pacific Fleet was thus reduced by a quarter.

There was a top-level discussion at the end of April 1941 about the transfer of even more vessels from the Pacific to the Atlantic. On the 29th two British officers, Rear Admiral Danckwerts and Captain Clark, just back from a visit to Pearl Harbor,

---

[35]CAB 69/8, DO(41)66, 20 October. Pound was arguing against Prime Minister Churchill here, proposing the despatch of a conventional balanced fleet of older ships, rather than the new battleship *Prince of Wales.* He was, nevertheless, repeating the strategic plan which had been in place since at least August 1941.

[36]CAB 81/105, JIC(41)449, 28 November. The assessment also assumed (fatally incorrectly) that for air support in Southeast Asia the Japanese would have to rely on *carrier-based* planes.

[37]ADM 116/4877, Danckwerts to COS, 29 April. Danckwerts made it clear that he was using "capital ships" as shorthand for the varied types of a balanced fleet.

were summoned by Secretary of the Navy Frank Knox to a meeting involving Secretary of War Stimson, Admiral Stark, and General Marshall (the U.S. Army Chief of Staff). Knox asked what the British thought would be the result if the U.S. "now moved almost the whole of their Pacific Fleet into the Atlantic, leaving Hawaii to defend itself. . . ." This would be done while the U.S. was still a non-belligerent. The intention would be to leave behind three to four battleships, nine light cruisers, thirty to forty destroyers, and twenty fleet submarines. Danckwerts was cool to the idea and replied that six to nine American battleships should be kept in the Pacific, in order to deter the Japanese.[38]

Throughout 1941, the two navies had to make difficult decisions. Fighting the Germans and Italians (for the British) or exercising sea power in the Atlantic (for the neutral Americans) had to be weighed against keeping sufficient forces in the Far East and the Pacific to deter the Japanese or to deal with the first weeks of a Japanese assault. The British COS put the dilemma clearly in their strategic review of 31 July 1941 (on the eve of the Argentia Conference): "The difficulty, of course, is to strike a balance between the essential needs of the theatres in which we are fighting, without neglecting the theatres where war may develop at any moment."[39]

The ABC-1 strategy envisaged Allied naval forces—whatever could be spared from the European war—being based on Hawaii and Singapore and boxing in the Japanese Combined Fleet. The eastern flank (the eastern Pacific) caused the British some concern. They were not sure the Americans were doing enough to deter the Japanese from advancing against Southeast Asia, or from stopping them if the attack really came about. As Vice Admiral Tom Phillips—Vice Chief of Naval Staff (VCNS) and future C-in-C of the ill-fated Eastern Fleet—minuted in March 1941, "a vast fleet at Hawaii making faces at Japan would be a waste."[40] The tasks of the U.S. Pacific Fleet under the WPL-46 war plan were indeed limited. The fleet was to raid enemy sea communications and positions and to "prepare" to capture the Marshall and the Caroline Islands in the Pacific (as well as the Azores and the Cape Verde Islands in the Atlantic). The details of the Pacific Fleet task implied operations (rather than just preparations) for "denial and capture of positions in the Marshalls," but to "[*p*]*repare* [my emphasis] to capture and establish control over the Caroline and Marshall Island area, and to [*prepare* to] establish an

[38]ADM 116/4877, Halifax to FO, 29 April. The American proposal to move the bulk of the U.S. Navy to the Atlantic was probably the result of what was seen in Washington as a worsening situation for the British. April 1941 was the time of the successful cruise of the battle cruisers *Scharnhorst* and *Gneisenau* in the Atlantic, and the collapse of the British position in the Balkans.

[39]AWP, Vol. 3, p. 337, "General Strategy," 31 July.

[40]ADM 116/4877, "Far East: Appreciation with US belligerent and Japan neutral," 10 March 1941

advance fleet base in Truk."[41] It was not clear if this would really worry the Tokyo government.

The Admiralty sent Rear Admiral Danckwerts on a flying visit to Pearl Harbor in April 1941, shortly after the completion of ABC-1. Danckwerts' chief aim was to impress on Admiral Husband Kimmel, C-in-C, Pacific Fleet (CINCPAC), and his staff the importance to Britain of the fleet in Hawaii undertaking offensive operations as soon as possible after the outbreak of war between America and Japan. This would allow it to achieve "maximum diversionary effect" when "the Associated [i.e. Allied] strength in the Far East would still be so far below the minimum necessary to maintain [the] position." In his official report to the COS, Danckwerts noted that Kimmel had not yet drafted war orders in response to ABC-1, and that he chose not to reveal existing Pacific Fleet plans to his British visitors. Kimmel had ruled out direct action against the home territory of Japan, in view of the distances involved and the limited value of isolated raids. "They have in mind," Danckwerts reported, "to undertake operations against Japanese bases in the Marshall and Caroline Islands as a kind of step-by-step offensive to the southwest, but they do not appear to have any definite plans for the purpose, nor detailed knowledge of Japanese defences and preparations there, nor the necessary means of overcoming Japanese shore resistance." Little thought had been given to a more direct threat to Japan: "Generally speaking they hold to a hope that the mere existence of the United States Pacific Fleet based on Hawaii will produce a containing effect on Japanese naval forces, but I do not think we can count on any active demonstrations in Japanese waters."[42]

Washington did contemplate naval activity to deter or divert Japan. In February 1941, the President had proposed a visit by a carrier, three or four cruisers, and a squadron of destroyers to Manila by way of the South Pacific. In April, Stark told Kimmel that FDR had the idea "to keep [ships] popping up here and there, and keep the Japs guessing." The Pacific Fleet war plan, WPPac-46, included a proposal to send two heavy cruisers on a raid west of the Nanpō Shotō, the island chain south of Japan which includes Iwo Jima; the cruisers would have refuelled at Midway.[43]

Both the British and the Americans considered the deterrent effect of potential carrier raids on Japan. The Admiralty's 1937 "War Memorandum (Eastern)" had stated that British forces "would be unlikely to take the initiative by embarking on

[41] RG 38, Box 147F, WPL-46, p. 28. On 12 August 1941, WPL-46 was modified to give the Pacific Fleet the additional task of defending Wake Atoll, a strategically important—if exposed—location in the Central Pacific.

[42] CAB 30/28, COS(41)308, 14 May. Danckwerts' first report, to the COS, was dated 17 April. His meetings in Hawaii were on 9 and 10 April. Any coolness on Kimmel's part might be explained by the fact that he had just (5 April) been informed by Washington that his fleet was to be reduced by a quarter to help the British in the Atlantic.

[43] RG 38, Box 147F, WPPac-46, p. I-4.

carrier-borne air attacks against open cities in Japan, although it is understood that Japan is particularly apprehensive of such action." Prime Minister Churchill himself in a letter to the Admiralty on 17 February 1941—during the period of the ABC talks—suggested that if war with Japan came, "some enterprise with aircraft-carriers strongly supported by fast ships against the Japanese homeland towns might be attempted."[44] In a briefing letter of 25 February 1941, Admiral Stark suggested to Admiral Kimmel at Pearl Harbor that he "study very carefully the matter of making aircraft raids on the inflammable Japanese cities (ostensibly on military objectives)." In April he returned to this idea of Kimmel sending a carrier and cruisers to the northwest (from Pearl Harbor): "[Y]ou can see what a striking force of the composition I gave you, and known to the Japs, would mean to them, in view of their unholy fear of bombing. This striking detachment would have been right in position for most [*sic*] anything." Twelve months later, of course, the Americans did indeed mount the famous—if physically ineffectual—carrier-based Doolittle raid on Japanese cities. Stark seems to have been ambivalent about the idea of such pre-war naval demonstrations; and he believed, as far as the State Department was concerned, that "it was a little too much for them."[45] Such operations (or overt preparations for them) might, however, have led the Japanese to rethink the commitment of large naval forces to Southeast Asia. As we will see below, the threat of carrier attacks on the Japanese homeland was not an insignificant factor in the development of the strategy of the Combined Fleet in 1941.

Nevertheless, forward deployments to the northwest did not feature in the actual plans of the U.S. Pacific Fleet. Two and a half months after the British visit, in July 1941, Kimmel and his staff completed the Pacific Fleet's operational plan (O-1 or WPPac-46), aimed to fulfil the tasks set out in Washington's WPL-46. This Pacific plan remained in effect through December 1941 and called for a movement of the whole Pacific Fleet to the southwest, into the Marshalls, with air strikes against Japanese installations there.[46] It has been suggested by the historian Edward Miller that Admiral Kimmel and his Chief of War Plans, Captain Charles H. McMorris, had developed in the course of the summer of 1941 a strategy to lure the Imperial Navy into a decisive battleship action in the Marshall Islands—and that this partly explains Kimmel's preoccupation with training and lack of attention to the defenses of Hawaii.[47]

---

[44]ADM 116/4393 (1937); ADM 116/4877, p. 172 verso (1941).

[45]PHA, part 16, p. 2149 (25 February), p. 2164 (19 April). On possible air attacks on Japan, see also Miller, *Orange*, pp. 320–21.

[46]RG 38, Box 147F, WPPac-46. For a summary of the planned operations, see Evan Mawdsley, *December 1941: Twelve Days that Began a World War* (New Haven: Yale University Press, 2011), pp. 106–08.

[47]Edward S. Miller, "Kimmel's Hidden Agenda," *Military History Quarterly* 4:1 (Autumn 1991), pp. 36–43; Miller, *Orange*, pp. 286–312. Miller admitted that his evidence for Kimmel's "battle design" was only "circumstantial."

Interesting as this theory is, Miller has not provided hard evidence to back it up. In reality, Kimmel was aware of the shortcomings of his fleet. In the spring of 1941 he had had to detach a quarter of his battleships to the Atlantic. The admiral also had to move a large proportion of his submarines westwards to the Philippines. He certainly had no Marine force or shipping to carry out offensive landing operations in the Marshalls. It seems much more likely that Kimmel was, as instructed, preparing to make a demonstration towards the Marshalls in connection with the general plan to distract the Japanese from the Malay barrier.

The British did not expect much to come of Kimmel's sortie into the Marshalls, although they were unable to obtain much information about it. In early October 1941, Rear Admiral R. Kelly Turner, Chief of the War Plans Division under Admiral Stark, refused "as a matter of secrecy" to reveal Pacific Fleet war plans (i.e. WPPac-46) to the British. Shortly afterwards—and only seven weeks before the actual outbreak of the war—the British Admiralty Delegation in Washington could only report in general about American plans for "harassing operations in the Mandates Area."[48]

For all the lack of information arriving in London about the use to which the nine battleships and three big carriers of the U.S. Pacific Fleet would be put, the western flank of the British-American pincer left even more to be desired. Until the very eve of the Pacific War, the British were unable to build up their naval strength based on Singapore. The Americans, meanwhile, did nothing to strengthen the surface-ship strength of the U.S. Asiatic Fleet in the Philippines.

As already mentioned, the ABC staff talks were the starting point for the new strategy. The situation, as secretly reported to major Royal Navy fleet commands on 13 May 1941, was that if the United States joined Britain in the war against Germany, and Japan entered the war, then a British fleet would be despatched to the Far East. It would be known as the "Eastern Fleet," and it would consist of six capital ships: five battleships (*Nelson, Rodney,* and three R class), and a battle cruiser (*Renown*). This would be a "balanced fleet," including the carrier *Ark Royal,* ten cruisers, thirty-two destroyers, and ten submarines.[49] Although not mentioned in

---

[48]CAB 122/8, Turner to Danckwerts, 3 October; CAB 122/8, BAD to Admiralty, 11 October. Turner noted that the situation would be different if both countries were at war in the Pacific; "we are never informed concerning the details of projected British Operations." This was a surprising remark given close existing American-British cooperation in the Atlantic.

[49]The surviving R class battleships were *Ramillies, Resolution, Revenge,* and *Royal Sovereign* (in later versions of the plan, all four vessels would be moved east). They entered service during the First World War and mounted 15-inch guns; a retrograde design compared to the Queen Elizabeth class, the R class could only make twenty-one knots. *Nelson* and *Rodney* were faster and armed with 16-inch guns. Commissioned in 1930 and 1927, respectively, they were regarded by the Admiralty as "modern" compared to ships completed or designed during the First World War. *Renown,* a sister ship of *Repulse,* dated from 1914–18, but she had been modernized. *Ark Royal* was a new fleet carrier.

the 13 May message, it was at this time that Vice Admiral Tom Phillips, then VCNS under Admiral Pound, was assigned as provisional commander of this formidable "Eastern Fleet." The U.S. Navy would meanwhile (on entering the war) have deployed in the Atlantic six battleships, three carriers, eight to thirteen heavy cruisers, four to eight light cruisers, sixty-two to seventy-seven destroyers, and twenty-two submarines (operating from Halifax, Bermuda, Gibraltar, and in UK home waters).[50]

The proposed Eastern Fleet was quite similar to the force the Admiralty had earmarked in August 1939 in the event Japan intervened when Britain was at war with Gemany and Italy (France would still have been an active British ally). At that time the ships to be sent from the Home Fleet and the Mediterranean consisted of seven battleships (*Nelson, Rodney,* two R class, and three Queen Elizabeth class), a carrier, eight cruisers, and thirty-five destroyers. The "China Station" already contained a considerable force: a carrier, four cruisers, fourteen destroyers, and fifteen submarines.[51]

The 13 May 1941 signal also explained that if Japan stayed neutral—despite America joining the war against Germany—it was probably only British Force H from Gibraltar (the battle cruiser *Renown* and the carrier *Ark Royal*) that would be sent east, and it would go to the Indian Ocean. The place of Force H at Gibraltar would be taken—on the American entry into the war with Germany—by three U.S. Navy battleships, a carrier, and four heavy cruisers.[52]

Admirals Pound and Phillips outlined the Admiralty's tactical view at meetings of the Cabinet's Defence Committee in late October 1941. The Japanese Navy, Phillips explained, like the Royal Navy, was "a mixture of old and modern ships." The four British R class were superior to the oldest Japanese battleships—by which Phillips must have meant *Ise, Hyūga, Fusō,* and *Yamashiro.* The R class, Pound stated, might have been used for convoy escort in the Indian Ocean, but "[t]he aim had always been to constitute a battle fleet with this as the nucleus."[53] (The R class were also actually comparable to seven of the American battleships at Pearl Harbor in December 1941.) Meanwhile, of larger battleships completed after the First

---

[50]ADM 116/4877, Admiralty to C-in-C, Home Fleet [etc.], Message 1904B/13, 13 May; ADM 199/1472B, Layton, "Supplementary Report . . . ," 25 April 1947.

[51]ADM 1/9767, "War Memorandum (Eastern)," August 1939, pp. 92–93. In 1939, the Admiralty believed it had to leave the battle cruisers and most of the aircraft carriers in the Atlantic to deal with fast German ships such as the battle cruisers *Scharnhorst* and *Gneisenau* and the "pocket battleships." In August 1939, there were also a considerable number of cruisers in the Australian and New Zealand navies, and in the Indian Ocean.

[52]ADM 116/4877, Admiralty to C-in-C, Home Fleet [etc.], Message 1904B/13, 13 May.

[53]CAB 69/2, DO(41)65, 17 October; CAB 69/8, DO(41)66, 20 October. The R class mounted eight 15-inch guns, and the four oldest Japanese ships twelve 14-inch guns. By 1941, they all had a similar full-load displacement of about 35,000 tons. The Japanese ships were four knots faster, and capable of twenty-five knots.

World War, *Nelson* and *Rodney* were believed to be superior to *Nagato* and *Mutsu*.[54] There was also probably an assumption that ship for ship, the British vessels were more effective.[55]

The next strategic developments were triggered by the events of three hectic weeks in the summer of 1941. Japanese forces moved into southern Indochina (24 July), severe American economic sanctions were imposed in reprisal (26 July), and President Roosevelt and Prime Minister Churchill met at the Atlantic Conference at Argentia, Newfoundland (9–12 August). Japan was taking more aggressive action and measures were required either to keep Tokyo's ambitions in check or to put forces in place if war began. For Whitehall, a special factor was that the Pacific Dominions felt more threatened and demanded tangible naval reinforcements.

We need to make a brief detour here—in July 1941—to outline the changing attitude of the Americans to the Philippines. Before this time, the attitude in Washington, including both armed services, was that the islands were indefensible—and that attitude was also essentially embodied in ABC-1 and RAINBOW 5. The July 1941 confrontation with Japan led the Administration and the U.S. Army to reconsider. General Douglas MacArthur was transferred back to the U.S. Army from the army of the Philippine Commonwealth, and preparations were made to send air-defense reinforcements and ground-force equipment to the islands. The U.S. Army Air Corps put forward a radical new strategy to use heavy bombers to defend the islands, deter the Japanese, and indeed dominate the western Pacific.[56] The U.S.

---

[54]The British ships mounted nine 16-inch guns, and the Japanese eight. They all displaced about 40,000 tons. The Japanese ships were four knots faster, and capable of twenty-seven knots.

[55]British intelligence regarding the Japanese Navy is a large and interesting topic which cannot be followed through here for reasons of space. The British had very limited knowledge of Japanese new construction of capital ships (let alone development of aircraft and torpedoes). They underrated their potential enemy in the Far East; the "War Memorandum (Eastern)" for August 1939 stated that allowance should be made for "the anticipated inferior efficiency of the Japanese Fleet (approximately 80 percent of our own)." See ADM 116/5757, "Naval Strength of Japan," also Christopher M. Bell, "The Royal Navy, War Planning and Intelligence between the Wars," in Peter Jackson and Jennifer Siegel, eds., *Intelligence and Statecraft: The Use and Limits of Intelligence in International Society* (Westport: Praeger, 2001), pp. 139–56, and Wesley K. Wark, "In Search of a Suitable Japan: British Naval Intelligence in the Pacific before the Second World War," *Intelligence and National Security* 1:2 (1986), pp. 189–211.

[56]American strategy in the Philippines is an important subject, but essentially outside the theme of *naval* strategy. Space does not permit a fuller treatment. See Russell F. Weigley, "The Role of the War Department and the Army," in Dorothy Borg and Okamoto Shumpei, eds., *Pearl Harbor as History: Japanese-American Relations, 1931–1941* (New York: Columbia University Press, 1973), pp. 165–88, and Louis Morton, *Fall of the Philippines* (Washington, DC: U.S. Army Center of Military History, 1953), pp. 14–50, 61–73. The best account of the air strategy is William H. Bartsch, *December 8, 1941: MacArthur's Pearl Harbor* (College Station: Texas A&M University Press, 2003).

Navy was cooler about a change of policy. Admiral Stark was the architect of the "Germany first" strategy, and the Pacific Fleet had just been reduced by a quarter. No preparations were made for the heavy ships of the U.S. Pacific Fleet to cooperate in the defense of the islands. The surface forces of the U.S. Asiatic Fleet were not increased beyond a token force (two cruisers), although there were also thirteen older destroyers. Despite the weakness of American naval forces in the Philippines, the Royal Navy, as we will see, took a growing interest in the islands.

The July 1941 crisis over Indochina, the demands of the Dominions for support, and the Atlantic Conference led to a British decision to send some naval forces to the Indian Ocean and the Far East (Phase I of the "Singapore strategy") when Britain and Japan were still at peace. The Prime Minister, who had just traveled to Argentia aboard the new battleship *Prince of Wales*, proposed in August that a similar ship be sent to the Far East, where its technical quality and speed would impress and deter the Japanese; Churchill had in mind replicating the impact on British strategy of the German battleship raiders in the Atlantic. Such a move, moreover, was one which could be effected almost immediately.

Admiral Pound and the professionals at the Admiralty had quite different ideas for a "peacetime" "Phase I."[57] They envisaged a strategy similar to the May 1941 ("wartime") plan, and effected gradually. What Pound proposed on 28 August was evidently the basic concept with which the Admiralty would be working over the next two months. Now all four R class battleships would be sent to the Indian Ocean as convoy escorts, starting with *Revenge* in September, and followed by the other three up to early January. The more powerful *Nelson* would arrive at Trincomalee (Ceylon) at the end of November, followed by *Rodney* at the end of January 1942; if a deterrent was necessary, these two big vessels might be moved forward to Singapore. The battle cruiser *Renown* would arrive in mid-January and the carrier *Ark Royal* in April 1942; the task of these two fast ships would be to hunt down raiding Japanese heavy cruisers.[58]

With these "peacetime" transfers (Phase I) planned for the coming six months in mind, the Royal Navy began in September 1941 to consider some remarkable operational deployments. A meeting of Admiral Tom Phillips (VCNS and C-in-C designate of the Eastern Fleet) and the chief planners was held in the Admiralty on 30 September 1941, and what was termed a "radical" new strategy for Phase II was discussed. Assuming sufficient screening destroyers (eight to twelve ships) could

---

[57]In the present context, "peacetime" means the period when Japan was not at war with Britain or the USA; "wartime" refers to a (hypothetical) situation in which Japan had entered the war. In the broader sense, of course, Britain had been at war—with Germany—since September 1939.

[58]Winston Churchill, *The Second World War*, Vol. 3 (London: Cassell, 1950), pp. 769–73, Pound to Churchill, 28 August. A signal was sent on this day to British fleet commanders informing them of the imminent movement of the R class (ADM 116/4877, Admiralty to C-in-C, Home Fleet [etc.], 1614A/28, 28 August).

be found to cover *Nelson* and *Rodney*, the Eastern Fleet might use Manila in the Philippines as an "operational base." Manila was a vital location north of the Malay barrier, the planners noted: "Its position on the flank of the Japanese communications to the China Sea makes it undoubtedly the best base from the strategical point of view." Some 450 miles from the nearest Japanese air base on Formosa, Manila was evidently not under danger of serious air attack.[59] Although this daring proposal must have been based on an awareness of the new American readiness to reinforce and defend the Philippines, nothing was said of the diplomatic complexities of basing a British fleet in a neutral American port. (This may well have been influenced by the growth of linked USN-RN operations in the Atlantic.)

Two days after Phillips' meeting, on 2 October, a revised memorandum was prepared for the British naval delegation (BAD) in Washington, as a non-binding "guide" to the Admiralty's view of Far Eastern naval strategy. This stated that since the U.S. Navy had already taken up its wartime deployment and tasks with respect to Germany, the Royal Navy had begun—at a time when Britain was not actually at war with Japan—the deployment of capital ships for the planned "Eastern Fleet" into the Indian Ocean; this move was to be completed by early 1942. Forces to be sent would include two "modern" capital ships (unnamed, but presumably *Nelson* and *Rodney*), an aircraft carrier, and a considerable number of destroyers. "[I]t may well be possible," the memorandum suggested, "to adopt more offensive courses of action in the early stages of the war than was previously considered desirable." The Philippines base was stressed: "The advantages of Manila as an advanced base for forces operating against the Japanese lines of communication are apparent and *consideration should consequently be given to its use as an advanced base by British forces including heavy ships.*"[60]

The plans circulating in the Admiralty in September and October provided the background to a better-known (and misunderstood) strategic development, the decision made by the Defence Committee on 20 October to send the new battleship *Prince of Wales* to the Far East.[61] The immediate and main cause of this event was—as in July/August 1941—a more threatening situation there. Prince Konoe

[59] ADM 116/4877, "Record of a meeting held in VCNS' Room," 30 September. The extent of the change in strategy is suggested by a note from Admiral Henry Harwood, the Assistant Chief of Naval Staff (Foreign), to the First Sea Lord, supporting Manila as an advanced base from which the Eastern Fleet would "probably" operate: "we have completely changed our outlook and come much nearer to the American viewpoint, i.e. to operate our ships North of the Malay Barrier" (ADM 116/4877, Harwood to Pound, 29 October 1941).

[60] ADM 116/4877, "Aide-Memoire for use by the British Representatives and Naval Planning Conference Singapore," 2 October. The underlined words were a pencil amendment to the original typescript passage which had read, in part, "and it is therefore desirable that as strong a British force as possible should be based on Manila as an advanced base. . . ."

[61] The best source on the whole tragedy of Force Z (*Prince of Wales* and *Repulse*) remains Marder, *Friends/Enemies*, pp. 213–41, 365–521.

resigned as Japanese Prime Minister on 16 October (his ministry was replaced, two days later, by a new one led by General Tōjō Hideki); there was also a fear that new Russian setbacks in the war with Germany might lead Japan to attack the Soviet Far East. Once again, the civilians desired to take some quick action to deter the Japanese, allay the fears of their Dominion partners, and demonstrate their strength and willingness to the Americans.

At the Defence Committee meeting on 20 October, Eden and Churchill had returned to the strategy of immediately sending a new battleship to the Far East as the core of a small deterrent squadron; the Admiralty persevered with the gradual build-up of a balanced fleet of older ships.[62] Eden argued that a "really modern ship" would have "a far greater effect politically than . . . a number of the last war's battleships." If that ship were to call publicly at Cape Town, "news of her movements would quickly reach Japan and the deterrent effect would begin from that date." The argument that the First Sea Lord and Vice Admiral Phillips made was that a force of six slower battleships would still act as a deterrent. The Japanese would have to commit their entire battle fleet to deal with it, and this in turn would uncover the home islands; therefore their hands would be tied.[63] Meanwhile, Pound, Phillips, and the Admiralty were reluctant to send one of the newest battleships to the Far East. This was not because they feared that the move would put such a ship at risk of Japanese attack, but because of the continuing threat of the German battleship *Tirpitz* in the Atlantic.

This time the politicians won the argument. Five days after the Defence Committee meeting, on 25 October, *Prince of Wales* departed the Clyde, with Vice Admiral Phillips aboard. *Repulse,* which was already on station in the Indian Ocean, would meet her in Ceylon.[64] In order to enhance the deterrent effect, the

[62] An Italian torpedo plane inflicted a torpedo hit on *Nelson* on 27 September 1941. This underrated event precluded the powerful battleship being sent, as originally proposed, to Ceylon in late November 1941, and weakened the proposals of the Admiralty. *Nelson* and *Rodney* together, with destroyers and a modern carrier, would have constituted a significant force. As Admiral Phillips pointed out, these six ships would have been a close match for the Japanese battle line in speed and gun power. Pound had only a very limited grasp of the potential of Japanese carrier air power, even less that of land-based naval air power, but he was not alone in this.

[63] According to the minutes of the meeting, Pound did not, in the discussion with the civilian members of the Defence Committee on 20 October, mention the possibility that this balanced fleet would use an advanced base in the Philippines. Ian Cowman has argued that this is indicative of an element of deception or conspiracy on the part of the 1st Sea Lord, who shared the Admiralty planners' obsession with "forward" Phase II operations; Cowman's case is not convincing (*Dominion or Decline,* pp. 247–55).

[64] The older battle cruiser *Repulse* was protecting convoys in the Indian Ocean. The force despatched to the Far East was originally supposed to include the new aircraft carrier *Indomitable* (commissioned as recently as 10 October 1941). This was consistent with the

movement of the British naval reinforcements was widely publicized, notably by the Prime Minister in a speech on 10 November. Owing to the improved naval situation in the Atlantic and the Mediterranean, Churchill declared, and "[owing] to the completion of our splendid new battleships and aircraft carriers of the largest size, . . . we now feel ourselves strong enough to provide a powerful naval force of heavy ships, with its necessary ancillary vessels, for service if needed in the Indian and Pacific Oceans . . . [T]his movement of our naval forces, in conjunction with the United States Main Fleet, may give a practical proof to all who have eyes to see that the forces of freedom and democracy have by no means reached the limits of their power."

The Times on 20 November reported that *Prince of Wales* was in Cape Town, and on 3 December the newspaper reported the battleship's arrival "with other units of the fleet" in Singapore and the creation of the "Eastern Fleet in Being" commanded by Vice Admiral Phillips.[65]

Despite the highly visible movement of *Prince of Wales,* the Admiralty plan in December was still to send a substantial fleet to the Far East. As recalled in 1942 by Admiral Layton, the C-in-C, China Station (until 3 December 1941), the expectation had still been "[t]hat by about 80 days after the outbreak of hostilities an eastern Fleet would be available which, although not capable of tackling the entire Japanese Fleet, should be able to force the enemy to discontinue a close investment of Singapore."[66] (One means of stopping—or deterring—the "close investment" would have been to threaten Japanese lines of communication from a forward base in Manila.)

On 5 November, Pound sent details to the Americans, conflating the two British strategies (Eden and Churchill's small deterrent squadron versus the Admiralty's larger balanced fleet), and indicating that the Royal Navy was forming a "capital ship force in Far Eastern waters before the outbreak of war with Japan." The ships included *Prince of Wales,* the old battleship *Revenge,* and the battle cruiser *Repulse* (the last two already in the Indian Ocean); they were to be be followed by *Royal Sovereign* and *Ramillies* (leaving the Atlantic in November) and *Resolution* (leaving in December). Pound also proposed that British ships would be based north of the Malay barrier: "In our review, the question has been raised of basing this force

---

original Admiralty plan of May 1941, although it was believed that the armored *Indomitable* was better suited to high-risk Mediterranean operations, and the unarmored *Ark Royal* better suited to the low-risk Far East. *Indomitable* was damaged by running aground on 3 November while working up in the West Indies. Even if this accident had not occurred, the sinking of *Ark Royal* by a U-boat on 13 November would probably—given the carrier shortage—have led to the retention of *Indomitable* in European waters.

[65]BBC Written Archives Centre, printed in Martin Gilbert, ed., *The Churchill War Papers,* Vol. 3 (London: W. W. Norton, 2000), p. 1426; *The Times,* 20 November, p. 4, 3 December, p. 4.

[66]ADM 199/1472A, "Remarks . . . [draft]."

on the Philippines with the object of attacking Japanese lines of communications. Experience [*sic*] seems to show that the air threat to Manila from Formosa may not be quite so serious as was previously felt."[67] The First Lord also requested that the Americans provide destroyers as escorts.

On 7 November, Admiral Stark's view of the situation was transmitted to London. He agreed "that the need exists for prompt action by both the British and the United States." He welcomed the creation of the Eastern Fleet under British command, but hoped it would establish its authority over Dominion naval forces. He stated that the island of Luzon (in the Philippines) would be available for use "if the United States were at war," but warned that Manila (on Luzon) might not be a safe base: "it is questionable . . . whether or not heavy naval surface forces of a strength markedly inferior to the Japanese should be initially stationed as far North as Manila, because of the danger to their lines of communication, and the lack of adequate base facilities and supplies." The CNO did agree that two divisions of American destroyers could be assigned to the British Eastern Fleet if the U.S. was at war with Japan.[68]

Pound relayed this message to Phillips two days later, reporting that arrangements with the Americans were now developing rapidly and stressing the need for an early meeting with the representatives of Australia and New Zealand to agree strategy. "I consider," he wrote, "that the sooner you can get to Singapore and hold the meeting . . . the better, as this is a necessary preliminary to finally clear up the Anglo-American plans. I am anxious not to lose any time in responding to Admiral Stark's gesture." Phillips arrived in Singapore on 29 November (he traveled by flying boat from Ceylon ahead of *Prince of Wales* and *Repulse*); four days later he flew off on a secret visit to consult with Admiral Thomas Hart (C-in-C, U.S. Asiatic Fleet) in Manila. The substance of the talks in Manila (in which General MacArthur also took part) involved not how to deal with an imminent attack by Japan—an unexpected event which would occur five days later—but rather the practicalities of basing British ships in the Philippines in about March 1942.[69]

Before leaving Allied naval strategy on the western flank, a further, and under-researched, subject should be mentioned: the large American submarine flotilla in the Philippines and the complete absence of any British submarines in the Far East. The British, in a stillborn planning document of August 1941, had suggested that "allied submarines . . . working in conjunction with our air forces, constitute our

[67]ADM 116/4877, Admiralty Message 1559A/5 to BAD, 5 November. Pound was presumably referring to "experience" of the Royal Navy using bases in European waters.

[68]ADM 116/4877, Ghormley to Pound, 7 November. Interestingly, Stark did not mention the danger of *air attack* on a fleet base at Manila. A USN destroyer division comprised four ships; three destroyer divisions were based in the Philippines.

[69]ADM 116/4877, Pound to Phillips, Message 1954A/9, 9 November; Naval Operational Archive, Hart Papers, Box 3, "Report of a Conference" [6 December]. The Japanese Navy evidently did not know about Phillips' visit to Manila until after 1945 (PWP, p. 266 (Chihaya)).

most important weapon for attacking Japanese seaborne forces." "Allied submarines" was presumably a reference to American submarines, as the sizeable British submarine flotilla had been withdrawn to the Mediterranean in 1940. (There were, however, also a number of Dutch submarines.) In his important message on joint strategy, transmitted to the First Sea Lord on 7 November 1941, Admiral Stark reported that this was an area where the Philippines were being strengthened: "The United States Navy is reinforcing the Asiatic Fleet with twelve modern submarines, eight of which departed Hawaii on the 24th [of October]. The remainder departed November 4th." Unfortunately, when war broke out the American submarines in the Philippines (and later in the NEI) had almost no success in attacking or even locating invading Japanese forces; only the Dutch boats had any effect.[70]

The most extraordinary—and most studied—new naval strategy developed by either side in the Pacific in the second half of 1941 was Admiral Yamamoto Isoroku's "Hawaiian Operation," a very long-range strike by six aircraft carriers against Pearl Harbor.[71]

Yamamoto had evidently begun to contemplate recasting the strategy of the Combined Fleet in the summer of 1940, at a time when the Navy leadership in Tokyo—the Navy Minister (Admiral Oikawa) and the Chief of the Naval General Staff (Prince Fushimi)—committed their support to a southward offensive that would take advantage of events in Europe. Shortly before 27 September 1940, the date of the signature of the Tripartite Pact (which Yamamoto opposed), he had his now well-known meeting with the Prime Minister, Prince Konoe, at the private residence of the latter outside Tokyo. The subject was the Navy's prospects in the event of war with the United States. This conversation is often misinterpreted. "If I am told to fight regardless of the consequences," the admiral reportedly said, "I shall run wild for the first six months or a year, but I have utterly no confidence for the second or third year."[72] This comment is often quoted, partly because it was so prescient—the Combined Fleet did indeed overrun the western and central Pacific in the first six months of the war, only to be smashed at Midway in June 1942. The

[70]CAB 122/8, ADB-2, p. 8 [August 1941]; ADM 116/4877, Ghormley to Pound, 7 November. In his 1942 draft "Report" to the Admiralty, Admiral Layton would be especially critical of the failure of the U.S. Asiatic Fleet submarines (ADM 199/1472A). The subject is discussed in Clay Blair, *Silent Victory: The U.S. Submarine War against Japan* (Philadelphia: Lippincott, 1975), pp. 127–62. See also: Hart Papers, Box 1, "War Activities Submarines, US Asiatic Fleet," 1 April 1942.

[71]The fullest account of Pearl Harbor planning, based in part on unique access to many of the Japanese principals, remains Prange, *Dawn.*

[72]Quoted in *Reports of General MacArthur,* Vol. II, Part I (Washington, DC: Department of the Army, 1966), p. 33, note 1. Agawa, *Reluctant Admiral,* p. 189, includes Yamamoto's account of the conversation, from a letter to Admiral Shimada Shigetarō; it is very similar. According to this version, the admiral was summoned by Prince Konoe and attended with some reluctance.

quote is also sometimes used—incorrectly, I would suggest—to establish Yamamoto's credentials as a realist.

In fact, what Admiral Yamamoto was opposing was the conventional strategy of the Imperial Navy. The Navy, he was arguing, would be able to make spectacular gains in the first six months against the small peacetime enemy forces committed to the Western Pacific. It would not be able to deal with the counterattack, when it came, by relying on the "attrition/decisive battle" strategy of the 1920s and 1930s. Yamamoto was more explicit in a letter to Navy Minister Oikawa in January 1941. The Combined Fleet was training for a set-piece battle, with each type of ship having a pre-determined role, but such a battle might well never occur in a war with the United States and Britain: "I think there may be no such case happening throughout the whole period of an expected war as all of the Combined Fleet closing in [*sic*] an enemy force, deploy, engage in gunnery and torpedo duel, and finally charge into the enemy force in as gallant a way as possible." Although exhaustively studied through (map-based) strategic exercises, the "attrition/decisive battle" strategy had never led in these simulations to "overwhelming victory" and indeed the map exercises were called off if it looked like they were leading to a situation where the Imperial Navy "might be dragged into gradual defeat." The Navy had instead, Yamamoto urged Oikawa, to prepare for the unexpected.[73]

Yamamoto was also, in his January 1941 letter to Oikawa, critical of ideas that were developing in Tokyo for concentration of an operation against British and American bases in Malaya and the Philippines (the "Southern Operation"):

> [When] we take a defensive stand toward the east [i.e. toward America] and await the enemy coming on . . . we cannot rule out the possibility that the enemy would [instead] dare to launch an [aircraft-carrier] attack upon our homeland to burn our capital city and other cities.
>
> If such happens, our Navy will be subject to fierce attacks by the public, even when we succeed in the southern operation. It is evidently clear that such a development will result in lowering the morale of the nation to such an extent that it cannot be recovered.[74]

[73] Donald J. Goldstein and Katherine V. Dillon, eds., *The Pearl Harbor Papers: Inside the Japanese Plans* (hereinafter PHP) (Washington, DC: Brassey's, 1993), pp. 115–17 (Yamamoto to Oikawa, 7 January 1941); Agawa, *Reluctant Admiral,* pp. 219–20; Koda Yoji, "A Commander's Dilemma: Admiral Yamamoto and the 'Gradual Attrition' Strategy," *Naval War College Review* 46:4 (Autumn 1993), pp. 63–74; Evans and Peattie, *Kaigun,* pp. 471–79.

[74] Yamamoto said that he thought the new C-in-C of the Pacific Fleet, Admiral Kimmel, would be active and offensively-minded. Although Yamamoto may have mentioned the question of carrier-based air raids as a means of convincing Oikawa, he may also have taken the threat of Japanese vulnerability seriously. He had mentioned the same threat in a letter to Baron Hirada on November 1940: "[I]f . . . Tokyo is completely destroyed by fire . . . and if I must witness it [from a base in the Inland Sea] while waiting for a strategically opportune

As we have seen, given discussions by American or British planners, this fear of a raid on Japanese cities was not completely unfounded.

Yamamoto had a radical strategic proposal to make to Oikawa. "The most important thing we have to do first of all in a war with the U.S., I firmly believe, is to fiercely attack and destroy the U.S. main fleet at the outset of the war, so that the morale of the U.S. Navy and her people [*sic*] goes down to such an extent that it cannot be recovered." Specifically, Yamamoto proposed to attack and physically block the fleet base at Pearl Harbor. The Imperial Navy, Yamamoto argued, should not be too concerned with the western Pacific, as American and British forces there were weak and poorly trained; the Philippines and Singapore should be attacked at the same time as Pearl Harbor, but they would be quickly overrun "if and when the U.S. main force is destroyed."[75]

Yamamoto came back to this justification ten months later, in an October 1941 letter to the new Navy Minister, Admiral Shimada. By this time he was making the case not so much for the Hawaiian operation (which was now far along in preparation and training), but for the commitment of maximum strength—all six fleet carriers—to ensure its success. "[T]he only way is to have a powerful air force strike deep at the enemy's heart at the very beginning of the war and thus to deal a blow, material and moral, from which it will not be able to recover for some time."[76] Yamamoto had somewhat shifted his argument since the January 1941 letter to Oikawa. Rather than stressing the impracticality of the conventional "attrition/decisive battle" strategy, he now dwelt on to the supposed high cost of the "southern operations," i.e. the attacks on Singapore, the Philippines, and the NEI.

---

time, I cannot remain still" (quoted in Prange, *Dawn*, p. 13). And he would repeat the argument in September 1941 in conversation with Kusaka and Ōnishi, two air officers who had become dubious about the Hawaiian operation: "[W]hat would you do if, while we were engaged in the South Pacific, the U.S. fleet launched air raids on Japan from the east? Are you suggesting that it's all right for Tokyo and Osaka to be burned to the ground so long as we get hold of oil?" (Agawa, *Reluctant Admiral*, p. 229).

[75]PHP, pp. 115–17 (Yamamoto to Oikawa, 7 January 1941). Yamamoto had apparently made the same point verbally to Oikawa in November 1940.

[76]Ibid., pp. 118–20 (Yamamoto to Shimada, 24 October 1941). Yamamoto did not specifically refer here to Pearl Harbor or Hawaii. It is not clear what point Yamamoto was making, as he had evidently already won the argument about using *all* the big carriers in the Hawaiian Operation by threatening to resign on 18 October.

"The resulting situation . . . ," Yamamoto stated, "will find the Navy's strength stretched beyond the limit of its capacity." In the letter, the admiral also made the point that "however smoothly the southern operations may progress, we are bound to suffer considerable losses in heavy cruisers [i.e. these were the ships of Admiral Kondō's 2nd Fleet] and aircraft." Especially due to the loss of aircraft, and the inability of Japanese industry to replace them, "it will be virtually impossible for the Navy to meet the needs of the major sea operation that is likely to follow."

Yamamoto complained in this letter that the argument had been made in the Naval General Staff that the Pearl Harbor raid was "nothing more than a secondary operation in which the chance of success was about fifty-fifty" and that therefore "the use of the entire air force [i.e. all six carriers of 1st Air Fleet] in such a venture was too risky to merit consideration." He returned to the danger of American carrier raids on Japanese cities ("Tokyo and Osaka").[77]

The main point to be made here is that what the C-in-C of the Combined Fleet was actually proposing was a decisive battle, staged farther east than had ever been considered in the inter-war plans. Alan Zimm has correctly described the Pearl Harbor attack, as seen by Yamamoto, as "an entirely new strategy to win the war."[78] However, the objective was not just, as Zimm has recently argued, "to cripple the Pacific Fleet sufficiently to prevent it from moving against the flank of the Japanese advance for at least six months."[79] Covering the eastern flank may well have been the intention of Admiral Nagano and the Naval General Staff, when they (belatedly) sanctioned a raid against Pearl Harbor. Indeed, that is what really happened between December 1941 and April 1942. But Yamamoto's view was that the Hawaiian Operation would have the decisive effect of breaking the will of the enemy ("to fiercely attack and destroy the U.S. main fleet at the outset of the war, so that the morale of the U.S. Navy and her people goes down to such an extent that it cannot

---

[77]This argument is puzzling. American, British, and Dutch naval forces in the Far East were still very weak in October 1941, and in reality Japan would lose very few ships in the three-month campaign. Yamamoto may have assumed that the Japanese Navy's land-based and carrier air groups would suffer heavy losses from land-based British and American aircraft and anti-aircraft fire in Malaya and the Philippines. At the Imperial Conference on 5 November, Admiral Nagano would predict that naval aircraft losses in the Southern Operation might reach one-third or one-half of aircraft committed, on top of considerable losses of merchant ships (JDW, p. 235). As it turned out, both Yamamoto and Nagano were too pessimistic.

[78]Alan D. Zimm, *The Attack on Pearl Harbor: Strategy, Combat, Myths, Deceptions* (Havertown, PA: Casemate, 2011), p. 366. Zimm argues that Yamamoto "sold" the Pearl Harbor concept to the NGS as a means of immobilizing the Pacific Fleet for six months, which is presumably what Zimm means when he refers to it as the admiral's "avowed [i.e. ostensible] objective" (p. 373).

[79]Zimm, *Attack,* p. 373. Zimm suggests that this "put the torch to conventional Japanese plans for a decisive battle between the fleets at odds that would allow a Japanese victory." His conclusion is that "[t]he most telling indictment against the Japanese strategists and intelligence service [*sic*] is that they did not need an attack on Pearl Harbor to get their needed six months." Logistic constraints and lack of oilers and other supply ships meant, Zimm suggests, that even an undamaged American fleet could not have begun major operations for six months. Zimm's logic breaks down when he suggests Yamamoto became aware of this and that was why he moved towards a decisive battle at Midway. My argument would be that Yamamoto got it disastrously wrong twice, in the unsuccessful decisive battle of December 1941 and the counter-productive decisive battle in June 1942.

be recovered").[80] It conflated the attrition and ambush phases of the classic strategy, and moved the site of the decisive battle to Hawaii.

To achieve decisive results, Admiral Yamamoto needed to concentrate all the big carriers of the Combined Fleet (four vessels—*Kaga, Akagi, Hiryū,* and *Sōryū*—in the initial planning—six when the new *Shōkaku* and *Zuikaku* joined the fleet, in August and September 1941 respectively). He also concentrated the largest part of his submarine force—some twenty-six big cruiser boats—in the Hawaiian area.[81] Even some of the high-speed "midget" submarines of the A class, conceived in the late 1930s and intended for open-seas attrition, were pressed into service for a penetration of the American naval base.

Not the least remarkable feature of the Hawaiian Operation was that it developed as a kind of "private venture," distinct from the strategy of the central authorities of the Imperial Navy. The leadership of the military in Imperial Japan was, as has been noted, badly fragmented.[82] Yamamoto flew his flag aboard the battleship *Nagato,* moored at the western end of the Inland Sea, 450 miles from naval headquarters in Tokyo. As C-in-C of the Combined Fleet, he regarded himself—especially in wartime—as directly subordinate to the Emperor, rather than to the Naval General Staff and Admiral Nagano.[83]

---

[80]On Pearl Harbor as a decisive battle, see also Marder, *Friend/Enemies,* p. 332, note 46, bringing in the Japanese official history, *Senshi sōsho.* A different view is taken by Evans and Peattie, who argue that "[f]or Yamamoto, as for the rest of the Japanese high command, the attack on the US Pacific Fleet at Hawaii was a secondary operation whose purpose was to ensure the security of the main Japanese objective: Southeast Asia" (*Kaigun,* p. 488).

[81]On submarine operations, see *Japanese Monograph* No. 102, *Submarine Operations: December 1941—April 1942,* reprinted in PWP, pp. 230–43. The most recent book on the subject is of limited value: Burl Burlingame, *Advance Force Pearl Harbor* (Annapolis: Naval Institute Press, 2002). Only a small number of submarines were required for the Southern Operation, deployed as a patrol line in the South China Sea between Singapore and the invasion area in northern Malaya and southern Thailand. The Hawaiian submarine concentration was not, evidently, a subject of a headquarters disagreement between the NGS and the Combined Fleet.

[82]On the breakdown of the naval establishment in the 1930s, see Asada, *Mahan,* pp. 161–86. As Asada Sadao described the situation in the 1930s: "in the navy the policymaking process came to be marked by diffusion, dispersion, and decentralization of authority and power" (Asada Sadao, "The Japanese Navy and the United States," in Borg and Okamoto, *Pearl Harbor,* p. 225).

[83]A not dissimilar situation existed in the U.S. Navy. When the British planners taking part in the ABC-1 talks requested information about Pacific Fleet operating plans they were told that Admiral Kimmel in Pearl Harbor would have to be consulted, not the Chief of Naval Operations in Washington. Admiral Danckwerts and Captain Clark traveled to Hawaii to do so (Cowman, *Dominion or Decline,* p. 216).

PWP, pp. 103–04. This account of Nagano's reaction is from a paper prepared by Ōmae Toshikazu, a member of the Naval Affairs Bureau of the Navy Ministry. Ōmae's account was apparently based on Fukudome's recollections, and it was supported by Rear Admiral

In the summer and early autumn of 1941, Admiral Nagano and the planners in Tokyo were concentrating on the Southern Operation and coordination with the Army. They were planning the advance against British and American bases in Malaya and the Philippines, and towards the resources of Southeast Asia. It is not entirely clear when Nagano became aware of what was in effect the rival Hawaii project. After the war, a number of former Japanese officers would recall a date as late as mid-October 1941. In a post-war account it was said that Rear Admiral Fukudome, Chief of the 1st Bureau of the NGS, gave a negative report on the Hawaiian Operation plan to Admiral Nagano, who replied only that he thought the operation "risky."[84] In any event, the Hawaiian operation was regarded in Tokyo as a "raid," an auxiliary strike intended to inhibit—for a certain period of time—the movement west of the U.S. Pacific Fleet. It was not something which justified

---

Tomioka, another of the surviving war planners. Fukudome's objections to the Hawaiian plan are detailed in his "Hawaii Operation" in O'Connor, *Japanese Navy*, pp. 6–8. (See note 23, above.)

It is very difficult to believe that before this time Combined Fleet headquarters could have kept its extensive planning and training for a mass carrier raid on Hawaii secret from the NGS, and from Admiral Nagano in particular. Two chiefs of staff of the Combined Fleet (who had to be fully aware of Pearl Harbor planning) were transferred to senior posts in the NGS in this period; Fukudome left in April 1941 to become chief of the NGS 1st Bureau, and Ito Seichi on 1 September 1941, to become Vice Chief of the NGS.

Rear Admiral Ōnishi, by then chief of staff of the 11th Air Fleet, presented a draft Hawaii plan to the newly transferred Fukudome at the NGS in April 1941 (Prange, *Dawn*, pp. 103–04). Captain Kuroshima, Yamamoto's senior staff officer and a key Pearl Harbor planner, visited the NGS in April and August 1941 and discussed the operation with Captain Tomioka, Chief of the Operations Section of the NGS (Prange, *Dawn*, pp. 104–05, 181–84; Agawa, *Reluctant Admiral*, pp. 223–24).

The creation in April 1941 of the 1st Air Fleet—the core of Admiral Nagumo's Mobile Force, the task force which attacked Pearl Harbor—was not in itself a giveaway as to the existence of the Hawaiian Operation. The air fleet's formation was linked to training and technical aviation matters, rather than operational ones (e.g. the Pearl Harbor raid). Formation of the 1st Air Fleet was apparently suggested on his own initiative by Rear Admiral Ozawa, commander of the 1st Carrier Division, and only belatedly approved by Yamamoto and Nagano (PWP, pp. 78–79 (Ozawa), p. 285 (Chihaya); O'Connor, *Japanese Navy* (Fukudome), p. 6).

[84] Michael Barnhart stated that "Yamamoto had submitted his bold plan to Nagano in May" ("Japanese Intelligence before the Second World War: 'Best Case' Analysis," in Ernest R. May, ed., *Knowing One's Enemy: Intelligence Assessment before the Two World Wars* (Princeton: Princeton University Press, 1986), p. 448). It would not be improbable that Nagano knew informally about the Pearl Harbor plan by early summer, but Barnhart did not provide hard evidence. Less convincing is Barnhart's suggestion that the NGS opposed the Hawaiian Operation because the elimination of the U.S. Pacific Fleet would have removed the Navy's justification for the attack on the Philippines. In reality, more important causes of opposition were the competing requirements of the Southern Operation and the high risk of the Pearl Harbor venture.

risking all the main carrier forces of the Combined Fleet. The overall war planning document discussed at the Imperial Conference on 15 November 1941, and presumably prepared with the involvement of Admiral Nagano, was still couched in traditional terms: "At the appropriate time we will endeavour by various means to lure the main fleet of the United States [near Japan] and destroy it."[85] Nagano had also said on 5 November that he expected a prolonged war, and even if the IJN succeeded in defeating the combined fleets of Britain and the United States, "the war will continue long after the Southern Operation."[86]

One factor which affected the attitude of the NGS in Tokyo were the weighty implications of the Southern Operation, first conceived in mid-1940. The requirements of Vice Admiral Kondō, C-in-C of the Southern Force, were becoming clearer. Kondō was a powerful figure in his own right; he had served as Vice Chief of the Naval General Staff, under Nagano, before moving to take command of the 2nd Fleet (the core of the future Southern Force) on 1 September 1941. The NGS wanted to assign him at least three of the best-trained large aircraft carriers, and half the tankers Yamamoto had earmarked for the Hawaiian Operation.[87] Such a disposition would have left Admiral Nagumo's Hawaii attack fleet, the Mobile Force, with two new and untrained big carriers, and perhaps one more experienced ship.

In the end, however, Admiral Yamamoto got his way and kept his mass striking force, although he had—in mid-October—to threaten resignation.[88] His letter of

---

[85] Evans and Peattie, *Kaigun,* p. 482; JDW, pp. 247–48.

[86] JDW, p. 233. It is not clear whether Nagano was thinking here about a defeat of local forces in the western Pacific (the reinforced fleets in the Philippines and Singapore), or the traditional decisive victory over the main force of the U.S. Pacific Fleet. The former case was more likely.

Prange, *Dawn*, pp. 223–31, 280–86, 295–300. The 1st Air Fleet was divided into three carrier divisions. The 1st Carrier Division consisted of *Kaga* and *Akagi*, and the 2nd Carrier Division consisted of *Hiryū* and *Sōryū*; all four ships had been in service for some years and had veteran air groups. The ships of the 3rd Division, *Shōkaku* and *Zuikaku*, were commissioned only in the summer of 1941, and their air groups had only limited training. In the map exercises held aboard the Combined Fleet flagship on 9–11 October, the Mobile Force included *Kaga*, *Shōkaku*, and *Zuikaku*, while *Akagi*, *Hiryū*, and *Sōryū* were assigned to the Southern Force.

[87] It needs to be kept in mind that the Southern Force was made up of ships from Yamamoto's Combined Fleet, and was subordinate to him. The 2nd Fleet was essentially the powerful heavy cruiser element of the Combined Fleet which, supported by two fast battleships and two small carriers (*Ryūjō* and *Zuihō*), eventually made up the Southern Force. Kondō's predecessor as C-in-C of the 2nd Fleet (Rear Admiral Koga Meneichi) had in the spring of 1941 also been "dead against" the concentration of all carriers in the 1st Air Fleet. Koga's opinion was that each of the two main fleets (i.e. the 1st Fleet and 2nd Fleet) should have its own carrier division—two vessels—to provide fighter cover (PWP, pp. 78–79 (Ozawa)).

[88] Captain Kuroshima, Yamamoto's first staff officer, presented this ultimatum to the NGS on 18 October. The fullest account is Prange, *Dawn,* pp. 296–300.

24 October to Admiral Shimada (who had just been appointed as Navy Minister in the new Tōjō cabinet) was part of a political campaign to ensure that all the big carriers would be committed to the Hawaii strike. The NGS did not feel ready to cope with such a crisis in the fleet command. A combination of Yamamoto's threat; perceived continued weakness of British, Dutch, and American forces in the Far East; and Japanese confidence in the capabilities of their land-based Navy attack planes, with long-range fighter escort, was sufficient to convince the naval authorities in Tokyo to accept the Combined Fleet's plans.[89] Yamamoto's ultimatum and the final commitment of all six carriers to the Hawaiian Operation occurred about three weeks before Churchill announced the commitment of modern ships to the Far East (on 10 November). This new development did not lead to a basic change in Japanese plans, although some land-based attack planes were redeployed from Formosa to Indochina (for use in the South China Sea, rather than in the Philippines), and Admiral Kondō's covering force of two fast battleships and a number of heavy cruisers was moved 1,000 miles south, from the Pescadores to Camranh Bay in Indochina.

Conclusion: "Strategical cloud cuckoo land"?

The British Vice Admiral K. G. B. Dewar condemned "British naval policy in Far Eastern waters [as] based on unrealistic threats and imaginary deterrents, conceived in Mr Churchill's strategical cloud cuckoo land."[90] Assigning all or even much of the blame to Churchill for the loss of *Prince of Wales* and *Repulse* is unfair; ultimately responsible were tactical decisions made by Admiral Phillips on 8–10 December.[91] The larger point, however, is that Pacific Ocean naval planning on all sides was lacking in realism.

Admiral Yamamoto believed that the balance of global naval and economic power made unfeasible Japan's 1920s/1930s strategy of fighting a prolonged naval war with the United States, based on attrition followed by decisive battle. The experience of 1942–44 was to prove him right. Yamamoto's "solution" to the strategic problem was, however, no more viable. The "decisive battle" attempted at Pearl Harbor by the aircraft and submarines of the Combined Fleet failed to destroy the U.S.

---

[89]Shimada Koichi, "Air Operations in the Philippines," in O'Connor, *Japanese Navy*, pp. 31–33.

[90]Quoted in Marder, *Friends/Enemies*, Vol. 1, p. 231; Marder provided no source. Dewar (1879–1964) was a rather controversial officer who had retired from active service in 1929.

[91]Ignorance of the the potential of the Japanese Navy's land-based torpedo bombers was a major factor, and was not unique to Admiral Phillips. He did in fact withdraw *Prince of Wales* and *Repulse* to the south as soon as he knew he had been sighted by Japanese scout planes (on 9 December). But Phillips had also, with a screen of only three destroyers, rashly ignored dangers from submarines and the powerful Japanese escort force of fast battleships, cruisers, and destroyers.

Pacific Fleet or break the American will to fight; it caused only limited destruction and unified the American public around a war of revenge.

The American government, for its part, was unable to develop a coherent Pacific strategy, or even a "joint" strategy which the Army and Navy could agree upon. President Roosevelt, against the advice of his military leaders, allowed the U.S. to be swept into a confrontation in the Far East for which it was unprepared. If anything deserves the description "strategical cloud cuckoo land" it was the U.S. Army's attempt to dominate the western Pacific with strategic air power. The U.S. Navy was generally more realistic; it did not in 1941 intend a deep drive across the Pacific on the lines of the extreme ORANGE war plans. But Admiral Kimmel still allowed limited operations against the Marshall Islands to take priority over securing his main base at Pearl Harbor.

The Royal Navy, equally, had a global strategy that did not correspond to its resources, and its planning contained a fatal underestimation of Japanese capabilities. Sending *Prince of Wales* and *Repulse* to Singapore was an example of "showing the flag" (or bluff) that in itself contained no element of extreme peril. The intention of the Admiralty—not of Churchill—to base a large fleet as far north as Manila in the spring of 1942 would have been considerably more risky.

The striking events of December 1941—the landing at Kota Bharu in northern Malaya, the carrier strike against Hawaii, the shattering of American air strength in the Philippines on 7/8 December, and the sinking of *Prince of Wales* and *Repulse* by Indochina-based torpedo bombers on 10 December—were a product partly of these confused strategies. The U.S. Army and U.S. Navy had incompatible strategic conceptions, and that led to fiasco in Hawaii and the Philippines. Churchill, the Admiralty, and local commanders in the Far East had different views about deterrence and action, and the result was the loss of major fleet units, and then of the whole "Malay barrier." The Japanese planners had succeeded in overturning the basic assumption of the "Associated Powers"—that the Imperial Navy could be boxed in between Hawaii and Singapore. The Japanese had two competing war-fighting plans in late 1941, and in the short term this ambiguity worked to their advantage. The British and Americans expected Japanese action focused on Southeast Asia (the Southern Operation) and they interpreted accordingly the extraordinary intelligence they received in late November and early December (mostly from codebreaking). They did not expect the raid against Hawaii or even the direct landings in northern Malaya. This contributed to sensational Japanese operational success in the first days and weeks of the war. But in the longer term, Tokyo's strategy, too, would prove to be fatally flawed.

3

The Perils of Technological Transformation

A Case Study of the Japanese Attack on Pearl Harbor

Alan D. Zimm

The modern military has a deeply instilled faith in technology. This is particularly true of aviators, who daily assume that technology trumps gravity. Carl Builder once commented that the U.S. Air Force worshiped on the altar of technology, while another observer went a step further and said that the Pentagon's demand for technological advances shares many of the characteristics of an addict's demand for cocaine—it is insatiable, the cost is high, the product is addictive, and it can lead to brain death. In short, the introduction of new technology does not guarantee improved effectiveness.

Many things can go wrong when introducing new technologies to a military force. There is often a failure to view a new technology holistically, in other words, to understand how the technology impacts the force as a whole. There may be new data requirements for tactical decisions. Command and control and combat doctrine must be adjusted to new realities. The effectiveness of a new technology may be viewed too optimistically. Immature technology can be unreliable. Faulty studies can either over- or under-estimate performance. Testing can be difficult, expensive, or time-consuming, leaving a great temptation to accept the results of a manufacturer's promises, or accept that a small number of operational tests under perfect conditions represent what can be expected under combat conditions.

In addition, organizations accustomed to operating in a certain way can fail to either recognize or to take advantage of new tactical possibilities offered by a new technology. Long-established assumptions and standard operating procedures may need adjustment, or the new technology might actually cause the effectiveness of the force as a whole to go down. Some leaders fixed in the old ways may distrust new technologies. Captain Robert Rubel, USN (Ret.) stated: "Besides facing the resistance of those who view it as a threat, the technology's full potential often remains unrealized because of a failure of imagination. Instead it is forced, at least initially, into existing functions and slotted into established intellectual 'pigeonholes.'"[1]

The Attack on Pearl Harbor: Unrealized Technical Potential

At the attack on Pearl Harbor on 7 December 1941, the Japanese introduced a number of technological and organization innovations to naval aviation. These innovations were either new to the Japanese Navy, new to the world, or employed for the first time against a peer competitor. The results shocked the world. Five battleships were sunk in a massive air attack by an underrated navy thousands of miles from their home base. Shock and amazement obscured a significant fact: the attack fell far short of its potential.

The idea that the attack did not achieve its potential may surprise those familiar with history's treatment of it, depicted by many authors (and even contemporary naval officers) as "brilliant," "textbook perfect," and "an unqualified success."[2] That it was a success is incontrovertible; to determine if it was brilliant or perfect or an unqualified success requires some analysis.

The attack's potential can be calculated by taking the expected number of hits by the various weapons and comparing it against the vulnerability of the targets. This calculation assumes perfect command and control, where the bomb and torpedo hits are perfectly distributed. All the hits are against targets that are vulnerable to the particular weapon, and each target is subjected to the right number of hits. No hits are wasted in "overkill," that is, hits in excess of the number needed to ensure a target's destruction. The hits should also be against targets that are the objective of the attack, not against targets of minor value when higher-priority targets remain.

The Japanese had sixteen priority targets: eight battleships and eight cruisers. They were to attack them with torpedoes, armor piercing (AP) bombs, and general purpose (GP) bombs.

Four battleships and four cruisers were moored where they were vulnerable to torpedo attack. The others were either shielded from torpedoes by ships moored

[1]Captain Robert C. Rubel, USN (Ret.), "Pigeonholes or Paradigm Shifts?," *Proceedings* 138:7 (July 2012), p. 42. The author would like to acknowledge contributions by Captain Chris Carlson, USN (Ret.) to the section on the design of the Japanese passive anti-torpedo system. This chapter was originally published in *Global War Studies* 12:2 (2015).

[2]Examples can be drawn from nearly any general history book on the Pacific War. Specific examples can be found in: Gordon W. Prange, *At Dawn We Slept: The Untold Story of Pearl Harbor* (New York: Penguin, 1981), p. 203; Michael Slackman, *Target: Pearl Harbor* (Honolulu: University of Hawaii Press, 1990), pp. 9, 76.

Table 3.1: Weapons Planned for the Attack and Expected Hits

Weapon	Planned	Expected Hits (modern)	Expected Hits (Japan)
Torpedoes	40	23	27
AP Bombs	50	unknown	5 to 8
GP Bombs	81	32	49

outboard of them or by shipyard piers, or in drydock. Four torpedo hits would be sufficient to sink a battleship; two to sink a cruiser.[3]

Battleships inaccessible to torpedoes could be most profitably attacked by level bombers dropping AP bombs. Each AP bomb hit had a 23% chance of detonating a magazine and destroying the ship, or a 20% chance of crippling the ship by a hit in the engineering spaces. Either would be a desirable outcome.[4]

Dive bombers brought 250kg GP bombs to the attack. GP bombs could not penetrate battleships' deck armor, and thus could be expected to only do superficial damage to topside spaces. One U.S. study estimated that it would take eighty-five hits by bombs in the 250kg GP bomb class to sink a battleship. The Japanese recognized this before the attack, and planned to use their GP bombs against cruisers, where four hits should be sufficient to sink the ship.[5]

In Table 3.1, the "Planned" column shows the number of weapons of each type that the planners allocated to attack fleet targets. "Expected Hits (modern)" shows a calculation of how many hits could be expected using a modern methodology, where the best average hit percentage achieved during training or rehearsals is reduced by one-third to account for actual battlefield conditions—adverse weather, deck aborts, smoke obscuration from battlefield fires, imperfect command and control, and the fact that human performance drops when being shot at—a correction colloquially called the "Fog of War" factor. This factor is a general rule of thumb used by many U.S. Department of Defense modelers based on historical studies. The resulting percentage is multiplied by the number of weapons allocated to obtain the expected number of hits, usually rounding down.

The best average performance by the Japanese torpedo bombers while in training was 87%, achieved in their last training sorties. Reducing this by one-third and multiplying by the forty weapons allocated yields twenty-three expected hits.

[3]Mark R. Peattie, *Sunburst: The Rise of Japanese Naval Air Power, 1909–1941* (Annapolis: Naval Institute Press, 2001), p. 36. U.S. Naval War College, *Maneuver Rules,* Issue of June 1941. Operations Evaluation Group, *Study No. 428, Number of Torpedo Hits Required to Sink a Ship,* 27 September 1950.

[4]Air Intelligence Group, Division of Naval Intelligence, *Striking Power of Air-Borne Weapons,* OPNAV-16-V A 43, July 1944.

[5]Prange, *At Dawn We Slept,* p. 259. *Striking Power of Air-Borne Weapons; Maneuver Rules.*

The expected number of AP bomb hits cannot be calculated using this methodology because the best Japanese training performance is not known (in English-language sources). We do know that one Japanese planner considered a 10% hit rate as "dismal," but at the same time, their chief planner appeared to consider a 5% hit rate as a case where "Lady Luck had been working overtime."[6]

"Expected hits (Japan)" shows the number of hits that the Japanese themselves expected, taken from the testimony of Japanese officers. The expectation of twenty-seven torpedo hits was recorded by Commander Genda Minoru, the lead planner. The expectation of forty-nine hits by dive bombers reflects Japanese achievement of 60% hits regularly attained during training, applied without the "Fog of War" correction, as apparently the Japanese expected their training ground performance would be duplicated in a surprise attack.

Working with the Japanese planners' expectation of twenty-seven torpedo hits, at least five AP bomb hits, and forty-nine GP bomb hits, and distributing those hits among the targets while observing proper weapon-to-target assignments, the calculation shows that the Japanese would expect sufficient potential hits to sink or cripple fourteen of their sixteen priority targets. The two survivors would be battleships shielded from torpedo attack, as the AP bombs would most likely only cripple or destroy two of the shielded four.[7]

After these fourteen ships were sunk by the requisite number of hits, there would remain eleven torpedo and seventeen GP bomb hits in excess. This was enough to sink or cripple another three battleships and four cruisers, also plenty to have sunk all the Pacific Fleet's relatively "soft" aircraft carriers, had they been present. In a perfect attack, these excess hits could be directed to tertiary targets such as destroyers and auxiliaries; in reality, they would more likely represent overkill applied to primary targets, a form of insurance that the planners would expect to make doubly certain their objective would be achieved, while they fervently hoped that no overkill hits would be applied while other primary targets still remained afloat.

Table 3.2 shows the results of the attack. "Actual Hits" shows the number of torpedoes and bombs that scored a hit on a floating hull. "Effective Hits" shows the number of hits that contributed to the objective of the attack, i.e., the destruction of the aforementioned sixteen targets. Effective torpedo hits are calculated by deducting hits on invalid targets and overkill from those hits achieved. Two hits on *Utah* were deducted, as she was a demilitarized target ship which the aviators were briefed to avoid. Of the five hits on *Oklahoma,* one was deducted as overkill, as were three of the seven hits on *West Virginia.* The remaining thirteen effective hits were less than half of the Japanese expectations.[8]

---

[6]Genda, quoted in Prange, *At Dawn We Slept,* p. 162.

[7]Alan D. Zimm, *Attack on Pearl Harbor: Strategy, Combat, Myths, Deceptions* (Philadelphia: Casemate, 2011), pp. 195–98.

[8]Ibid., p. 229.

Table 3.2: Weapons Planned for the Attack and Expected Hits vs. Actual and Effective Hits

Weapon	Planned	Expected Hits (modern)	Expected Hits (Japan)	Actual Hits	Effective Hits
Torpedoes	40	23	27	19	13
AP Bombs	50	unknown	5 to 8	10	3
GP Bombs	81	32	49	15	1

Of the AP bomb hits, the aviators exceeded expectations by scoring ten hits. Two were on the repair ship *Vestal,* caught in the pattern of bombs intended for *Arizona.* Seven of the bombs malfunctioned, that is, either failed to explode, broke apart upon impact, passed through the ship without exploding, or exploded with a reduced yield. Only three could be judged as effective, and of those, only one penetrated to a magazine or an engineering space.

Overall, instead of scoring killing hits on fourteen ships with plenty of overkill available to make sure of the task, the Japanese placed killing hits on only three ships: *Oklahoma,* five torpedo hits; *West Virginia,* seven torpedo hits; and *Arizona,* the AP bomb hit which detonated her magazine.

The dive bombers were ineffective. They scored only fifteen hits instead of the expected forty-nine. Most of these hits were on destroyers or auxiliaries that probably were hit by bombs that missed their intended targets so badly they hit these adjacent, low value ships. Seven hits were scored on battleships, hits that could be expected to cause only superficial topside damage.

Two additional battleships sank due to circumstances that cannot be credited to Japanese weapons delivery accuracy. *California* sank from two torpedo hits, half of the calculated requirement. Neither torpedo warhead breached the ship's internal torpedo defense holding bulkhead. However, access covers to some anti-torpedo voids were open or loosened for an inspection to be conducted on the following day, allowing internal flooding. Before this flooding was contained, she was abandoned when burning oil, floating down from the destroyed *Arizona,* threatened to engulf the ship. The crew neglected to shut watertight doors as they departed, allowing progressive flooding to spread throughout the ship. When the crew returned about an hour later, it was too late.

The other unexpected loss was that of *Nevada,* succumbing to one torpedo hit and five GP bomb hits. The torpedo hit, as in *California's* case, did not breach the anti-torpedo holding bulkhead, although there was some manageable leakage at strained seams and deformed butts. *Nevada* actually succumbed to a damage control error and a ship design error, along with her poor material condition associated with an older ship soon to be retired. Bomb hits started a fire forward. As a precaution, the order was issued to flood the forward magazine; in a communications blunder, someone flooded the aft magazine as well. This mass settled the ship deep in the water, enough to bring the second deck under the ship's new waterline. The second deck proved not to be watertight, and water spread through the deck and leaked into compartments below. At the same time, water poured into a poorly designed ventilation system. Aged clapper valves did not interrupt the flow of water, and the ventilation system distributed water throughout the ship. *Nevada* sank, but her sinking should not be attributed to the efficiency of the Japanese attack.

Overall, the Japanese only directly sank three of their priority targets compared to their capacity to sink fourteen, or 20% of the available vulnerable targets. If the unrealized potential of the overkill hits are included in the calculation, they achieved only 14% of their potential. This falls far short of an attack that has been depicted as "brilliant," "textbook perfect," and "an unqualified success."

Much of the shortfall can be attributed to poor integration and poor organizational adaptation to new technologies. Some of these new technologies: a) were not well assimilated, or b) required changes that were in conflict with Japanese naval aviation organizational culture, or c) were inadequate due to miscalculations, or d) were not properly tested, or e) were only effective under the best of conditions and were intolerant of changing environmental conditions.

Certainly there were additional reasons (other than technological stumbles) that contributed to this shortfall. For example, many of the aviators were green, barely carrier qualified. Many, perhaps most, were very young. For example, one of the more experienced B5N Kate torpedo bomber pilots assigned to the carrier *Zuikaku* was only nineteen years old, and his gunner and observer were sixteen or seventeen years old. As another aviator serving on *Zuikaku* related, "Most of our men were not that experienced as pilots."[9] Most were seeing combat for the first time; only some of the flight leaders had combat experience, as most of the naval contribution to the air war in China came from land-based Nell twin-engine bombers, not carrier aviation. There were a few A6M Zero fighter units committed to operate from shore bases for a few months; carriers contributed a few coastal strikes, but only a handful.

The Japanese expected to achieve surprise. Instead, they were shocked at the speed with which the American ships employed defensive anti-aircraft fire. The first torpedo bombers initiating runs against Battleship Row were immediately taken under frightening levels of .50-caliber machine gun and 5"/25 gunfire. Command and control broke down at the outset of the attack, thrusting huge responsibilities of target selection and attack coordination onto individual, young aircrews.

---

[9]Testimony of Ensign Honma Hideo, in Ron Werneth, *Beyond Pearl Harbor: The Untold Stories of Japan's Naval Airmen* (Atglen, PA: Schiffer, 2008), pp. 199–200.

While these are some of the reasons that assuredly contributed to the Japanese underperformance, technology also played a part—a critical part.

Technological Innovations at Pearl Harbor

Over the years between the First and Second World Wars, the Japanese naval air forces developed a number of technological innovations that were employed by their naval air forces at Pearl Harbor. These technologies were both material (improvements or innovations in equipment) and social (improvements or innovations in the way forces were employed, and improvements to doctrine). In many ways, the attack on Pearl Harbor was as novel to the Japanese as it was to the world.

Among the technological and organizational innovations seen at Pearl Harbor (and associated issues): bombsights on precision attack dive bombers and high altitude level bombers (accuracy); armor-piercing (AP) bombs (penetration of armor, testing, fuzing, manufacturing); general purpose (GP) bombs (testing, fuzing, manufacturing); mass employment of carriers (command and control, and aircraft radios); and aerial torpedo warheads (defeating battleship underwater protection systems).

A Low-Tech Success Story

New technology does not have to be "high-tech" to be effective. At Pearl Harbor, a low-tech innovation actually exceeded expectations, and it was two years or more before its performance was eclipsed by contemporary high-tech systems.

Before the attack, level bombing was falling out of favor in the world's navies, primarily because of low hit percentages. With hit percentages hovering below 10%, the return on investment—both in money and in carrier deck capacity—seemed too low to justify including level bombing in a carrier's repertoire. The Americans and British at times had like opinions: the U.S. carrier *Ranger* was designed to operate only fighters and dive bombers, while British carriers concentrated on torpedo bombers for strikes against warships.[10]

For the Pearl Harbor attack, the Japanese were faced with a dilemma: they expected many of the battleships to be shielded from torpedo attack by ships moored outboard of them, or possibly all the battleships would be protected by anti-torpedo nets. They knew that their dive bombers' weapon could not penetrate a battleship's deck armor. So, they desperately needed some capability to attack battleships with heavy armor-piercing bombs. These weapons had to be released at a sufficiently high altitude to give them enough striking velocity to penetrate deck armor. But, with the low hit percentages demonstrated by high-altitude bombing

[10]Norman Friedman, *U.S. Aircraft Carriers: An Illustrated Design History* (Annapolis: Naval Institute Press, 1983), p. 69.

before the attack planning began, level bombing alone could not expect to attain sufficient hits to achieve the objective of the attack.

The Japanese used a domestically upgraded version of the Course Setting Bomb Sight (CSBC), obtained from the British after World War I. In contrast to this was the American Norden Bomb Sight, a very high-tech tachometric bomb sight with a mechanical computer, gyroscopes, and an autopilot to eliminate pilot error. It was reputed to be able to "put a bomb in a pickle barrel from 30,000 feet." As a simplistic comparison, the CSBC was geometry, the Norden was calculus. If the CSBC was an abacus, the Norden was an electronic computer.

The Japanese assigned some skilled chief petty officer aviators and a recent officer honors graduate from their level bombing training course to improve the hit percentage using the CSBC. They found that the most significant problem was the coordination between the bombardier and the pilot. While the Norden placed the aircraft under the control of an autopilot linked to the bombardier, in the Japanese system the bombardier directed the pilot onto the target by voice commands. The pilot was responsible for keeping the aircraft level in pitch and roll and in following the course corrections smoothly and accurately. Their coordination had to be precise. For example, using the CSBC from 6,000 feet at a speed of 100 miles per hour, a pitch error of only two degrees would cause the bomb to miss by nearly 200 feet; a roll error of two degrees from 10,000 feet could cause the bomb to miss by nearly 350 feet.[11] The Japanese would bomb at a speed considerably higher than 100 miles per hour (about 125 knots, or 144 miles per hour), which would exacerbate the problem.

Through intense practice and enhanced coordination between the bombardier and pilot, select aircrews achieved great improvements in bombing accuracy, regularly scoring 15% hits against a battleship-size target. Rather than trying to attain proficiency with all the pilots and bombardiers and have each aircrew aim their bombs, tight formations were used, with all the aircraft toggling their bombs when they saw the lead bombardier release his weapon. This technique was adopted later in the war by U.S. formations. During the attack, witnesses on the ground testified that the Japanese AP bombs maintained a perfect, tight formation all the way to the target; a photograph of a bomb salvo hitting *Vestal* and *Arizona* shows a tight salvo grouping.

The Japanese planners expected "four out of five salvoes to hit," meaning that four out of five of the five-plane formations would score at least one hit.[12] Overall, the Japanese expected five to eight AP bomb hits from the ten formations in the attack.[13]

---

[11]Stephen L. McFarland, *America's Pursuit of Precision Bombing, 1910–1945* (Washington, DC: Smithsonian Institution Press, 1995), pp. 16–17.

[12]Werneth, *Beyond Pearl Harbor*, pp. 160–61.

[13]Donald M. Goldstein and Katherine V. Dillon, eds., *The Pearl Harbor Papers: Inside the Japanese Plans* (Washington, DC: Brassey's, 1993), pp. 283–84.

The Japanese bombsight technology and technique was a clear winner early in the war. At Pearl Harbor, the Japanese CBSC scored ten to fourteen hits out of forty-nine bombs dropped, a remarkable 20% hits (or 16% if you exclude the two "collateral damage" hits against *Vestal*). Elsewhere, the same system devastated the Cavite Shipyard and Clark Field in the Philippines, and sank the aircraft transport *Langley* while she was underway during the Dutch East Indies campaign. In contrast, 291 bombs were dropped using the Norden sight during the Battle of Midway for zero hits.

This is a case from the attack on Pearl Harbor where a less technically advanced system clearly delivered the required performance, and actually exceeded expectations. The difference was in the dedication of the personnel. Level bombing had advocates who were determined to make it work.

Investigating Bomb Performance

At Pearl Harbor, the D3A Val dive bomber employed a 250kg general purpose (GP) bomb with a 0.2-second delay fuze. Dive bombing was a significant technological advance which had the potential to contribute greatly to the success of the attack. The bomber under normal conditions was accurate enough to attack "point" targets, in contrast to level bombers that were more suited to large or "area" targets. The Japanese had every reason to expect that a surprise attack would achieve a clean sweep of the eight cruisers assigned to the dive bombers. Instead, the dive bombers achieved practically nothing.

Something went wrong with the D3A Val's accuracy. It happened to all the dive bombers, not some subset of pilots or units, so it was a systems' problem. In addition to an accuracy problem, the command and control of the attack fell apart. Only seventeen bombs were directed against cruisers, the remainder against misidentified targets or targets that could not be seriously harmed by the 250kg GP bomb. These issues will be discussed later.

To add to their woes, many of the bombs dropped by the dive bombers were faulty, with an inordinately high number of duds and low-order detonations. As a sample, about 30% of the bombs dropped on Ford Island malfunctioned. One sailor found an unexploded bomb in a hangar and rolled it by hand across the taxiways and dumped it into the harbor. Other accounts describe bombs that dispersed a large cloud of toxic yellow particulate. Witnesses also described bright red "fireballs" associated with hits on shore and against two ships in drydock. These signatures were so different from usual bomb detonations that one officer formally reported his suspicion that the Japanese were employing incendiaries.[14]

A red fireball or a cloud of yellow particulate is a sign of improper detonation, traceable to manufacturing defects. The process of assembling a bomb involves

[14]Commanding Officer, USS *Downes* (DD375), Report of Pearl Harbor Attack, DD375/A16-3, 17 December 1941.

building the hollow case, softening the explosive charge (the "fill") by heating, and then pouring or packing the charge into the bomb case. The explosive should be about the consistency of pancake batter, and well mixed.

As can be imagined, this is a dangerous process. The danger can be reduced by lowering the temperature of the fill, making it more the consistency of bread dough. But explosives at this consistency do not pack uniformly, often capturing bubbles of air and leaving creases and layering in the fill. When the bomb goes off, these bubbles and creases break the coupling of the reaction, so portions of the fill deflagrate rather than explode. The blast is visually impressive, as burning fill produces a bright red, lovely fireball suitable to a Hollywood movie. If the malfunction is more significant, part of the fill does not even ignite, and the result is a cloud of yellow unburned atomized explosive. In contrast, an efficient explosion leaves only small amounts of thin grey smoke (the glorious Hollywood red fireballs that seem to be the result of every car crash are created by propping a few gallons of gasoline mixed with a bit of strontium on top of the explosive). The efficient explosion transmits a much greater shock wave and generates much faster (and more destructive) casing fragments. The Hollywood explosion is more spectacular, but much less lethal. When this happens to a large proportion of the bombs, anything more than about 5%, it is an indication of poor manufacturing practices, poor testing, poor quality control, and perhaps a hint that Japan was operating at the ragged edge of its technology resources.

The Japanese staff decided to use the same GP bomb with the same fuze delay on dive bombers assigned to hit ground targets as well as ship targets. The long fuze delay was appropriate for attacking ships, since it allowed the bomb to penetrate deep into the hull before exploding. However, this time delay made the bombs ineffective against ground targets. There is a photograph showing the crater made by a 250kg GP bomb that landed just outside the Ford Island dispensary within about ten feet of the exterior wall—the crater actually slightly undermines one wall. But the damage to the building was minimal, and even the undermined portion of the wall was intact. There were only a few chips of paint off the walls. Doors and windows within a few feet of the lip of the crater still retain some of their glass. All of the force of the explosive was expended in digging a hole and sending dirt straight up, rather than blasting the building flat. Another bomb landed just outside the Wheeler Field hangar door. The crater here was not as deep because of the harder surface of the taxiway, but there was no significant damage or marks on the concrete pillar next to the crater, and much of the glass remains in the hangar door a few feet away.

In comparison, a U.S. 500-pound bomb with a superquick fuze was rated to destroy hard targets (like cinder block walls) out to a radius of thirty to sixty feet. Soft targets such as personnel and aircraft (the intended targets at Pearl Harbor) were at risk up to 300 feet away. While the crater of the Japanese bomb that hit outside the Ford Island dispensary was about six feet deep with a diameter of about 20–24 feet, a comparable U.S. 600-pound (272kg) Mark 1M1 Demolition bomb would dig a crater about twelve feet deep and thirty-eight feet in diameter.[15] So, the Japanese bomb achieved something like 6% to 25% of the damage potential of a comparable weapon with the proper fuze.

Staff officers are responsible for selecting the appropriate ordnance to match the target. The 0.2-second fuze was a poor selection for ground targets, indicating faulty staff work. Better bombs might not have been available, but for an attack where the planning started nine months in advance, this still indicates faulty staff work. Looked at another way, the shore ordnance and testing establishments were responsible for the overall performance of the bomb, and clearly they failed, too. The 250kg GP bombs used at Pearl Harbor were not up to international standards of effectiveness and reliability.

The performance of the Japanese 800kg AP bombs dropped against Battleship Row can be compared with the performance of their 250kg GP bombs to see if there is any corroboration of the indictment against the Japanese shore munitions establishment.

In early 1941, the Japanese Navy did not have any suitable AP bombs. They decided to modify superseded 41-centimeter armor piercing shells of the type once fired by the battleship *Nagato.* The modifications included removing the explosive charge, enlarging the explosive charge cavity by milling, adding dual fuzes into the base plate, attaching flight stabilization fins, and reinstalling the fill. Most significantly, since the shell was too heavy to be carried by the available aircraft (the B5N Kate carrier attack bomber), about 400 pounds of metal was shaved off the after end of the shell, thinning the side wall and tapering the bomb from about the midpoint to the rear.[16]

The level bombers scored ten confirmed hits,[17] a magnificent performance by the aircrews. But of the ten, no less than seven were duds, malfunctions, or low-order detonations.

One bomb that hit *West Virginia* struck a main battery turret. The nose of the bomb penetrated the turret's top armor, but the bomb's forward momentum caused the afterbody—the section that had been stripped of 400 pounds of metal—to slap against the forward edge of the hole. This impact shattered the bomb. The fuze failed and the explosive charge scattered into the turret, burning instead of detonating. The damage was not much worse than if the turret had been penetrated by

---

[15]United States Army Air Forces, "Craters of Demolition Bombs," chart reproduced in Stephen L. McFarland, *America's Pursuit of Precision Bombing, 1910–1945* (Tuscaloosa: University of Alabama Press, 2008), pp. 164–65.

[16]U.S. Technical Mission to Japan, *Japanese Bombs,* Ordnance Data Page for 250-kg GP bomb, December 1945.

[17]Another four possible hits may have been discovered by Bill Jurens while examining reports from the team salvaging *Arizona.* Bill Jurens, letter to the author, 18 November 2010.

an inert cannon ball. Repairs involved replacing a loading slide, renewing the electrical wiring, and cleanup, a few weeks' work.

A 70% malfunction rate is horrible. That failure rate in a fielded weapon is the result of bad engineering and inadequate testing.

The Japanese tested this bomb by dropping sample bombs against a mock-up of a section of *West Virginia's* deck armor. Bombs dropped from 4,000 meters penetrated the plating. At nearly the last minute, they considered reducing the attack altitude to 3,000 meters to improve accuracy, but they were unsure whether the bombs would penetrate the ships' armor from the lower altitude. Bomb after bomb was dropped from 3,000 meters until, on the verge of giving up, a hit was obtained and the bomb smashed through the plate. The Japanese made the decision to bomb from 3,000 meters based on that single successful trial.

A more robust testing program was obviously warranted. The usual practice in the United States and Great Britain was to test bombs by firing them out of a howitzer that was placed only a few yards away from a vertically-mounted section of armor plate. The gun's powder charge was adjusted to obtain the desired impact velocity. The plate could be tilted to simulate different impact angles. Many, many trials could be run to gain a complete understanding of the bomb's performance under various conditions. However, the Japanese did not have the ordnance establishment infrastructure to quickly perform tests in this manner, or they did not allocate the funding to suitably modify a howitzer, or they simply were content with the single successful trial. Considering that the order for AP bombs was placed at least nine months before the required delivery, there was sufficient time to build a suitable test rig and perform comprehensive tests.

This analysis provides an interesting snapshot of the readiness of the Japanese war establishment and their capability to sustain the war effort. It suggests that Japan, as a relatively new industrial power, still had significant problems with the development and manufacturing of even relatively simple weapons, and was operating on the ragged edge of her industrial capability.

Employment of Massed Carriers: Problems of Command and Control

The Japanese at Pearl Harbor were the first to form a six-carrier striking force with the ability to combine aircraft from all the carriers into a single strike. Control of such a massive strike in the approach to the target, and coordination of their targeting assignments, were significant issues. How were the attackers to ensure killing levels of ordnance were placed on each priority target and ordnance not wasted in overkill, while still optimizing the overall attack effort? The command and control innovations involve both changes in equipment (the introduction of aircraft radios to carrier aircraft) and doctrine (where command of aircraft from one carrier might be placed in the hands of an officer from another carrier).

Much of the blame for the poor attack performance by the torpedo bombers at Pearl Harbor can be attributed to poor planning and poor command and control at the point of attack. The command and control deficiencies can be traced back to doctrine that was not suitable for a massive employment of aircraft, and an organizational culture that placed too much reliance on the individual air crews and not enough authority in the hands of senior aviators. There were communications systems that were introduced that could have mitigated these problems, such as aircraft-to-aircraft FM voice radios, which the Japanese introduced into their carrier aircraft over the year before the attack. But there were flaws in the technology that limited its effectiveness, and when those deficiencies came up against the prevailing organizational culture, the potential benefits from the technology were negated.

The aviator culture was influenced by the *samurai* ethic and Japanese cultural values. The sociology of the Japanese military is a subject that deserves more extensive analysis, but a few characteristics can be noted: a propensity to accept high-risk operations coupled with the planning assumptions that the high-risk actions would inevitably be successful; individual responsibility for actions with extreme consequences of failure; a lack of a combined arms or cooperative forces mentality; an intolerance of error; and staff work that was inflexible and accepted assumptions as nearly inviolate.[18]

A few examples illustrate this behavior.

During the pre-war period, Japanese training was intense. Bombers practiced night attacks against targets that were protected by searchlights. The glare from the searchlights blinded some pilots, resulting in air-to-air collisions with loss of aircraft and lives. These losses were accepted, and the training continued under the same conditions.

The fleet practiced an exercise where they would enter an anchorage at night without navigation lights. One of the Japanese bridge officers related that if a collision occurred with any significant damage to His Majesty's ships, the officer responsible would be expected to commit ritual suicide.

The staff did not provide any planning for the eventuality that there might be torpedo nets protecting the battleship anchorage at Pearl Harbor. Torpedo nets (also called torpedo baffles) could shield battleships from damage from torpedoes. A logical and effective solution would have been for the staff to develop a contingency plan redirecting some of the dive bombers to blast open the nets with their general purpose bombs. However, the Japanese staffs did not have a combined arms mindset, where aircraft types would cooperate to improve the overall effectiveness of the attack. Consequently, the problem of torpedo nets was left to the torpedo bomber aircrews to solve during the outbound voyage. Their solution was to have some of the torpedo bomber pilots crash their B5N Kate carrier attack bombers into the nets. Volunteers came forward for this task.

---

[18]See Ruth Benedict, *The Chrysanthemum and the Sword: Patterns of Japanese Culture* (New York: New American Library, 1946).

These characteristics resulted in an attack style that placed a great deal of responsibility on the individual aviators. Commanders were mostly responsible for delivering their charges to the attack point, whereupon the aircraft were released to execute their attacks. Commanders had only hand signals, visual signals like flares, and aircraft movements like waggling their wings to communicate with their subordinates. "Follow me" leadership was the norm for *shotai* (groups of three aircraft) or *chotai* (groups of nine aircraft) commanders. While this might work for small numbers of aircraft associated with single carrier strikes, it could be potentially disastrous with many carriers' aircraft all attempting to attack at the same time.

The Japanese planned for the torpedo bombers to strike the fleet targets first, before the defenses could be aroused. They would make landfall at the northern shore of Oahu and head for an initial point (IP) north of the harbor. En route to the IP they would shift from cruise formation to attack formation, and the torpedo bombers would consolidate into four groups. Separating at the IP, two groups from *Soryu* and *Hiryu* would circle around to approach from the west to attack moorings where the Japanese expected carriers would be anchored, while two other groups from *Akagi* and *Kaga* approached from the east to attack Battleship Row. With this scheme, all four groups could initiate their attacks "nearly simultaneously," and the total duration of the torpedo attack would be about ninety seconds. All this was to be done without communications between the various groups. So, if the IP was missed, or one or the other of the groups delayed, there was no way to coordinate the attack. There was no flexibility built into the system.

Attack routes against the likely battleship, carrier, and cruiser anchorages were briefed to the torpedo bomber aircrews, taking into account the shallow waters in the harbor and the distance needed to stabilize the aircraft in the correct altitude, attitude, and airspeed and drop their weapon with sufficient distance to allow the torpedo to arm. Specific attack paths were not assigned to specific aircraft, as the aviators were expected to evaluate the situation on their own and select their targets independently.

This approach was fraught with dangers. Many of the attack routes crossed, between groups and even within groups. If the attacks were conducted simultaneously, as would be preferred, there would be the potential for collisions and other forms of mutual interference. In modern terminology, the attack routes were not "deconflicted."

Deconfliction is usually done by path, by altitude, or by time. The constrained dimensions and shallow water of Pearl Harbor meant that most of the routes could not be moved, making path deconfliction impossible. The torpedo bombers all had to attack at the same altitude, making altitude deconfliction impossible. The only available method was by time. To do this, the planners decided to form the bombers into four long lines with each aircraft separated by about seven seconds, or 200 meters. Each aircraft would attack singly, after the aircraft ahead of it went into the attack. While they would follow their leader in the approach, each aircrew was responsible for identifying their own target and selecting their own attack route.

This formation choice was only appropriate if complete surprise was achieved. Without surprise, the strung-out torpedo bombers would be too far separated to provide mutual support and could have been picked off individually by defending fighters, even the obsolete American P-26 "Peashooter." A better solution would have been for the torpedo bombers to stay in larger formations, and be dispatched one at a time for their attacks by the formation leader using radio communications. However, the Japanese chose not to employ their FM voice radios in the command and control of the attack.

The Japanese first established their own radio factory in 1940, with a planned capacity of over 1,000 sets per month. Full fleet installation on all aircraft of *Kido Butai* (the carrier striking force) was to be completed by late 1941, so it is likely that all the Japanese aircraft in the attack were equipped with radios, and had time to practice with them.

The early production radios were of good quality, but the installation standards were horrendous. There was no coordination between the people who built the radios and the airframe manufacturer. The result was that radios were often jammed into whatever space was available, with little consideration for ergonomics. Operators often had difficulty reading the instruments and switch settings and manipulating the controls.[19]

In a well-designed cockpit, all the instrument dials are readable, and there are no loose wires that might be snagged or pulled loose. Switches and controls are accessible, and labeled so as to be readable from the operator's normal position in his seat.

A photograph exists of the interior of a Japanese A6M Zero's cockpit, taken of an aircraft that crashed at Fort Kamehameha during the Pearl Harbor attack. The entire cockpit shows little attention to pilot comfort or accessibility of controls. For example, the seat back was about eighteen inches high, reaching only up to the small of the pilot's back. This could not have been comfortable on long flights, contributing to pilot fatigue. It also did not allow shoulder straps, so the pilots' head and shoulders were free to bang about the cockpit during the "yankin' and bankin'" of air combat maneuvering.

The radio was obviously retroactively fitted into the cockpit. The radio installation consists of the two boxes mounted in the interior of the cockpit on the side of the airframe below the pilot's thigh. The transmitter and receiver boxes were "shock mounted" by hanging the boxes from criss-crossed bungee cords, which meant that the sets themselves could move, making adjusting the controls difficult. Loose wiring is everywhere evident, held in place only by a few notch clips. There was a bundle of excess cord between the boxes—the installer did not bother to trim the cord to the proper length. This is bad practice, as this excess wire could serve as an antenna and pick up static and power surges. Little attention was given to proper

[19]Military History Section, Headquarters, Army Forces Far East, *Operational History of Naval Communications, December 1941–August 1945,* Japanese Monograph No. 118.

shielding and grounding of the wiring and equipment, so that the receiver would often pick up the static generated when the aircraft engine spark plugs fired, sounding like a continuous blast of machine gun fire in the pilot's earphones. Blasts of static discharges would also be generated by the airframe as it flew through the atmosphere; all the excess wiring would pick that up as well. Loose wiring even ran into the rudder wells next to the pilot's feet, making it a wonder that the pilot did not kick out some connection in the course of getting into or out of the aircraft.

There are two jacks coming out of the face of the receiver just aft of a selector knob—the jacks stuck out about two inches and would block the pilot's access to the selector knob. The jacks could be banged by the pilot's legs and knees during air combat maneuvering, which could bend or dislodge the jacks. The pilot could not see the ammeter on the transmitter face (or, for that matter, any of the other meters or control labels) from his normal seated position, so it would be nearly impossible to tune the radio to a new frequency while aloft, or to restore a setting if the pilot brushed against the controls or kicked the dials off their original positions. The location and orientation of the dials and switches forced the pilot to operate the set by touch. The voice radio required two hands for operation, one on the microphone and one operating a "transmit-receive" switch, forcing the pilot to control the aircraft by holding the stick between his knees. The radio's range was at best fifty nautical miles, often much less—when anything could be heard over the static.

It is no wonder that this radio was disliked by the aviators. Fighter pilots did not see this as an improvement over the old "tried and true" methods of hand signals and flares. Some considered it an infringement on the autonomy given to fighter pilots during a mission. Pilots were known to rip the sets out of their aircraft, showing that they valued the improved maneuverability due to the weight reduction over the ability to communicate with their fellows.[20]

Given all these problems, it is not a surprise that the aviators had little use for their voice radio. Other than one signal releasing the formations to begin their attack (over high frequency keyed radio broadcast from a B5N Kate), radio technology was not incorporated into the tactical planning or employed by the formation commanders. The results of this decision can be seen by looking at the execution of the torpedo bombers' attack.

Command and Control of the Torpedo Attack

The Japanese had several alternative plans for the attack. In the case that surprise was achieved, the General Commander of the Strike, Commander Fuchida, would fire a single flare, and the torpedo bombers would immediately attack, while the dive bombers scheduled to attack the Ford Island airfield and Hickam Field would

[20]Saburo Sakai with Martin Caidin and Fred Saito, *Samurai!* (New York: Ballantine Books, 1957), p. 79.

be held back as a deconfliction measure. If surprise was not achieved, two flares would be fired and the dive bombers would lead the attack to distract anti-aircraft (AA) gunners away from the slower and more vulnerable torpedo bombers.

This plan fell apart from the outset. First, when landfall was made at the northern tip of Oahu, clouds obscured the mountains. Fuchida decided to divert from the intended track to fly along the west side of the island, out of the clouds. Considering the size of his formation, this was undoubtedly the right decision, but he failed to adjust to this decision. He never attempted to regain the planned track or achieve the planned IP. The torpedo bombers would all be released west of the harbor, eliminating the chance of delivering the attacks against Battleship Row and the carrier moorings simultaneously, since the Battleship Row attackers now had to fly past the harbor and double back. A short conversation on FM voice radio prior to releasing the attack between the leaders of the Battleship Row attack force and the carrier moorings attack force could have overcome this issue—the aircraft assigned to the carrier moorings could have agreed to orbit for the six minutes required for the aircraft assigned to attack Battleship Row to get into position. A word on the radio could then initiate all four groups' attacks. But there was no provision for such communications in the plan, and evidently the leaders did not think to use their radios in this way.

While still west of the island, Fuchida then fired one flare, activating the "surprise achieved" plan. This also ordered a shift into attack formation. The dive bombers and torpedo bombers began to reorient themselves, but the fighters did not react. Fuchida thought that the fighters' commander had missed the signal, so he fired another Black Dragon flare.

This was an incredible blunder. The dive bombers counted two flares, so they firewalled their throttles and charged directly toward the harbor and airfields. The torpedo bombers' leadership interpreted the flares as a duplicate single flare signal, but seeing the dive bombers lunge forward, and not wanting to wait for the dive bombers to awaken the defenses, the torpedo bombers also broke for the harbor, before their formation change was complete. It was like firing the starting gun at a horse race, and devil take the hindmost.

The torpedo bombers' formation change was thrown into disarray. Intervals between aircraft stretched to 500 meters. Intervals between formations stretched to as much as 1,200 meters. Aircraft at the rear lost sight of their leaders, and at least one group of four torpedo bombers ended up orbiting Ewa Town about ten miles to the west of the harbor, trying to sort out what they were to do.

The leader of one of the formations aiming for the carrier moorings, Lieutenant Commander Matsumura, could not identify a target with the morning sun in his eyes, so he turned to the south to delay the approach. The leader of the other carrier moorings attack group, Lieutenant Nagai, pulled up alongside and made hand gestures imploring Matsumura to speed up the attack. Matsumura released Nagai, who promptly initiated an attack against *Utah,* a demilitarized battleship being used as a target auxiliary, a ship that the aviators had been specifically briefed to avoid. This could have been corrected with effective voice radio communications and better command and control.

Utah was moored close to two cruisers, *Raleigh* and *Detroit,* both legitimate targets (for dive bombers), but a waste of a precious torpedo. Nagai's wingman, Petty Officer Mori, thought to himself, "How silly. Can't they see that two of the ships are nothing but cruisers?" Then, the two aircraft following him broke off to join Nagai and attack *Utah.* Mori had no way to recall them.

Nagai eventually recognized *Utah* and aborted his attack, but the aircraft following him were not as observant. In all, six torpedoes were wasted in this attack, which was conducted at about 07.55. Two of the six hit *Utah,* while one torpedo missed so badly that it hit *Raleigh* in the adjacent mooring.

Nagai circled south of Ford Island looking for a target. He misidentified the backlit silhouettes of *Oglala* moored outboard of *Helena* as a battleship, and led four other torpedo bombers into the attack. Three torpedoes rammed into the harbor bottom mud, and one missed and exploded against the side of the quay. One torpedo, likely Nagai's, passed under the hull of *Oglala* and hit *Helena* amidships. The detonation was so close to *Oglala* that it opened the seams of the World War I-era minelayer, and she eventually rolled over and sank, to be forever known as "the ship that was frightened to death."

The torpedo bombers from *Akagi* and *Kaga* circled around south of the harbor and initiated their attack against Battleship Row about six minutes after the beginning of the attack on *Utah,* and about nine minutes after the first bombs from dive bombers struck Ford Island, waking up the harbor. To put this in context, the usual standard for setting General Quarters (and thus achieving battle standards of interior watertight integrity) on a battleship was eight minutes.

Consequently, most battleships were well on their way to setting General Quarters before the first torpedo attack aircraft approached. Originally planned to take less than ninety seconds, Fuchida's blunders caused the attack on Battleship Row to stretch out for over ten minutes. Most of the available ships' anti-aircraft batteries were firing by the time the last of the torpedo bombers initiated their attacks, as witnessed by the fact that five of the last seven torpedo bombers were shot down. Even the first aircraft to attack the battleships was greeted by heavy fire that was both surprising and dismaying. The intense and unexpected fire influenced target selection: faced with the fear that they would be shot down before delivering their attack, over half of the torpedo bombers attacked along the easiest and most direct approach leading to *Oklahoma* and *West Virginia.* Four precious torpedoes, 10% of those brought to the attack, were wasted in overkill. While all the accessible battleships were hit by at least one torpedo, only two were hit by four or more. Not one torpedo bomber intentionally attacked a cruiser.

The bombers following Nagai in his attack against *Helena* crossed the paths of the stream of bombers attacking Battleship Row. Other *Soryu* and *Hiryu* aircraft tried to merge into the line of *Akagi* and *Kaga* bombers while they were trying to set up their aircraft within the narrow airspeed, altitude, and attitude parameters required for a successful torpedo launch in a shallow harbor. Visualize piloting a heavily-loaded aircraft two wingspans above the water and just a few knots above stall speed, and then having another aircraft suddenly cut in front of you, restricting your line of sight and buffeting you with turbulence. There were several near collisions. At least one attack was aborted when the prop wash from an interfering aircraft nearly caused another to lose control. A torpedo was jettisoned over land after a near collision. Other aircraft had to cancel their attacks and go around and repeat their runs against increasingly effective anti-aircraft fire. All this could have been prevented by some firm leadership over the FM voice radio.

There are some lessons that can be learned from the Japanese experience.

First, introducing new technology does not guarantee that it will be properly employed—it does not even guarantee that people will even think about how to best use it. The old "tried and true" ways have a great deal of appeal.

Second, new technologies may require, or may facilitate, new approaches. The "old ways" may have to be abandoned or modified.

Third, a lesson that technologists throughout the world must repeatedly relearn: human proclivities trump technology. It does not matter how good the technology if it is not accepted at the deck plate level (or, as in this case, the cockpit level). If the performance of the new system is flawed, there will be excuses not to use it.

Overall, no effective command and control was exerted over the torpedo attack. Had it not been for individual aircrews boring in for individual attacks against a fierce and unexpected resistance, the attack could have been a complete failure.

Torpedo and Anti-torpedo Technology

American passive anti-torpedo defenses were among the best in the world in 1940, even in those older battleships that completed construction during and just after World War I. For example, in a New Mexico-class battleship the anti-torpedo defenses consisted of an outer hull, four intermediate anti-torpedo bulkheads (three of which consisted of 0.75-inch (19mm) armor plate), and an inner holding bulkhead, with a total depth of about eighteen feet. The intermediate voids between these bulkheads could be filled with water or fuel oil to stop fragments and absorb the blast effect.

Like all navies, the Japanese wanted torpedoes powerful enough to breach these defenses. When the terms of the Washington Naval Treaty called for the disposal of the hull of the partially completed battlecruiser *Tosa,* they used the opportunity to experiment. They detonated a 200-kilogram charge against the ship's hull below the waterline. The charge blew open the hull and created a hole about twenty-four feet wide; most importantly, it also severely breached the inner holding bulkhead. The Japanese also built models of the defenses of Colorado-class battleships, and performed a large number of subscale explosive tests. From these experiments, the Japanese concluded that their Type 91 Mod. 2 aerial torpedo would breach any existing U.S. passive anti-torpedo defense system.[21]

At Pearl Harbor torpedoes hit *Oklahoma, Nevada, West Virginia,* and *California.* The effectiveness of the anti-torpedo defenses in *Oklahoma* cannot be properly evaluated as they were crushed as the ship rolled over. None of the torpedoes that hit *Nevada, West Virginia,* or *California* breached the innermost holding bulkhead.

Nevada was one of the oldest of the battleships, nearing retirement and in poor material condition. She was hit in a location where the torpedo defense system was the weakest, between the forward main battery turrets. Her holding bulkhead was deformed with some leaks at the butts and joints, but the bulkhead otherwise held.

None of these three ships experienced the complete defeat of the anti-torpedo system seen in *Tosa.*

What went wrong with the Japanese testing program? A major part of the answer comes from within the assumptions that the Japanese engineers made while conducting their tests. In designing their underwater protection systems, the Japanese were more concerned with defeating "diving shells," that is, gun shells that hit short of the ship and traveled underwater to hit the ship below its belt armor. They found that empty anti-torpedo voids performed best against diving shells. They made the assumption that American naval architects would come to the same conclusion. So, they sized their aerial torpedo warheads to be powerful enough to penetrate a "dry" system.

However, U.S. naval architects were most concerned with the damage effects caused by torpedoes. Against torpedoes it was better to have several of the intermediate anti-torpedo voids filled with either water or fuel oil. The Japanese aerial torpedoes had insufficient explosive force to breach this type of system.

For the historian, the lesson is: just because there is a technical report in the archives that claims that a particular weapon is the scourge of the seas, it does not necessarily make it so. Period technical reports have to be verified by later physical tests or, better, by combat experience. Assumptions must always be examined. Assumptions are often buried in some appendix of a report, or more often not reported at all. Yet, as seen again in the next section, assumptions can make the difference between an effective and ineffective system.

Bombsight Technology: The D3A Val Dive Bomber over Pearl Harbor

The dive bomber was one of the most effective new technological systems seen at the beginning of World War II. The Japanese D3A Val dive bomber was extremely accurate when employed as intended. For example, it is reputed to have scored on

[21] U.S. Naval Technical Mission to Japan, *Characteristics of Japanese Naval Vessels, Article 9, Underwater Protection,* January 1946.

the order of 90% hits and damaging near misses in sinking the British heavy cruisers *Cornwall* and *Dorsetshire* in the Indian Ocean in early 1942.

The Japanese Type 96 dive bomber bombsight was mounted above the engine cowling forward of the cockpit. It was a long telescope that protruded out of the front of the cockpit. The sight picture consisted of grid lines like that seen on graph paper, with a central dot.

The Japanese had developed a unique method of team dive bombing. Attacks were made in *shotai* of three aircraft attacking from 3,000 meters altitude at about thirty-second intervals (as opposed to ten-second intervals used by U.S. dive bombers). The lead bomber would center his sight picture central dot on a standard place on the target, usually on the top of a ship's foremast. He would attack in a sixty-degree dive and release his bomb at an altitude of 400 to 600 meters. Where the bomb would hit relative to the aim point was assumed to be the result of wind drift and target motion. The following dive bomber would then adjust his aim point—if the bomb hit four squares up and three to the right of the aim point on his bombsight, then the pilot would shift his aim point to four down and three left, which should compensate for the wind and target motion. If either the second or third dive bomber in the *shotai* scored a hit, the *shotai's* attack was considered a success. Using this method, the D3A Val regularly achieved 60% hits in practice sessions.

Based on this hit rate, the Japanese could expect to achieve forty-nine hits out of the eighty-one dive bombers planned against fleet targets. They achieved far less than this. Only nine bombs hit targets at which they were aimed; another six hits came from bombs so badly off target that they hit some other ship. Not one of the dive bombers' hits was of strategic importance. Only one hit was achieved against their primary target, cruisers.

What happened? It was an issue of assumptions, environmental conditions, and failures in command and control.

There is a very famous photograph by a *Time* photographer at the beginning of the second wave dive bombers' attack on the harbor. Taken from a mountain slope to the north of the harbor, the photograph is a bit overexposed, but it gives the impression of a bright and sunny day, marred only by the black dots of AA fire and the ugly smoke from the oil fires coming from *Nevada,* Hickam Field, and the destroyers *Cassin, Downes,* and *Shaw.*

A nearly duplicate photograph was taken by a Navy photographer's mate from the same location at nearly the same time. Taken with a different exposure time, this photograph shows the clouds, a feature washed out by overexposure in the *Time* photograph. The Navy photograph shows a harbor covered by nearly 100% cloud cover between 3,000 to 5,000 feet, with the morning sun breaking through the clouds leaving a streak of light across the Submarine Base and Battleship Row, illuminating the ships below.

Other photographs taken just before the arrival of the second wave attack also show heavy cloud cover. The Pearl Harbor weather forecast for the day broadcast by a civilian radio station called for solid cloud cover between 3,500 and 5,000 feet.

The Japanese technique for dive bombing required clear visibility of the target from their roll-over altitude of 3,000 meters. All members of the *shotai* needed to be able to see the target before the attack was initiated, and needed to see the impact point of the leader's bomb just after they dove into their attack. With the cloud base down to 3,500 feet, this technique was impossible.

Observers on the ground noted that during the attack some Japanese dive bombers would break through a hole in the cloud deck and then attempt to line up on a target. Many delivered their attack from very steep dives, greater than sixty degrees, while some "glide bombed" with angles of descent of thirty degrees or less. As their bomb sights were calibrated assuming a sixty-degree dive, the pilots had to drop their bombs by eye, unassisted, something they had not practiced. The dive bombers' sight and technique was effective only under certain conditions; it was not robust enough to be effective under the less optimal conditions experienced during the attack. Thus, accuracy was a fraction of expectations.

With the cloud conditions and no radio communications, there was no way to direct the attack against the best targets. While being subjected to heavy AA fire and with their visibility blocked by clouds, the dive bombers had difficulty identifying valid targets and delivering their attacks. Considering that the dive bombers' primary targets were cruisers, most instead attacked whatever target of opportunity they noticed as targets became visible as they plummeted through the clouds. Destroyers were attacked and the aviators reported that they had bombed cruisers; tenders were attacked and the aviators reported that they had struck battleships. Some intentionally attacked battleships due to a miscommunication on the flight deck prior to the second wave's takeoff which they took as new directions; without radio communications, they had no way of clarifying their target assignments while aloft. As recorded in U.S. logs and reports, only seventeen of the bombs were released against cruisers, which should have been their objective; attacks otherwise were scattered willy-nilly over the entire harbor, with even some attacks against auxiliaries and destroyers outside the harbor.

The technological lesson is obvious: ensure that any new system can operate in all the expected environmental conditions. If it cannot, there must be some kind of effective backup.

Conclusion

This study suggests a number of points regarding the effectiveness of new technology in the Japanese attack on Pearl Harbor: 1) Introducing new technology does not guarantee improved performance; 2) Introducing new technology does not guarantee that it will be employed in the most effective manner. The new technology may require new approaches, may require technological developments in supporting areas, and must be coupled with appropriate tactics; 3) Culture trumps technology. The technology will be ineffective if you cannot convince people on the deckplate or cockpit level to use it, and to use it properly. A new technology can be stymied by the organizational or military culture of the force attempting to employ it; 4) A military force must constantly check its assumptions, and reconfirm them to be valid for all possible conditions. Assumptions must not be allowed to dominate the answer to any problem. Historians must be aware of these assumptions, and understand the place of these assumptions in the historical picture. When analyzing why a particular force failed, check the assumptions under which they were operating—there may be a causal linkage; 5) Technologists must perform sensitivity analyses to validate their assumptions. Calculations must be revisited under varied conditions. Care must be taken regarding reuse of previous work—the assumptions may not hold in the new environment; 6) There is a significant gap between test and training results and what can be expected to be achieved on the battlefield; 7) Training trumps technology.

In the case of the attack on Pearl Harbor, the Japanese were introducing a large number of technological and organizational changes into their naval aviation community. Their attack fell far short of its potential. Their greatest error was a failure to anticipate.

They did not understand the complexities that were involved in strikes involving a large number of carriers and massive numbers of aircraft. It was a significant achievement just to launch from six carriers simultaneously and marshal nearly 200 aircraft, a most impressive, world-leading achievement, but they did not go the next step to ensure that this mass of aircraft was effective over the target. Telling leaders of small sections of three to nine aircraft to independently pick their target will work for a strike of twenty aircraft, but not with one of 200. They needed procedures to prevent collisions, deconflict attack runs, and identify and kill the best targets while preventing wasting ordnance by overkilling the easiest targets. They needed cooperation between the aircraft types. In short, they needed positive command-and-control, not individual *samurai* heroism.

In the case of the torpedo bombers, their cultural values and established techniques led them to retain the old ways of assigning targets, placing the responsibilities on the individual aviators. There was no effective command and control vested in the more experienced commanders. They had a potential solution for this problem in the FM voice radio, but technological problems with installation of these systems caused the Japanese aviators to discard that prospective approach.

This brief study also has detected significant failings in the Japanese ashore support services. An extensive effort to determine the best form of anti-torpedo underwater protection was undone by some calculation errors and one faulty assumption, which shaped both the design of their own ships' anti-torpedo protection as well as the determination of the weight of explosives required in their own aerial torpedoes. Improper manufacturing techniques, as well as a clearly inadequate testing program, resulted in bombs that were far less effective than they should have been.

Lastly, a dive bombing system was developed that was dependent upon clear skies and good visibility for proper execution. This system was not robust enough to handle poor weather conditions; again, the Japanese failed to anticipate, in this case the problem of low-lying clouds, and did not have a backup plan in case that eventuality came to the fore.

A part of a good leader's responsibility lies not only in developing new innovations, but also in understanding the ramifications of a technological advance. The extraordinary leader anticipates the future, and understands fully the strengths and weaknesses of his instruments, and their implications. The Japanese attack on Pearl Harbor serves as an excellent example of the perils of technological innovations.

4

The Dominions and British Maritime Power in the Second World War

Iain E. Johnston-White

There has been increasing acknowledgement in recent World War II historiography of the UK's reliance on its empire to first survive and then take the offensive against the Axis powers.[1] While national historians outside the UK have long explored their respective dominions' or colonies' contributions to the cause—through war histories or in narratives on the development of nationhood—war historians for some time largely considered the "British effort" as an umbrella term, incorporating the empire in its entirety.[2] If the imperial contribution is highlighted now more than ever before, there has still been limited attention paid to one group in particular: the British Commonwealth of Nations.

The UK and its self-governing dominions—Canada, Australia, New Zealand, and South Africa[3]—formed an imperial sub-group, a voluntary alliance within and yet also curiously aloof from the empire. And while the question of whether the entire empire was a source of strategic strength or weakness to the UK as war approached has been the source of considerable debate,[4] there can be little

[1]Examples include, David Edgerton, *Britain's War Machine: Weapons, Resources and Experts in the Second World War* (London: Penguin, 2011), and Ashley Jackson, *The British Empire and the Second World War* (London: Hambledon Continuum, 2006). This chapter was originally published in *Global War Studies* 11:1 (2014).

[2]Several examples include, Martin Kitchen, *A World in Flames: A Short History of the Second World War in Europe and Asia, 1939–1945* (London: Longman, 1990); Williamson Murray and Allan R. Millett, *A War to be Won: Fighting the Second World War* (London: The Belknap Press of Harvard University Press, 2000); Richard Overy, *Why the Allies Won* (London: Jonathan Cape, 1995).

[3]Newfoundland's independent dominion status had lapsed due to the effects of the Great Depression and its subsequent inability to service its own debt, while the Irish Free State determinedly maintained neutrality throughout the war.

[4]This debate continued for some time, see for instance, Correlli Barnett, *The Collapse of British Power* (Gloucester: Alan Sutton, 1972), p. 566; Michael Howard, *Continental Commitment: The Dilemma of British Defence Policy in the Era of the Two World Wars*

argument about the value of the dominions. Nevertheless, the relatively seamless nature of the Commonwealth alliance can sometimes obscure the scale of its success. An initial look at the figures, for instance, shows that the RAF had 487 squadrons by 1945, and the dominions were providing 100 of them;[5] delving deeper, however, it becomes apparent that over 60% of all imperial aircrews were trained in the dominions.[6] Indeed, the extent of the dominions' contribution in several areas was nothing short of essential to British strategy—a point made startlingly apparent by considering the war at sea.

On the eve of the Second World War, the UK's worldwide empire was dependent upon the sea to as great an extent as ever before. The growth in air travel had provided an alternative and quicker mode of transport, but global sea lanes were the arteries of the British world system, moving men and bulk trade around the Commonwealth and colonies on an unrivalled scale. The UK remained the predominant sea power after the First World War, but the interwar period challenged this position of primacy. The end of the Anglo-Japanese alliance in 1922 weakened the UK's position in the Far East, while the naval treaties of 1922 and 1930 limited British freedom in expanding the Royal Navy (RN). The growing naval strength of a number of other powers—the United States, Germany, Italy, and Japan—meant that by September 1939 Britannia could no longer claim to rule the waves.

Nevertheless, British war plans were still primarily based upon naval dominance. It did not take long, however, in 1939–40 for the shortcomings of a strategy based around the naval blockade of Germany to become apparent, even before the Nazi conquest of continental Europe neutralized its effect; nor did the enemy waste any time in exposing the vulnerability of the Commonwealth's reliance on its merchant fleet. All combatants had to face the problem that the Second World War was a highly mobile conflict, requiring the global movement of fighting forces to an unprecedented degree. Heavily dependent upon imports and sea communications with its allies,[7] however, the UK relied on merchant shipping not only to conduct war, but also for the fundamental purpose of subsistence at home. German Admiral Karl Dönitz, senior submarine officer in—and later head of—the

---

(London: Maurice Temple Smith, 1972), p. 100; G. C. Peden, "The Burden of Imperial Defence and the Continental Commitment Reconsidered," *The Historical Journal* 27:2 (June 1984), pp. 405–23.

[5]Jackson, *British Empire*, pp. 38–39.

[6]The National Archives, London (hereinafter TNA), AIR 10/551 Air Ministry, *Second World War, 1939–1945: RAF flying training, policy and planning* (London, 1952), Appendix No. 3, Analysis, by Air Forces, of the Total Output of Qualified Aircrew, pp. 279–80.

[7]For all of its oil and more than 50% of its food (in calories), Lizzie Collingham, *The Taste of War: World War Two and the Battle for Food* (London: Penguin, 2011), p. 105.

German Navy, noted that "England[8] was in every respect dependent on sea-borne supply for food and import of raw materials, as well as for development of every type of military power. The single task of the German Navy was, therefore, to interrupt or cut these sea communications."[9]

While the UK could boast of the largest pool of merchant shipping in the world in 1939,[10] it also had the largest empire to service, including the dominions.[11] During the war, conflicting demands meant that the strain on shipping became acute. Substantial tonnage was required not just to begin any sizeable offensive, but also to maintain supplies for active theaters. Evicted from continental Europe, the UK found its main land theaters in North Africa and the Middle East—a steady stream of men, munitions, and materiel needed to be transported there from the UK, the dominions, and colonies, demanding vast amounts of shipping traveling long distances. The experience of Operation Torch,[12] which coincided with the heaviest month of shipping losses for the Allies, was a clear example of just how intimately shipping was tied to active operations.[13] The movement around the Commonwealth of men for the air training schemes, the transportation of muni-

---

[8]Used by Dönitz (and the German military) as a synonym for the UK. Dönitz also regularly used "England" to describe the whole imperial effort. Canada, for instance, is not mentioned at all in his description of the Battle of the Atlantic. Dönitz, "The Conduct of the War at Sea." Published by the Division of Naval Intelligence in 1946, and reproduced at http://www.ibiblio.org/hyperwar/Germany/KM/WarAtSea/index.html#A.

[9]Ibid.

[10]This was the combined ocean-faring shipping resources available to the Commonwealth, the allocation of which was controlled from London by the Ministry of Shipping (from 1941, this role was assumed by the Ministry of War Transport). The USA's entry into war meant that allocations of tonnage by London and Washington were overseen by the Combined Shipping Adjustment Board, to avoid wastage or duplication of effort. This was particularly important for Australia and New Zealand as they provided a base for American forces and thus it was not always entirely clear which source of shipping, American or Commonwealth, should provide particular imports. TNA, MT 59/378, Australia and New Zealand Shipping Requirements, Letter from MWT to British Shipping Mission in Washington, 30 October 1942.

[11]In September 1939, the UK possessed about one third of the world total of merchant tonnage. In 1914, the figure had been 48%—again reflecting the general decline of the British naval position. C. B. A. Behrens, *Merchant Shipping and the Demands of War* (London: HMSO, 1955), pp. 2–3. Kennedy puts these figures at 42% in 1914 and 26% in 1938, Paul Kennedy, *The Rise and Fall of British Naval Mastery* (London: Fontana, 1991), p. 318.

[12]The Allied invasion of French North Africa in 1942.

[13]Even more so because defeating the Axis in North Africa was partly justified by the fact it would re-open the Mediterranean to Allied shipping and thus ease the strain on resources. Shipping diverted for the preparations for Operation Torch additionally cost the UK around a 30% reduction in imports during 1942. Collingham, *Taste of War*, p. 111.

tions from the major supply sources in the UK and North America to allies,[14] and the requirements of any new theater that the UK hoped to fight in all had to be met from the shipping pool, with enemies all the while sinking as much British tonnage as possible to deplete shipping resources.

In sharing a resource as greatly strained as shipping, the Commonwealth was compelled to work—and suffer—together to meet the challenges the Axis and global strategy posed. In the North Atlantic, the UK increasingly turned to Canadian assistance to combat the submarine threat. At the Cape of Good Hope, it was left to South Africa to facilitate the flow of supplies to and from the eastern empire once Italian belligerency had severed the line through the Mediterranean to Egypt and beyond. Geography and Japanese belligerency meanwhile meant that deference to Allied strategic plans, dependence on America, and a willingness to go without were often the most valuable contributions the Antipodes[15] could offer to the shipping cause. Nevertheless, the Commonwealth's special place and importance in British strategy meant that it did not suffer like other parts of the empire due to the shipping shortage and it had a uniquely important role in the maintenance of Britain's seaborne communications network. Having acknowledged the UK's place at the center of "a maritime Commonwealth," Churchill claimed shipping was "the . . . sole foundation of our war strategy."[16] The dominions played a vital role in ensuring its safe passage worldwide.

British Naval Strategy to 1945

British strategy had long been built around the strength of the RN. The importance of naval power was highlighted by defense expenditure in the interwar years, with the Admiralty annually receiving the majority of British defense funds down to 1938. These figures reflected the reality of imperial defense as much as the protection of the British Isles—if each portion of the empire was ostensibly responsible for its local defense, the pieces looked to the RN to keep the whole connected. Even though the dominions of Canada and Australia had opted against contributing to the RN in favor of establishing their own navies and the Union formed its own

[14]In particular, supplying the Soviet Union was a significant drain on shipping from 1942. Library and Archives Canada (hereinafter LAC), United Nations Shipping Position, August 1942—C. D. Howe Papers, MG27 III B20 Vol. 57, S.23-4, "Combined Shipping Adjustment Board 1942–43," United Nations Shipping Position Report, August 1942. The Allies did not start supplying Russia until August 1941 and by the end of that year sixty-four ships had sailed. During 1942, that number increased to 256 and by the end of the war 813 vessels had attempted the journey. S. W. Roskill, *The War at Sea, 1939–45*, Vol. 3, *The Offensive*, Part 2 (London: HMSO, 1961), Appendix R, pp. 432–33.

[15]Australia and New Zealand.

[16]J. R. M. Butler, *Grand Strategy*, Vol. III, *June 1941—August 1942*, Part II (London: HMSO, 1964), p. 548.

independent South African Naval Service (SANS), the RN remained the major component of every dominion's maritime defense. The psychological impact on Australia of the failure of the Singapore strategy in 1942, for instance, displayed how deeply the faith in the RN had been held in areas as distant from the Home Waters as the South Pacific.

Favoring international treaties which facilitated global disarmament and limitations on naval forces over alliances, London developed a one-power standard in the interwar period—effectively a pledge to remain at least as strong as the largest navy in the world. This policy was combined with a strategy to keep that one power, the USA, within naval bounds that the UK could afford to compete with while simultaneously keeping the biggest threat—Japanese naval power in the Pacific—at manageable proportions. Moves in this direction coincided with the lapse of the Anglo-Japanese alliance in 1922 and were further enshrined through the 1930 London Naval Treaty. If consigning the RN to treaty limits for the first time was a sign that the era of British "naval mastery," to borrow Kennedy's phrase, had ended, this was not the intention of the architects of these agreements. At the Admiralty, the realization that treaties could placate rival powers and influence their building programs favorably, exactly when British naval mastery was most open to challenge for over a century, meant that naval agreements were considered to be a method to preserve the UK's position until it was ready for another all-out naval arms race.[17]

It was not until after the 1930 treaty lapsed on 31 December 1936 that the RN could follow the rearmament trend. Despite the growing threat of Germany and the unclear intentions of Italy and Japan, the British Government continued to place its faith in treaties, such as the Anglo-German Naval Treaty of 1935,[18] and in the one-power standard—going against the ambitious hopes of the Committee of Imperial Defence (CID), which recommended the creation of a two-power standard aimed towards Germany and Japan. In any case, the British shipbuilding industry, having deteriorated in the interwar period and just recovering from the grave effects of the Great Depression, could not have kept pace with the larger scheme and was at full stretch just meeting the requirements of the one-power standard as war approached. During the interwar period, the expense of warship tonnage rapidly increased while the building capacity of British yards had halved by 1935.[19] In the face of growing challenges and threats, these were both limiting factors as the RN rearmed.

[17]Kennedy, *Naval Mastery,* pp. 325, 339; Joe Maiolo, *Cry Havoc: The Arms Race and the Second World War, 1931–1941* (London: John Murray, 2010), pp. 102–23.

[18]The 1935 Anglo-German Naval Treaty permitted Germany to build its Navy up to 35% the strength of the RN. This was in excess of the limit placed on Germany's Navy by the Treaty of Versailles, and additionally undermined the Stresa Front of the UK, France, and Italy due to a lack of consultation with the latter two powers.

[19]Maiolo, *Cry Havoc,* pp. 105–06.

The fall of France quickly cast doubt on two pre-war naval assumptions: that the blockade could break the Reich's economy and that the major threat to British merchant shipping would be surface vessels. Instead, it was the UK's economy which had to face the challenge of sustaining imports and it was U-boats which presented the greatest threat.[20] Furthermore, technological developments meant that this conflict was very different from the Great War. Air power transformed naval warfare in a way few had predicted before the outbreak of hostilities, not least through the effectiveness of direct assaults on battleships and the importance of reconnaissance aircraft in locating threats. Air power therefore represented both a new threat to the RN and a new element to be incorporated into British sea power. The importance of the latter factor was demonstrated in the way focal areas for submarine attacks were moved from Allied coasts to the mid-Atlantic air gap as aircraft were more successfully employed. Intelligence, too, became central to the outcome of the shipping war, as dominance in the Atlantic could swing from one adversary to the other through adroit utilization of Ultra or as Enigma cipher codes were broken or altered.[21]

For submarines, convoys were the UK's answer. While pre-war anti-submarine preparations had been underwhelming, due to the belief that Germany would not wage unrestricted submarine warfare for fear of antagonizing the Americans, the early loss of a British ship was enough to ensure that work towards a considerable convoy system began in earnest.[22] The success of the convoy system was fundamental to British survival and Allied victory. Despite suffering a very similar average monthly loss of tonnage in both wars,[23] the number of ships sunk was forty-one per month during the Second World War, compared with ninety-five in the Great War—and in the latter war this was achieved largely without assistance from the French Fleet.[24] Convoys alone were by no means a panacea for the U-boat threat—in September 1940, for instance, forty out of fifty-nine sinkings occurred in convoys—but they were a vital element in securing the safe passage of shipping across the oceans.[25] Anomalous statistics such as the aforementioned must be read in light of German code-breaking of Admiralty messages, improved German technology and naval tactics, weather conditions, bad luck, and an

[20]Although they did operate on the surface intermittingly, the challenge of defeating them was very different to regular surface vessels due to their ability to rapidly submerge.

[21]See F. H. Hinsley et al., *British Intelligence in the Second World War*, Abridged Version (London: HMSO, 1993).

[22]The SS *Athenia*. S. W. Roskill, *The Navy at War, 1939–1945* (London: Collins, 1960), p. 35; Roskill, *A Merchant Fleet in War: Alfred Holt & Co., 1939–1945* (London: Collins, 1962), p. 22. Within a month, convoys were established for all of the most valuable shipping routes.

[23]Approximately 215,000 tons per month. TNA, ADM 1/19387 Battle of the Atlantic.

[24]Ibid.

[25]ADM 223/220 Battle of the Atlantic, 1939–44.

increasing percentage of shipping being placed in convoy as opposed to independent sailing. The effectiveness of convoys began early—only seven of the first 164 ships sunk were in convoy—and remained relatively consistent through the war.[26] Of 5,756 ships which sailed in convoy to the end of 1939, only four were sunk by submarines.[27]

The Royal Navy therefore had to meet a mammoth challenge—all waters that were navigable to British merchant vessels were potentially areas where protection would be required.[28] Utilizing the convoy system to meet this threat, even in just the most crucial areas, required a large number of escorts, consistently above prewar assumptions.[29] Even at the outbreak of war, when the optimistically low number of seventy cruisers was presumed necessary, the Admiralty had only around fifty at its disposal—including those contributed by the dominions.[30] As the war progressed, commitments for British shipping continued to grow and requirements for escorts increased concurrently. The RN simply did not possess the capability to meet these demands alone.

The Dominions and Naval Power

The dominions provided a significant amount of assistance. Dominion naval power in 1939, however, was inauspicious at best. Despite the dominions maintaining separate navies,[31] the entire Commonwealth was united by its reliance on the RN to provide true naval security. South Africa and New Zealand were incapable of producing major vessels,[32] while even Australia and Canada, each with their own Royal navies, had relatively limited shipbuilding capacities, which required the stimulus of wartime expenditure to facilitate an effective contribution. Furthermore, training methods, institutional structures, and vessels were British or local derivations of RN models, so that the naval forces of the dominions were operationally little more than an addition of the RN itself. The relationship

[26]Ibid.

[27]J. R. M. Butler, *Grand Strategy*, Vol. II, *September 1939—June 1941* (London: HMSO, 1956), p. 84.

[28]LAC, W. C. Clark Papers, RG19 Vol. 3988, "Estimates Navy," Appreciation: Canada's Naval War Effort During 1941.

[29]G. C. Peden, *Arms, Economics and British Strategy: From Dreadnoughts to Hydrogen Bombs* (Cambridge: Cambridge University Press, 2007), p. 179.

[30]Table 4.2 shows that on paper the Admiralty in fact had more than fifty, but, just as with the American destroyers exchanged in return for bases later in the war, many needed to be refitted and updated before they could see action or simply were not suitable for working with fleets. Butler, *Grand Strategy*, Vol. II, p. 25.

[31]With the exception of New Zealand at the outbreak of war.

[32]South Africa did build minesweepers and convert trawlers.

Table 4.1: Strength of the Royal Navy and dominion navies at the outbreak of war—personnel numbers

Navy	RN	RCN	RAN	RNZN	SANS
Permanent Force	118,932	1,990	5,440	1,339	6
Reserves	69,754	1,700	4,819	670	600
Total	118,686	3,690	10,259	2,009	606

Note: Throughout the war, thousands of RN personnel were of dominion nationality.

was such that interwar naval treaties designed to limit RN expansion also tied dominion hands, too—in the eyes of the world the dominion forces were local extensions of the RN.

In practice, this proved largely to be the case. New Zealand's naval forces had this status officially until October 1941; the Royal Australian Navy (RAN) did not take long to fall entirely under Admiralty control once war began; South Africa's forces, except for its Royal Naval Volunteer Reserve (RNVR SA), were negligible; and Canada, ever the ostensibly independent dominion under Mackenzie King, nevertheless instructed Royal Canadian Navy (RCN) forces to cooperate with the Admiralty to the fullest, largely making it a *de facto* extension of the RN.[33] The Admiralty therefore exerted significant control over the direction of the naval forces of the entire Commonwealth.

How great was this addition to RN strength? Table 4.1 highlights the relatively small addition the dominion navies made to British naval personnel on the outbreak of war. The combined personnel numbers of 16,564, of which the RAN alone made up almost two-thirds, were dwarfed by the UK's figure of almost 190,000. As a total Commonwealth force of just over 200,000, the dominion element of naval personnel constituted only a little over 8%. Table 4.2 shows that these few men were responsible for manning an equally small number of ships.

The RAN once more made up more than half of the available dominion forces, yet even this number was completely overshadowed by the strength of the RN. At the other end of the scale, the SANS, comprising just six men (three officers and three ratings) and having retired its final two vessels of the interwar period, was effectively a non-entity.

The operational strength of the RN was therefore not vastly enhanced by the dominion forces of September 1939. Nevertheless, the small increment gained even at this stage was of elevated importance given that the Admiralty was faced

[33]C. P. Stacey, *Arms, Men and Governments: The War Policies of Canada, 1939–1945* (Ottawa: Queen's Printer, 1970), pp. 308–09. Stacey also described the RCN as "the most British of the Canadian services."

Table 4.2: Strength of the Royal Navy and dominion navies at the outbreak of war—ships

Navy	RN	RCN	RAN	RNZN	SANS
Cruisers	58	–	6	2	–
Destroyers	173	6	5	–	–
Sloops, Escorts, And Patrol Vessels	45	–	2	2	–
Capital Ships	15	–	–	–	–
Aircraft Carriers	7	–	–	–	–

with shortages in almost every area, particularly escort vessels. While the RN could not rely on a great deal of assistance from the dominions upon the outbreak of war, it was also the case that the small size and nature of the dominion forces meant that they were incapable of pursuing independent strategies and were therefore inclined to fall in line with Admiralty direction.

In the manning of merchant shipping, the UK was also central. While in Canada it was accepted that "the provision of merchant seamen, their training, care and protection" was "essential to the proper conduct of war," the UK was nevertheless still providing some 40% of the officers and engineers in Canada's merchant fleet by November 1943.[34] This provision of skilled seamen to the dominions was an important service which the UK was able to provide even under the strain of wartime conditions, a hegemonic burden that was necessarily undertaken to buttress the contribution of the UK's junior partners.

The chief task for which the dominion forces could be employed early in the war was in escorting convoys. Despite their small numbers, dominion vessels were largely suited to this task. The dominions, however, did more than just protect the movement of merchant shipping—they provided their own on a larger scale than ever before. While British merchant tonnage as a proportion of global supply had dropped compared to 1914, the overall tonnage remained at around 20.5 million gross tons and the 9,488 ships of which this was composed were dispersed globally on their usual duties when war broke out.[35] The contribution of the dominions, however, had risen by over 80% since 1914—from just over 1.6 million gross tons

[34]W. C. Clark Papers, RG19 Vol. 3988, "Advisory Committee on Merchant Shipping Policy," Order-in-Council quoted in "Canadian Merchant Seamen" Report, 10 October 1944; W. C. Clark Papers, RG19 Vol. 3988, "Advisory Committee Merchant Shipping Meetings," Minutes of a meeting of the Advisory Committee on Merchant Shipping Policy, 12 November 1943.

[35]Roskill, *Merchant Fleet in War;* Behrens, *Merchant Shipping,* Appendix 1 and p. 3.

to just under three million. The main contributors were Australia[36] and Canada;[37] South Africa had only 22,000 tons to its name.[38]

Australia and Canada continued to build merchant shipping during the war, although the latter to a much more significant extent than the former. Some 3.6 million tons of shipping emerged from Canadian yards; much less than the UK itself (8.3 million tons), but still comfortably more than the ten ocean-going merchant ships eventually produced in Australia.[39] The dominions could have produced a more significant effort in this field with London's approval, but the production of merchant shipping was stifled by other demands and a reliance on the USA to fill the gap in British shipping.[40] American production, however, did not come into full effect until the latter years of the war and the production of Commonwealth tonnage in the dominions was extremely useful before this occurred. Looking ahead to 1942, for instance, the Allies predicted that they would make a net gain of 500,000 tons when comparing new building with losses, while in the same period Canada and Australia were expected to build at least 450,000 tons. This seemingly small amount emerging from the dominions was therefore vital, because the margins between losses and production were so fine that without it the Allies could not expect to increase carrying capacity, with all the implications

[36]Australia had 489,000 deadweight tons of ocean-going dry-cargo shipping. Behrens, *Merchant Shipping,* pp. 203–04.

[37]This was in spite of a disappointing attempt at maintaining a government-operated merchant fleet between the wars. Despite initial success from 1919, this costly failure was incrementally run down and ultimately abandoned when the last vessels, running from Canada to the Antipodes, were relinquished to a private company in April 1936. W. C. Clark Papers, RG19 Vol. 3988, "Advisory Committee on Merchant Shipping Policy," Brief for Shipbuilding, publication by the Canadian Shipbuilding and Ship Repairing Association, pp. 55–56.

[38]The government owned three deep-sea freighters, totalling 15,000 tons, which carried trade to and from Australia. There was an additional 7,000 tons of privately owned shipping. Department of Economics, University of Natal, *The Port of Durban* (Durban: Department of Economics, University of Natal, 1969), p. 22.

[39]H. Duncan Hall, *North American Supply* (London: HMSO, 1955), p. 425. Australia failed to offer more because extensive or large shipbuilding required "both heavy capital investment and an advanced stage of industrial development which had not been achieved in Australia." S. J. Butlin, *Australia in the War of 1939–45: War Economy, 1939–1942* (Canberra: The Griffin Press, 1955), p. 168.

[40]The Admiralty, for instance, placed orders for merchant ships with Hong Kong, but not Australia, instead encouraging the building of warships in Australian yards. The demands of ship repairs, felt across the Commonwealth, were also a major factor in the Australian Government's decision not to pursue its most ambitious plans (of sixty vessels) for merchant shipbuilding. H. Duncan Hall and C. C. Wrigley, *Studies of Overseas Supply* (London: HMSO, 1956), pp. 483–84; Butlin, *War Economy,* pp. 172, 177.

this had for global strategy.[41] Despite being a small expense on UK tonnage as war approached—the estimated 17.7% of British tonnage that was employed in inter-dominion or dominion-foreign trade was fractionally more than what the dominions brought to the pool—the dominions provided a useful and somewhat unexpected addition to Commonwealth merchant ship production during the war.[42] It was, however, in protecting and maintaining rather than providing shipping where the dominions made their most significant contribution.

North Atlantic Convoys: Canada's Special Role

Thrown into the fight unprepared and suffering through a testing period once the submarine threat emerged in the western Atlantic, the RCN reached a low ebb of heavy losses and redeployment in the winter of 1942–43, before re-emerging as a distinct success in Atlantic operations. Indeed, no other dominion came close to matching the effort at sea which Canada produced. The RCN grew immensely from its humble interwar status and finished the war as a strong force in its own right. In 1935, Sir Maurice Hankey, Chairman of the Committee of Imperial Defence, had considered the Canadian Navy to represent the nadir not only of the Canadian military services, but possibly even of all the dominions;[43] by the end of the war it was the third largest Allied navy[44] and had made a contribution that was fundamental to Allied victory.

---

[41]TNA, CAB 123/86 The UK Import Programme: Reports by the Shipping Committee.

[42]Behrens, *Merchant Shipping*, p. 5—based on 1936 statistics. The four dominions combined by December 1940 were estimated to cause a net loss of only 500,000 tons of shipping to the UK, significantly less than the 1.4 million tons they had required from the UK in 1918. CAB 21/1240 Shipping Committee General Correspondence.

[43]CAB 63/81 Briefing of the defence situation in Canada, August 1934—March 1935, Impressions of Canadian Defence Policy, December 1934, paragraph 7. Even by 1938, when comparing naval expenditure in relation to every $1000 of external trade, Canada ($2.50) was spending significantly less than not only the UK ($75.22) and the USA ($93.22), but also Australia ($13.47). Naval expenditure per capita also compared unfavorably in 1938–39, with the Canadian Navy ($0.58) faring worse than the RN ($14.80), the United States Navy ($5.18), as well as Australian ($3.23) and New Zealand forces ($2.46). South Africa was not deemed worthy of comparison, having retired all of its warships. W. C. Clark Papers, RG19 Vol. 3988, "Estimates Navy," Figures for Naval Appreciation.

[44]In numerical strength, although this was a highly specialized naval force largely comprised of escorts. The main striking power the RCN possessed was destroyers, with the RN congratulating the RCN for becoming a "Big Ship" Navy only in December 1943 after the acquisition of two modern destroyers and a further two cruisers from the RN, which came into service in 1944. W. C. Clark Papers, RG19 Vol. 3988, "Naval Service Programme," Message from First Lord of the Admiralty, 24 December 1943.

RCN personnel doubled between 1936 and 1939, but it remained a painfully small contingent when war arrived, especially in light of the tasks it was soon to face. During the war, however, an impressive 106,552 personnel enlisted for the service—the force peaked in strength in January 1945, with 92,441 men and women serving.[45] This enlistment, over fifty-three times the number of personnel available in September 1939, was mirrored by a staggering growth in vessels. The six-destroyer RCN of 1939 was transformed into a force of 365 vessels. The weight of this vastly enhanced establishment was felt in the brutal Battle of the Atlantic.

The Atlantic theater was exceptional in its importance to British survival and was the setting for one of the most enduring and fiercely contested battles of the entire war. Key munitions, personnel, and exports were being sent to North America from the UK; more importantly, North America was an increasingly vital source of munitions and raw materials for the UK, especially as other areas fell under enemy control.[46] The shipping shortage additionally encouraged imports which required the absolute minimum of British tonnage, namely manufactured goods—of which North America was also a major producer. Although inter-imperial purchases often meant saving crucial dollars, the loss of shipping capacity involved on such long voyages further encouraged London to draw from North America's resources, which were relatively close at hand. Canada alone sent 70% of the armaments it produced to its allies, mainly the UK.[47] It is little wonder the Chiefs of Staff wrote, "our life continues so long as we . . . do not lose the Battle of the Atlantic."[48]

Table 4.3 shows how increasingly important munitions imports alone from North America were. In 1943 and 1944, the Commonwealth received over a third of its munitions from North America. For the UK, ensuring these imports arrived safely was a priority.

RCN involvement with escort duties began as two of Canada's six destroyers accompanied the same number of RN cruisers in sailing with the first convoy to depart from Halifax on 16 September 1939. With London aware of the large gap in escort vessels available to the Commonwealth and Ottawa equally awake to the massive role it could play in the Atlantic, the latter quickly took steps to improve

[45]See Canada entry in I. C. B. Dear and M. R. D. Foot, eds., *The Oxford Companion to World War II* (Oxford: Oxford University Press, 2001, *Oxford Reference* online edition).

[46]For instance, raw materials imported from the United States rose from 1,874,500 tons (8% of total raw materials imported) in 1939 to 5,614,800 tons (37%) in 1941. From Canada, the proportion rose from 10.4% to 15% in the same years. Over half of the UK's raw materials were therefore coming across the Atlantic by this time. Joel Hurstfield, *The Control of Raw Materials* (London: HMSO, 1953), pp. 160, 161, 164.

[47]The UK received 60% of the tanks, 67% of the artillery, and 70% of the rifles produced in Canada. Hall and Wrigley, *Overseas Supplies*, p. 46; Jackson, *British Empire*, p. 63.

[48]Butler, *Grand Strategy*, Vol. II, p. 579.

Table 4.3: Total British Commonwealth supplies of munitions and percentage from each source. See, H. Duncan Hall, *North American Supply* (London: HMSO, 1955), p. 428

	1939 (Sept. - Dec.) and 1940	1941	1942	1943	1944	1945 (1st half)	Total
Total Supplies ($ million)	9,200	13,000	19,000	24,800	24,700	9,300	100,900
Percent from:	%	%	%	%	%	%	%
United Kingdom	90.7	81.8	72.6	62.4	61.2	66.1	69.5
Canada	2.6	5.2	8.6	8.8	8.9	10.0	7.0
Eastern Group	1.1	1.5	1.9	1.9	1.2	1.7	1.6
Purchase in USA	5.6	9.1	4.7	2.4	1.5	1.2	3.7
USA Lend-Lease	–	2.4	12.2	24.5	27.2	21.0	17.3

RCN strength. Indeed, the nature of the RCN's growth became just as important as its extent. As early as February 1940, contracts were placed by the Canadian government for sixty-four corvettes—small escort ships which were to form a crucial part in the protection of convoys throughout the war. Eventually, Canada built 122 of the vessels.[49] The ships produced domestically and acquired from abroad displayed Ottawa's commitment to protecting shipping as its primary and most important contribution at sea.

Despite the enthusiastic build-up of strength, however, the RCN encountered numerous difficulties early in the war. With the RN at full stretch, it provided little

[49] Canada, Dear and Foot, eds., *Oxford Companion to World War II.* Mackenzie King was a keen proponent of shipbuilding in Canada from an early stage, writing in his diary that he had "stressed the necessity of shipbuilding time and again," as early as 29 January 1940. W. A. B. Douglas et al., *No Higher Purpose: The Official Operational History of the Royal Canadian Navy in the Second World War, 1939–1943,* Vol. II, Part I (St. Catharines: Vanwell, 2002), p. 81.

assistance to the small Canadian navy which was forced to expand rapidly to meet its new daunting role—a situation which resulted in Canadian crews largely learning on the job. Drawn from civilian occupations and thrown hastily into the fray, Canadian crews lacked experience and lengthy training when they were forced to face the established U-boat force.[50] Nevertheless, the growing RCN continued to shoulder as much of the North Atlantic burden as it could, freeing the RN for action elsewhere.

During the winter of 1940–41, the U-boat campaign was at one of its two most critical levels. Yet despite the battle being largely fought off the coasts of the British Isles, potential breathing space for Ottawa to train and develop its force did not materialize. The grievous situation in the east merely exerted more pressure on the RCN to rapidly transform from its inadequate pre-war size to an establishment capable of relieving the RN of as much escort responsibility in the western Atlantic as possible, thus allowing the Admiralty to focus on the most threatened area.

By mid-1941, the growing RCN had taken over all convoy escort duty in Newfoundland's territory. This move coincided with the first real submarine threat in Canadian waters, as U-boats were now pushed away from British coasts by a combination of increased convoy protection—in part aided by escort vessels relieved by the RCN—and the diminished possibility of German invasion, as well as an increasingly effective system of air protection and anti-submarine work by the RAF.[51] From spring 1941 to the end of the year, the U-boats mainly focused on the area under North American air cover and even more determinedly in the mid-Atlantic air gap.[52] This shift, combined with the dominant successes of British forces in sinking U-boats when compared to their RCN counterparts, were bad portents for Canada in 1942.

Indeed, the first six months of 1942 were catastrophic for Allied shipping losses, with U-boats enjoying rampant successes on the east coast of the USA and in the Caribbean. Compared to these losses, Canadian forces got off lightly; set against the areas under RN control, they fared very badly. U-boats operating west of Newfoundland near Canadian ports, as well as under Canadian air cover, took a toll of shipping well above that in the eastern Atlantic. The RCN responded by continuing its dramatic expansion to counter the threat—by September 1942, the force had 188 warships and was providing just under half the surface escorts available for the crucial North America to UK route.[53]

---

[50]Canada, Dear and Foot, eds., *Oxford Companion to World War II.* Dönitz makes a similar case about the imbalance in experience being important to the success of the U-boats when they started operating in the waters around the USA. Dönitz, "The Conduct of War at Sea."

[51]Dönitz, "The Conduct of War at Sea."

[52]This was between the areas which North American and British anti-submarine aircraft could operate.

[53]Jackson, *British Empire,* p. 69.

This unceasing expansion continued to come at the expense of operational effectiveness, however, as Ottawa continually chose to push for adequate escort numbers over a smaller, but better equipped and trained, force[54]—mainly because the latter would not have the capacity to cope with escorting the quantity of vessels that the UK desperately needed to travel east. It was an unenviable choice. In November and December 1942, the toll such a massive task was taking on the over-extended and under-prepared RCN became apparent, as Admiralty assessments showed that four out of every five merchant ships sunk in convoy during those two months were being escorted by the Canadians.[55] These statistics were due to much more than Canadian inefficiency: the RN zones under RAF air cover had the most experienced and effective anti-submarine defenses, necessitated by the poor British performance early in the war, and the USA had only relatively recently forced the U-boats back from its coasts given its own harsh experiences at the start of the year.

Canada was now bearing the brunt of the U-boat offensive and feeling the consequences, an experience the UK and USA had already suffered through. Nevertheless, the scale of the disaster prompted action, especially when November 1942 saw 712,000 tons of merchant shipping lost.[56] In light of these set-backs, it was agreed that the RCN should take responsibility for the less dangerous Gibraltar escort route, to gain experience and confidence without the intense pressure it was facing on the North Atlantic routes.

This decision, however, did not solve the problem in the North Atlantic. Losses were lower in January and February 1943, but still at an unacceptable rate. The RN's failure to resolve the issues its RCN partners had faced in the mid-Atlantic area highlighted that even the more efficient and trained force could do little except reduce the gravity of the situation in this area. Nevertheless, a combination of better air-support,[57] technological advancements in weapons and location

[54]The "backwardness" of Canadian warships when compared to the RN was constantly lamented by Canadian officers and Ottawa fought a continuous battle to rectify the problems. Canadian Lieutenant Commander Piers complained as late as June 1943 that "R. C. N. personnel do not get the chance they deserve. Compared to R. N. Ships doing the same job, H. M. C. Ships are almost invariably outdated by 12 to 18 months in the matter of new equipment." W. A. B. Douglas et al., *A Blue Water Navy: The Official Operational History of the Royal Canadian Navy in the Second World War, 1943–1945,* Vol. 2, Part II (St. Catharines: Vanwell, 2007), pp. 55–56.

[55]Canada, Dear and Foot, eds., *Oxford Companion to World War II.*

[56]A number almost equivalent to the tonnage lost in the first *nine* months of the war (790,817 tons). ADM 1/19387 Battle of the Atlantic.

[57]The RCAF played a major role in supporting the RCN in protecting shipping off the east coast of Canada. Canadian shore-based aircraft sunk more than forty percent of all U-boats that Canadian forces accounted for, Jackson, *British Empire,* p. 69. The decision at the Washington Convoy Conference of March 1943 to provide the RCAF with twenty very-

equipment, as well as successes by the intelligence services, meant that the Allies turned the battle around in the final days of March. In this same month, it was decided that the RCN should return with sole responsibility for convoy duty in the newly-formed North-West Atlantic Command. With a new confidence and operational capability, the RCN quickly established dominance in the area.

Yet losses had been such under RN control during the first days of the month that March was in fact the worst month for convoys in the war—68% of tonnage lost was being escorted.[58] Even though factors such as adverse weather and the small proportion of non-escorted ships helped create this unfortunate statistic, the fact that the RCN was not the consistent factor in convoy casualties in the western and mid-Atlantic areas had become obvious to all concerned. Furthermore, it was no coincidence that when the deeper problems of the anti-submarine war—such as broken British codes and inadequate air cover in the mid-ocean area—had been solved, the Allies enjoyed a superb three months against the U-boats (June to August 1943) in the North Atlantic and merchant losses were reduced and displaced to other theaters. Meeting the technological challenges of this new war had taken time, but the successful and widespread application of aircraft in the shipping war became increasingly decisive. Similarly, the breaking of the new Enigma key employed by the German Navy, from which subsequent Ultra intelligence provided the foundation for a steady rise in operational successes, was a critical element in this turnaround in the spring of 1943. The RCN, returning to the North Atlantic routes, played a large part in the implementation of these accomplishments at sea.

The Canadian experience was not therefore straightforward or easy, yet rapid expansion of a tiny force was achieved alongside eventual dominance in the north-west Atlantic theater. For a considerable period in 1943 and 1944, over 80% of North Atlantic convoy escort duties were the responsibility of the RCN. In fact, two thirds of the submarines sunk by the RCN came after March 1943, showing how difficult the early war experience had been—but also the great level of strength the RCN achieved.[59] In 1944, the RCN took sole responsibility for mid-ocean escorts and, despite some losses, this area became a zone where U-boats were being sunk faster than merchant vessels for the duration of the war. From April to September 1944, including the critical Normandy invasion period, all trade convoys between North America and the UK were under RCN protection.[60]

---

long-range aircraft to aid the escort carriers in closing the Atlantic air-gap was also important to the air coverage provided. Washington Convoy Conference, Dear and Foot, eds., *Oxford Companion to World War II.*

[58] Indeed, March 1943 marked the lowest level that was reached by dry-cargo merchant shipping in the Commonwealth pool during the war, at 18,449,000 tons. Behrens, *Merchant Shipping,* Appendix VIII.

[59] Canada, Dear and Foot, eds., *Oxford Companion to World War II.*

[60] TNA, DO 35/1207 UK and Dominion roles in the Battle of the Atlantic.

The threat presented to British survival by the offensive against Allied shipping was very real. While the winter of 1940–41 was the closest the Axis came to critically blockading the UK, the cost of the offensive continued to have a real impact on what the Allies could hope to achieve operationally as shipping and escorts were a consistent limiting factor. Indeed, the First Sea Lord noted that "the shipping shortage will, and does indeed already restrict our whole offensive strategy."[61] Sixty-four percent of all shipping losses occurred in the Atlantic and the British bore the brunt of the Axis' global effort—shouldering 54% of Allied and neutral shipping losses. The U-boat threat was the gravest, as they alone accounted for 69% of losses, sinking some 2,765 vessels.[62] In this frightening environment, the value of convoys was extremely apparent: 85,775 ships sailed in 2,889 trade convoys to and from the UK; 654 were sunk, the rate of loss being roughly 0.7%.[63]

In the Atlantic, some 75,000 ships sailed in 2,200 convoys during the war. The RCN was responsible for escorting over one-third of this total—since shipping resources were so strained and the UK desperately relied on North American imports, this was a massive contribution. While some notable achievements, such as single-handedly escorting a 187-ship convoy in the summer of 1944—the largest collective to make the journey during the war—without losing a single ship, prove great examples of the successes which the RCN enjoyed, the fundamental statistic is that Canada escorted 25,343 ships to the UK.[64] The RN could only have escorted these vessels and their crucial cargoes without RCN help at the expense of crucial offensive strategy elsewhere. Put simply, Canada played an intrinsic part in the UK's capacity to survive and fight—this was Canada's "irreplaceable" role in the war at sea.[65]

Shipping on the Cape Route: South Africa's Special Role

Since the Battle of the Atlantic threatened the UK's very survival, it quite rightly has a special and often comprehensive place in the history books of the Second World War. The global nature of the war, however, meant that the effects of one theater often rippled through many others. The Cape Route, so important once British shipping could not safely sail through the Mediterranean, was an example of an area remote from, but crucially intertwined with, many other theaters. Following Italian entry into the war, the Cape Route swiftly became a focus for Axis attacks: it linked the Atlantic to the Indian Ocean; it was the quickest route to the North African and Middle East land theaters in which the UK was conducting most of its fighting; and it had become the primary link between the whole empire

---

[61] CAB 101/121 JGG Wootton: Shipping.

[62] ADM 1/19387 Battle of the Atlantic.

[63] Roskill, *Navy at War,* p. 451.

[64] DO 35/1207 UK and Dominion roles in the Battle of the Atlantic.

[65] Douglas et al., *No Higher Purpose,* p. 27.

east of South Africa and the Mother Country. With American ships also increasingly rounding the Cape of Good Hope from 1942, the resources of British ports around Africa were tested to an unprecedented degree.

In the early phase of war in particular, the UK looked to its African and Asian colonies as well as the southern dominions for numerous supplies.[66] While the southern dominions and India were the only significant imperial contributors of munitions outside the UK and Canada, albeit still on a relatively small scale, the raw materials and food resources which they provided were of great importance to the UK.[67] Following the closure of the Mediterranean, two trends increasingly took hold for London—looking for replacement supplies from North America and sending increasing numbers of ships in convoys around the Cape of Good Hope. Immediately a plethora of facilities which were either inadequate or plainly non-existent were required from the British ports on African coasts.

South Africa formed the hub of the Cape Route—the point where Atlantic Ocean met Indian Ocean. While it was not the only area suddenly flooded with shipping,[68] it was the only dominion along the whole route and it was also home to a significant RN naval base at Simon's Town. Among the deluge of ships it was now to accommodate were those of the WS convoys.[69] The first sailing of this series of convoys came towards the end of June 1940 and was followed by convoys traveling on average once a month, carrying vital supplies and troops, first for the Suez military base, but later for the Far East and India, too. In addition to the movement of military supplies, inter-imperial trade and the transfer of raw materials around the empire were now largely conducted along this route.

Shipping on the Cape Route needed to be as expeditious as possible for several reasons. The two- to three-month journey around the Cape to Egypt already placed the Commonwealth at a significant disadvantage against the Axis, whose supplies for their North African forces were making the much shorter journey from Italian ports directly across the Mediterranean. Furthermore, shipping tied to this route and delayed or sunk for whatever reason was a disproportionately large drain on the whole shipping pool,[70] since the journeys involved were so much

[66]The UK had, for instance, imported in 1938 over 75% of its iron ore, ferro-alloys, bauxite, and flax from what became occupied areas of Europe (and their territories); from colonies subsequently lost to Japan, it had imported in 1941 over 75% of its rubber, raw silk, and tungsten ore. Hurstfield, *Raw Materials,* pp. 158, 167.

[67]See Table 4.3 above.

[68]Freetown was the first African port visited by most convoys on the Cape Route and similarly experienced problems with a lack of preparation and facilities for such a huge volume of shipping, in addition to sustained Axis attention.

[69]Named after Winston Churchill.

[70]Roskill, *Merchant Navy,* pp. 250–51. The eventual re-opening of the Mediterranean to Allied shipping was estimated to save the Allies one million tons of shipping.

longer and planning was required further in advance. Resources were so strained by the extra burden of traveling the Cape Route that British ships could not be found to transport 20,000 troops to the Far East in November 1941.[71] Compounding these problems was the inevitable need to accommodate the RN escort vessels which accompanied the heavy flow of convoys.[72]

The South African experience with shipping proved to be a microcosm of the Commonwealth's as a whole. As with most areas in the war at sea, submarines provided a central threat to shipping in South African waters. Unexpectedly, however, the enemy was initially of relatively little hindrance in this area. For the Axis, an assault on the Cape Route appeared somewhat unprofitable, given the distance and traveling time involved, problems with refueling, the abundance of still-exposed shipping north of the equator, and the lack of adequate submarine numbers even for the Atlantic.[73] For the Union and the UK this was extremely fortunate, as the virtually non-existent SANS was in no position to deal with a serious challenge early in the war, while the RN was immediately spread thin by wartime requirements.

As time progressed, however, South Africa steadily improved its forces. Lacking a single vessel at the beginning of September 1939, by the end of the year the SANS had fifteen ships in service, all whalers and trawlers converted into minesweepers and anti-submarine vessels. In January 1940, a new Seaward Defence Force (SDF) was created under a retired RN officer, which became responsible for the Union's naval defense. Having looked to the RN and the fledgling South African Air Force (SAAF)[74] to adopt the role of protecting South Africa's coast alongside its land-based coastal defenses, the Union now increasingly took responsibility for its own coastline. The close cooperation between Commonwealth forces—of the SDF, the SAAF, and the RN—was a key feature in protecting shipping on the Cape Route.

Early Union operations were mainly limited to minesweeping in the coastal waters of the Union. Despite being largely successful in this role, the freedom of movement of the German raider *Atlantis* between the Indian Ocean and the Atlantic displayed the weakness and lack of balance of South African and RN strength on the Cape. Nevertheless, Union naval strength continued to develop. In January

---

[71] Eventually the USA agreed to transport the men, despite not yet being a belligerent. Convoys, Dear and Foot, eds., *Oxford Companion to World War II.*

[72] Roskill, *Navy at War,* p. 33.

[73] L. C. Turner, H. R. Gordon-Cumming, and J. E. Betzler, *War in the Southern Oceans, 1939–45* (Cape Town: Oxford University Press, 1961), pp. 97–98.

[74] The SAAF made an impressive effort in the South African naval war. It attacked over 70% of enemy submarines operating within 1,000 sea miles of South African coasts, flying over 15,000 wartime coastal patrols. Andre Wessels, "The First Two Years of War: The Development of the Union Defence Forces (UDF) September 1939 to September 1941," *Military History Journal* 11:5 (June 2000). The importance of air power to the war at sea was again apparent.

1941, the Admiralty was able to request four of the SDF's minesweeping vessels for the Mediterranean. Another nine were subsequently sent, some not returning to the Union's waters until December 1945. The Union's token pre-war naval force, of half a dozen men, expanded to the size of 1,643 men and fifty ships by September 1941.[75] Growing from strength to strength, the SDF and RNVR (SA) merged in August 1942 to form the South African Naval Forces (SANF). The Union was now taking care of protecting its own coastal waters.

Like with the other dominions, South Africans regularly served in the RN (mainly RNVR (SA) volunteers) as well as alongside British forces in theaters outside South Africa. To maintain an individual identity, South Africans kept distinctive orange shoulder flashes on their uniforms and Union vessels flew the country's flag. Complications were raised by South Africa's "Red Oath"—essentially a pledge by servicemen to only serve anywhere in Africa, in the name of regional security. In the case of the minesweepers sent to the Mediterranean, this was bypassed by basing all South African forces at the naval base at Alexandria.[76]

The establishment of the SANF came just in time for the most challenging phase of the war in proximity of the Cape, between August 1942 and May 1943. The main threats to South Africa's coastal waters were U-boats and mines. For the latter, South Africa developed something of a speciality in dealing with the threat. The former, however, reaped a much greater harvest for the Axis. German tactics outside the main Atlantic theater involved moving small "wolf-packs," often numbering around six U-boats, to peripheral theaters for short periods of time. This meant catching weaker convoys or unprotected shipping until reinforcements were sent, at which point the U-boats were relocated to a newly exposed area. Despite an assault around the Cape having long been predicted, the U-boats successfully sunk twenty-four ships when they arrived in South African waters in October 1942, totalling 161,000 tons and including three troopships of a WS convoy.[77] When the U-boats returned in May 1943, they again found some easy short-term targets, before being forced to move on once more.

The Union therefore enjoyed some mixed results—strong and successful minesweeping efforts balanced by a chequered record against U-boats. The fact that U-boats failed to wage a sustained campaign in Union waters, combined with the relatively small loss of tonnage when compared to the vast movement of ships through its waters, however, was a testament to the significant defenses the Union developed in a relatively short period of time alongside the RN. The biggest shipping challenge for the Union, therefore, was not the protection of convoys which already arrived in the company of RN vessels, or indeed the anti-submarine and

[75]While these were not major warships, they were fit for purpose in South African waters, consisting of fifteen anti-submarine and thirty-five minesweeping vessels.

[76]J. C. Goosen, *South Africa's Navy: The First Fifty Years* (Cape Town: W. J. Flesch, 1973), p. 52.

[77]Roskill, *Navy at War*, p. 227.

anti-minesweeping tasks which its SANS and SAAF performed with significant effectiveness in combination with the RN.

In fact, the most intractable and persistent difficulties experienced by the Union came in the ostensibly basic area of providing supply and repair facilities for the vast tonnage of exhausted Allied shipping. These vessels looked to the African dominion as a haven on their lengthy Cape Route journeys, yet the Union's facilities often fell short of the task designated to them. Furthermore, South Africa was plagued by many of the same issues which every member of the Commonwealth shared, most notably in the rate at which over-worked tonnage was being repaired. Only Freetown, alone as a notable British port on the west coast of Africa, experienced comparable problems—but this was without the added expectations of dominion status.

Union ports, dealing with a level of tonnage never envisioned to be passing through them, needed to be efficient enough to allow the quick turnover of ships carrying—and sometimes loading and unloading—large cargoes and crews, while simultaneously requiring supplies such as coal and water. Furthermore, repairs required for this shipping placed a constant drain on both the space available at docks and, when not conducted at an adequate speed, to the shipping pool itself. There were no easy solutions to the backlogs created.

The extent of repairs required was deemed entirely unprecedented, even measured against the standards of the Great War.[78] The problem loomed large in London and the Union alike: during the critical winter of 1940–41 it was estimated in January that almost 13% of British shipping was unavailable pending repairs;[79] while it was thought at one point the Union was dealing with a situation in which over 60% of ships which docked in its ports required repairs of some nature.[80] That ships leaving the Union were headed either into the Atlantic or Indian Ocean, both areas consistently threatened by Axis submarines, only exacerbated the need to ensure they were adequately maintained while docked in South Africa. The whole situation reflected one cost of the strain that war placed on shipping, because merchant tonnage was constantly employed when not awaiting repairs or transferring cargo and subsequently suffered the effects of this continual use. The value of quality maintenance was also not in doubt—"maritime casualties," including the loss of worn-out or inadequately repaired ships, took their own considerable toll on

---

[78]W. C. Clark Papers, RG19 Vol. 3988, "Advisory Committee on Merchant Shipping Policy," Brief for Shipbuilding, publication by the Canadian Shipbuilding and Ship Repairing Association, p. 57.

[79]A problem aggravated by the loss of European repair yards, Butler, *Grand Strategy*, Vol. II, pp. 398, 238.

[80]TNA, CO 852/475/4 Cape route: report by C. G. Wurtzburg and R. H. Toleston, 16 December 1942.

British shipping.[81] The problem was exacerbated by the age of much of the tonnage. Ships originally released from the non-belligerent USA for the Allies to purchase, for instance, were all at least twenty years old.[82]

Yet the South African provision of services appeared to be failing to meet the requirements asked of it. At the ports, repairs were not taking place with enough speed,[83] while behind the front line the Union lacked sufficient railway capacity to move goods to and from the harbors with adequate haste to alleviate cargo congestion.[84] As Behrens has argued, the Union had "not enough dry-docks, not enough skilled labour, not enough railway wagons; not enough, indeed, of most of the necessary facilities and provisions, and as the shortages increased so did the sources of confusion." The tailbacks this caused, including up to twenty-five ships at a time just waiting to dock outside Union harbors, were calculated to cost the UK alone roughly 500,000 tons of imports a year.[85] Neither was confusion the only issue—the British mission sent to help the Union deal with these persistent problems noted that "nobody visiting South Africa for the first time from the United Kingdom . . . can fail to be struck with the general absence of that sense of 'urgency' so dominant at home."[86]

The effects of these difficulties were furthermore reciprocal. With at one time as many as seventy-eight ships idle in or outside Union ports,[87] inevitably the rate

---

[81]Such losses were quite substantial, usually exceeding the combined sinkings of aircraft, surface ships, and mines. Gerhard L. Weinberg, *A World at Arms: A Global History of World War II* (Cambridge: Cambridge University Press, 1994), p. 370.

[82]W. C. Clark Papers, RG19 Vol. 3988, "Canadian Shipping Board," Letter to Mr Heeney, Clerk of the Privy Council, Re: The Canadian Shipping Board, 15 May 1940.

[83]This was a problem which plagued the entire Commonwealth. South Africa further struggled because it relied on imports for almost all of its industrial machinery and war conditions made imports of most kinds hard to come by. Hall and Wrigley, *Overseas Supply,* p. 423.

[84]Behrens, *Merchant Shipping,* p. 226. The vast majority of tonnage was directed to Durban or Cape Town because alternative ports, such as Port Elizabeth and East London, had rail line and fuel provision shortages.

[85]Ibid., pp. 259–60. Congestion also existed in other Commonwealth ports, for Australia, see National Archives of Australia (hereinafter NAA), A5954, 508/3, "Capital Ship Dock, Sydney—Priority. Report on Ship Repair Facilities."

[86]CO 852/475/4 Cape route: report by C. G. Wurtzburg and R. H. Toleston, 16 December 1942.

[87]Behrens notes that the average for the first nine months of 1942 was about forty, adding that the problems the Union experienced at ports in the summer "reached proportions . . . never equalled anywhere else." Behrens, *Merchant Shipping,* p. 260. It has in fact been suggested that the peak at the Port of Durban reached a high of over eighty ships, with an average above fifty. At Cape Town, the average for June 1942 was forty-two, with a peak of sixty-two vessels outside berthage space. Even the latter figure could be an underestimation of the problem since vessels deemed surplus—at one point as many as eighty—were required to avoid queuing at Cape Town and temporarily anchor in Table Bay. Goosen, *South Africa's Navy,* p. 93; Turner et al., *Southern Oceans,* p. 96.

of imports from the Mother Country to the dominion fell. The volume had already decreased considerably to 40,000 tons a month by 1941, but with the escalation of port congestion it fell further to a paltry 5,000 tons a month in the spring of 1942. American deliveries, which had been at a rate of twelve to fifteen ships a month, were completely suspended in April 1942.[88] Washington was soon demanding that the UK instruct Pretoria to place fewer orders in the USA, because the unmoved Union cargo was causing further congestion in American ports as it waited for collection.[89] The situation was not relieved until Churchill's decision in early 1943 to cut shipping traveling to the Indian Ocean Area by 60% and finally by the subsequent reopening of the Mediterranean to British shipping.

The Union, like Canada, was therefore a dominion wholly unprepared for the drastic nature of the task which British strategy demanded of it.[90] Yet it emerged to provide a service which, just as Canada was facilitating the survival of the UK, kept the link between western and eastern empire open at a time when this bond was essential to imperial armies and, indeed, London's entire war strategy. Despite the criticism the Union's effort has received,[91] it is difficult to see how South Africa could have coped with such a magnitude of shipping arriving into its limited docking space, especially since at times over half of it required repairs in addition to the usual provisions expected at port.

The fall of France dramatically changed the whole conduct of the war at sea. Dealing with this situation in the Union would have required initiatives (alongside large investment) before the war began in order to have had a significant impact in 1942 when the problems were most serious. Aside from such a development being largely unwarranted before the crisis of June 1940, the UK had been reticent when dealing with requests for financial assistance to expand Union harbor defenses in the 1930s. With prevailing political conditions in South Africa equally unlikely to facilitate large funds being spent on anything that even hinted at imperial-minded

[88] Behrens, *Merchant Shipping,* p. 298.

[89] For instance, in June 1942 the USA complained to London that the Union had a backlog of 100,000 deadweight tons of supplies to be moved as soon as shipping could be made available. MT 59/398 Australia and New Zealand: Shipping Requirements Policy; MT 59/382 South Africa: Tonnage Requirements.

[90] The mission sent to South Africa acknowledged that the volume of shipping the Union was being asked to deal with was known to be well above its capacity. CO 852/475/4 Cape route: report by C. G. Wurtzburg and R. H. Toleston, 16 December 1942.

[91] Behrens, in particular, is critical of the cost to British imports due to the delays in South Africa. Equally it is not entirely fair to claim, as one history of the Port of Durban does, that the Durban port (or the others) was "equal to the formidable task which awaited it." Behrens, *Merchant Shipping,* p. 260; University of Natal, *Port of Durban,* p. 24. In response to Behrens' criticisms, the *Official Yearbook of the Union of South Africa,* No. 29 (1958), contains a summary of the sheer scale of the task that was pressed on the Union during the war. See also footnote 28 in Turner et al., *Southern Oceans,* p. 188.

expansion, funding would have presented a major stumbling block. Furthermore, even when action was taken, such as at the key port of Durban,[92] the results were not significant until after the war had concluded due to the time required for construction and the wartime scarcity of skilled labor and construction materials.[93] Therefore, even if expansion had been set in motion when France fell, it is questionable whether additional facilities would have been available in time to deal with the most difficult period for the Union's ports, not to mention the detrimental effect dock construction work could have had in the interim on already stretched resources.[94]

Moreover, problems with the speed of repair and turnover of ships were not exclusively South African—they were prevalent throughout the Commonwealth.[95] In the UK, merchant ship repairs were a constant preoccupation in the early years of the war and new initiatives were sought and implemented to get maximum efficiency from British dock workers.[96] Even New Zealand, always relied upon to

[92]Sometimes referred to as the Port of Natal.

[93]Work was started in 1939 on plans to provide more efficient incoming channels for shipping at Durban, the key port of the Union (which handled more than 50% of tonnage passing through South Africa's ports). Yet it was not until 1950 that this scheme was completed. A previous expansion, building the Prince Edward graving dock, had taken some six years to complete (1919–25). It is therefore doubtful whether quick action on the Union's part would have compensated for the lack of a major pre-war expansion of port facilities. University of Natal, *Port of Durban,* pp. 28–29, 36. Other dominions, such as New Zealand, suffered similar problems in wartime expansion of facilities, Sydney D. Waters, *The Royal New Zealand Navy* (Wellington: War History Branch, Department of Internal Affairs, 1956), p. 12.

[94]New Zealand, for instance, explicitly stated that it worried that expanding its ports would come at the expense of immediate efficiency and that persistent extension work would cause facilities to operate below maximum capacity. See the report from Archives New Zealand (hereinafter ANZ), N1, Box 142, 6/35/4, "Ship Building—Ship Building Committee and Controller of Repairs—Construction of Vessels for US Forces," 1941–1942.

[95]This situation was exacerbated by the UK, which pleaded with the dominions to relieve the British Isles of as much of this burden as possible. Repairs were deemed to be a vital element in the successful prosecution of the Battle of the Atlantic, too, leading to skilled laborers in Halifax working sixty-eight hour weeks while labor shortages could not be met. W. C. Clark Papers, RG19 Vol. 3988, "Ship Repair Facilities—Halifax and St. John"; C. D. Howe Papers, MG27 III B20 Vol. 42, S.9–26, "Ship Repairs," Letter to Mr. Robertson, Acting Under-Secretary of State for External Affairs, 24 May 1941; In New Zealand, schoolboys were drafted in for basic labor in some yards to keep work proceeding apace. ANZ, N1, Box 408, 13/28/51, "Personnel—War Service—Trained Tradesmen for Ship Repair Work."

[96]CAB 21/1499 Merchant Ship Repairs; CAB 86/1 Battle of the Atlantic Committee—measures to quicken the rate of repairs on merchant shipping in the UK were frequently at, or very near, the top of the agenda for the Committee; CAB 120/409 Battle of the Atlantic: Shipping.

provide the utmost effort for the Mother Country, found that its inadequate provision at ports attracted London's criticism. The message from the Metropole, that the "Ministry of Shipping wish everything possible to be done to ensure prompt despatch and would welcome any reasonable scheme to improve waterfront conditions in New Zealand which are now notorious for inefficiency and expense," brought home these problems to Wellington.[97] Canada and Australia similarly failed to escape the inherent problems of dealing with a merchant fleet ravaged by attacks and overuse.[98] Where the most extensive repair facilities were eventually created, in Canada, the question eventually arose over what to do with these resources—that so far exceeded peacetime requirements and had been created through large capital expenditure—when the war ended.[99]

In light of this, criticism of South Africa's effort must be tempered. Drastic action from Pretoria regarding ports could hardly have been expected before the fall of France, however, the Union was certainly slow to deal with the new circumstances which emerged and obviously took limited action as a crisis emerged in its ports. Indeed, it was not until the British mission arrived that measures were taken to rectify the most fundamental structural issues which hampered the Union's efforts.[100] On the other hand, Smuts authorized complete cooperation with British representatives and the actions subsequently recommended were often implemented.

Furthermore, the Union's shortcomings should not obscure the successes it enjoyed. From having just 100 workers undertaking shipping repairs in September 1939, the Union had increased its workforce thirty-fold in that field to over 3,000 by the second half of 1942.[101] In light of a critical lack of workmen experienced

---

[97] J. V. T. Baker, *The New Zealand People at War: War Economy* (Wellington: Historical Publications Branch, Department of Internal Affairs, 1965), p. 395. For problems with the rate of repairs in New Zealand, see ANZ, N1, Box 92, 6/10/2, "HM Ships—Repairs to ships—At Wellington."

[98] For Australia, see Hall and Wrigley, *Overseas Supply,* p. 484; for Churchill requesting improved facilities and speed of repairs in Canadian ports, see CAB 120/409 Battle of the Atlantic: Shipping; for more problems with the rate of Canadian repairs and Canada's stretched facilities, see Douglas et al., *Blue Water Navy,* pp. 47–48; on delays at the ports of Halifax and St. Johns, see W. C. Clark Papers, RG19 Vol. 3988, "Wartime Requirements Board—Naval and Shipping Problems."

[99] By 1944, $10,000,000 had been spent on creating and improving repair facilities in Canada. W. C. Clark Papers, RG19 Vol. 3988, "Advisory Committee on Merchant Shipping Policy," Brief for Shipbuilding, publication by the Canadian Shipbuilding and Ship Repairing Association, p. 57.

[100] The mission departed the UK on 17 July 1942 and returned on 7 December. Their report was published before the end of the year. CO 853/475/4 Cape route: report by C. G. Wurtzburg and R. H. Toleston.

[101] South Africa, Dear and Foot, eds., *Oxford Companion to World War II.* By way of comparison, Canada *began* the war with 4,000 workers and eventually employed some 75,000 at

with marine repairs and the difficulty in obtaining British help given the dire situation in UK ports, this was no mean feat. Even more impressively, Union shipyards repaired over 13,000 ships during the war[102]—that the entire British merchant fleet, the largest in the world in 1939, numbered less than 10,000 vessels demonstrates the magnitude of this achievement. Furthermore, the Union's ports serviced some 400 convoys, carrying six million troops.[103] South Africa truly was a Commonwealth shipping hub and the effort produced was foundational to British strategy. It was therefore no exaggeration when the British mission reported that shipping services formed the most valuable involvement the Union could have in the war, with repairs in particular deemed to be "the greatest single contribution which South Africa can make to the Allied war effort."[104]

The Antipodes and the Shipping War

The island dominions of Australia and New Zealand both had a special interest in the protection of shipping and despite being so far from the early theaters of conflict, they were quickly called upon to provide support for the UK at sea. Five RAN destroyers joined the RN in the Battle for the Mediterranean, while RAN vessels served as escorts to the troopships transporting Antipodean soldiers to North Africa. The RAN became deeply engaged in Mediterranean operations, involved in everything from supplying Malta, escort duties, fighting the German and Italian fleets, and bombarding shore-based enemy positions.

Japan's aggressive attitude, however, caused RAN naval forces to return to the Pacific, so that when HMAS *Sydney* was sunk off the coast of Australia by a German cruiser in November 1941, most RAN vessels were in Australia or Singapore.[105] From then on, Australian naval forces would largely feature in the Pacific theater, for

---

its peak. Even here, however, there were problems with a shortage of skilled labor. W. C. Clark Papers, RG19 Vol. 3988, "Advisory Committee on Merchant Shipping Policy," Order-in-Council quoted in "Canadian Merchant Seamen" Report, 10 October 1944; for New Zealand lacking staff for repairs at Wellington, see N1, Box 92, 6/10/2, "HM Ships—Repairs to ships—At Wellington."

[102]South Africa, Dear and Foot, eds., *Oxford Companion to World War II.* Canada was the major contributor outside the UK in this field, repairing approximately 25,000 merchant and naval vessels in a four year period. W. C. Clark Papers, RG19 Vol. 3988, "Advisory Committee on Merchant Shipping Policy," Brief for Shipbuilding, publication by the Canadian Shipbuilding and Ship Repairing Association, p. 57.

[103]South Africa, Dear and Foot, eds., *Oxford Companion to World War II.*

[104]CO 852/475/4 Cape route: report by C. G. Wurtzburg and R. H. Toleston, 16 December 1942.

[105]The German auxiliary cruiser involved, *Kormoran,* was so badly damaged by the engagement that its crew scuttled the vessel shortly afterwards. Australia, Dear and Foot, eds., *Oxford Companion to World War II.*

a long period under the overall operational control of the USA. While the RAN was the strongest dominion navy on the outbreak of war—unsurprising given its unique combination of resources and insecurity[106]—it was superseded during the conflict by the RCN's dramatic growth. Nevertheless, the RAN was an impressive size for a country of Australia's means when its strength peaked on 30 June 1945: 39,650 personnel and 337 vessels.[107]

For New Zealand, the priorities of defense were also initially imperial. As a division of the RN, New Zealand's naval forces were at the disposal of the Admiralty and additionally involved in protecting Pacific convoys when not called on by the RN. When war arrived, the New Zealand Division already had four vessels engaged in war missions, as the navy provided New Zealand's most organized contribution to the early British war effort.[108] The imperial nature of the force was striking in numbers as well as in action: 43% of its ratings were on loan from the RN, while New Zealanders made up less than 7% of the officers.[109]

Despite the imperial obligations of the force, which became officially independent as the Royal New Zealand Navy (RNZN) on 1 October 1941, it was also required closer to home. The German raiders *Orion* and *Komet,*[110] which accounted for seventeen merchant ships of 114,118 tons, mainly in the Pacific, sunk four of these vessels in New Zealand's waters. The RNZN eventually peaked in strength in July 1945 with 10,649 personnel.[111] In addition to minor craft, the RNZN then had thirty vessels, predominantly minesweepers and anti-submarine boats.[112]

Like the RAN, the RNZN in fact operated in the American zone of control for the majority of the war. Once informed by London in June 1940 that a fleet would probably not arrive to reinforce Singapore in the case of Japanese belligerency, the USA was an essential component in Antipodean naval plans. As one New Zealand

---

[106]While not possessing the industrial maturity of Canada, it lacked the safety that the North American dominion possessed through proximity to the RN and the USA.

[107]G. Hermon Gill, *Australia in the War of 1939–1945,* Series Two, *Navy,* Part II, *Royal Australian Navy, 1942–1945* (Canberra: The Griffin Press, 1968), Appendix 3, p. 710.

[108]W. David McIntyre, *New Zealand Prepares for War: Defence Policy, 1919–1939* (Christchurch: University of Canterbury Press, 1988), pp. 219, 223.

[109]This was largely because the New Zealand naval forces formed a division of the RN. Reciprocal support, for instance in the form of hundreds of New Zealanders who served directly with the RN in the Battle of the Atlantic, was provided by the dominion. Waters, RNZN, p. 480.

[110]*Orion* arrived in New Zealand waters in June 1940, uniting with *Komet* in October. The two ships subsequently operated together until the following year.

[111]3,790 were serving with, and 570 were on loan from, the RN. This is one example of how the imperial nature of New Zealand's effort persisted despite the USA's operational control over the South Pacific.

[112]Like with the SANS, the force was built around the task of coastal protection.

historian has noted, "from 1942 in practice, and from 1940, in anticipation, New Zealand stood or fell with the USA, rather than Britain."[113] Nevertheless, the imperial element in the Antipodes' naval efforts persisted throughout the war. An Anzac squadron was formed to operate from Suva; Australia, New Zealand, and Canada coordinated their efforts in ensuring the safety of air cadets traveling to Vancouver and of some British shipping in the Pacific; and the Antipodes first pressed for, and then contributed men and vessels to, the British Pacific Fleet, which formed in November 1944.

Allied merchant shipping fared well in the Pacific—a relatively small amount of tonnage was lost during the war, although this achievement has been attributed mainly to the success of evasive routing rather than dominion naval strength.[114] Besides those ships sunk around New Zealand, nine merchant ships were lost due to German mines near Australia by the end of 1941, while Japanese assaults sunk twenty ships in this area during 1942. Disguised merchant raiders also posed a consistent threat to the Antipodes, as the Admiralty refused to press neutral tonnage for fear of antagonizing neutrals, particularly in the early phase of war.[115] By late 1943, however, these attacks had tailed off. Australia and New Zealand nevertheless relied heavily on British merchant vessels due to the necessity of overseas trade and London controlled the allocation of shipping to the dominions. Furthermore, aside from overseas trade, both dominions were reliant on their coastal vessels—the vast territory of Australia, for instance, conducted some 85–90% of its interstate trade by this method.[116] Tasman Sea convoys were initiated between the two Pacific dominions, although the adequate number of escorts was rarely found.[117]

Indeed, it is difficult to underestimate the importance of shipping to the island dominions of Australia and New Zealand, largely isolated in the South Pacific and

[113]Given the imperial nature of the Antipodes' contributions even after 1942, however, it might be fairer to say that these dominions stood beside London and Washington, but would fall with the latter. McIntyre, *New Zealand Prepares,* p. 242.

[114]Evasive routing was crucial, too, in the Battle of the Atlantic and was one reason why winning the intelligence war was so important. Inevitably in the larger Pacific Ocean, with much smaller forces dedicated to destroying merchant shipping, this method enjoyed more consistent success. David Stevens, "South-West Pacific Sea Frontiers: Sea-power in the Australian Context," in David Stevens, ed., *The Royal Australian Navy in World War II* (St. Leonards: Allen & Unwin, 1996), p. 93.

[115]NAA, A1608, B17/2/1, "War Emergency 1939. Instructions to Air Force re Merchant Shipping."

[116]MT 59/429, Australian Shipping Requirements; NAA, A5954, 621/3, "Prime Ministers Visit Abroad 1941. Australian Shipping Position."

[117]NAA, A2670, 233/1940, "War Cabinet Agendum—No 233/1940—Co-ordination of operations of Australian and New Zealand forces in the Tasman Sea."

so far from the main supply base of the UK.[118] Though tonnage levels were now deemed satisfactory, Canberra was rarely satisfied with what London allocated it.[119] Australian insecurity was born from its Great War experience and early warnings emanating from London—even in November 1939 Canberra was told that wheat purchases by the UK from the dominion resolved itself primarily as a question of shipping.[120] This prompted attempts to create an Australian-controlled pool of shipping[121]—yet despite building merchant vessels and establishing a Commonwealth Ships Chartering Committee,[122] eventually responsible for nearly forty vessels on interstate and international trade, Canberra remained heavily reliant on

[118]Australia informed the UK that adequate shipping constituted a "matter of economic and national strength" for the dominion. NAA, A1608, F61/2/2 PART 1, "Merchant Shipping—Organisation of for Essential Commodities, Part I."

[119]This situation was largely unique to Canberra. An interim report by the South Africa Shipping Commission, for instance, in January 1944 considered that "Union shipping services to and from all quarters of the globe, preponderantly under the British flag, have been adequate to the country's need and efficiently operated." Yet the UK specifically cited a comparison between Australia and South Africa to show that the UK had taken more non-essential goods such as wine from the former than the latter, and that in general British imports from Australia had increased over the three years of war to that date. For the South African Shipping Commission, see W. C. Clark Papers, RG19 Vol. 3988, "Advisory Committee on Merchant Shipping Policy—Documentation," Report from the Canadian High Commissioner in South Africa on South African Post-war Merchant Marine, 25 January 1944; for the UK's claims on Australian imports, see A1608, F61/2/2 PART 1, "Merchant Shipping—Organisation of for Essential Commodities, Part I"; see also A1608, F61/2/2 PART 3, "War Section. Merchant Shipping—Organisation of for Essential Commodities, Part 3," for how well Australia continued to do when comparing backlogs of exports with South Africa by the autumn of 1942.

[120]See A1608, F61/2/2 PART 1, "Merchant Shipping—Organisation of for Essential Commodities, Part I." As early as March 1940 the Australian High Commissioner in London was informing Canberra that he found the Australian allocation of shipping "disquieting." During the Great War, Australian Prime Minister Billy Hughes personally purchased fifteen ships on the private market for his government to force London's hand over what he deemed to be an unfair distribution of imperial shipping.

[121]Australia wished to work through the Ministry of Shipping for acquiring vessels to avoid competing and raising prices. The Ministry, however, informed Canberra that it would only act as any dominion's agent if the "vessels acquired would be operated by the Governments concerned in the interest of the joint war effort"—in other words, they would have to be ultimately controlled from London. Even when competing, Canberra complained that "everywhere we turned the Ministry of Shipping appeared to have got in first." Butlin, *War Economy*, pp. 163, 166; A1608, H61/2/2, "War 1939 Neutral Shipping—Use of Britain during the War."

[122]In 1942 this body, which had worked under the Australian Department of Commerce, was absorbed into the Department of Supply.

British and American ships. Overseas shipping vessels entering Australia fell from 1,905 (of 7,128,404 tons) in 1937–38, to 1,276 (of 5,174,118 tons) in 1941–42.[123] Merchant shipping shortages were alleviated to some degree by the USA's entry into the war;[124] however, they still persisted as late as 1945 when the dominion was struggling to maintain adequate supplies for its forces fighting in the South-West Pacific Area (SWPA).[125]

New Zealand also felt the shipping strain. With an economy largely dependent on the export of a narrow range of agricultural products and with 97% of its food exports going to the UK before the war, shipping was a necessary lifeline. Imports furthermore made up 40% of all goods in use in the country.[126] Combined with a shortage of funds and the unavailability of many products due to the onset of war, however, the shipping shortage soon caused imports to the dominion to fall. Measured against the 1937 level, they were at 75% in 1940 and fell further to two-thirds of the 1937 standard in 1941.[127] Shipping tonnage in New Zealand's ports, meanwhile, was down 17% by 1940 from the 1938 level, and continued to fall in the subsequent year.[128] UK imports fell away sharply and New Zealand turned to Canberra for support; however, given Australia's own pressing needs, only limited help could be provided. Wellington was forced into immediate action in many cases in the hope of preventing the shortage of supplies becoming untenable—petrol rationing, not introduced in Australia until October 1940, was implemented from the first week of war by Wellington.[129] Domestic producers were by 1941 providing their country with over 70% of the goods in use, up from 60% before the war began.[130]

---

[123]Butlin, *War Economy,* p. 193.

[124]The changing nature of Australian supply is well highlighted by the origin of the overseas ships on commercial services entering and clearing Australian ports. In the year ending 30 June 1939, British Empire ships made up roughly 73% of these vessels—by the year ending 30 June 1944, this proportion had fallen to 43%. In the same period, the proportion from the USA rose from 3% to 33%. Behrens, *Merchant Shipping,* Appendix XL.

[125]Australia, Dear and Foot, eds., *Oxford Companion to World War II.* Likewise, Australia's facilities remained below what was deemed desirable. By September 1944, for instance, Sydney was still considered to be inadequate for use by imperial forces in the predicted last phase of the war against Japan. NAA, A5954, 517/39, "Delays to Merchant Shipping."

[126]Butlin, *War Economy,* p. 107.

[127]Ibid.

[128]This loss of imports, and those occurring across the Tasman Sea, did provoke concern in New Zealand. See ANZ, N1, Box 139, 6/34, "Merchant ships—Loss of and damage to—General."

[129]By January 1940, New Zealand was facing a shortage of fuel oil to meet necessary requirements. DO 35/1059/1 New Zealand Shipping.

[130]Butlin, *War Economy,* pp. 177, 123.

From 1942, the arrival of American troops and the initiation of lend-lease and Canadian mutual aid, however, began to transform the picture. In 1943, imports subsequently rose to 21% above the 1937 level. That the effects of the USA's entry into the war were not largely felt until the latter half of 1943 were partly due to the general Allied shortage of shipping.[131] The expansion of Wellington's port facilities from 1942 provided further relief for the import situation.[132] By 1945, New Zealand was finally considered to be providing repairs of high quality at adequate speed.[133]

The New Zealand naval effort also severely disrupted coastal shipping around the dominion because vessels were requisitioned by the RNZN. In June 1942, all coastal ships over 350 tons were required to work continuously and even this measure could not prevent supply being inadequate for the majority of the war. Despite the difficult years before 1942, however, New Zealand avoided what would have been a hugely destructive total disruption of shipping.[134] While New Zealand played its part in this success,[135] this outcome was largely thanks to the ongoing efforts in the North Atlantic and around the Cape Route of Africa.

---

[131]The USA remained unhappy for some time with the slow turnaround of ships in New Zealand ports. ANZ, EA1, Box 764, 95/4/11 Part 1, "Transport—Supply of Shipping—Shipping construction and repairs in NZ."

[132]ANZ, N1, Box 92, 6/10/2, "HM Ships—Repairs to ships—At Wellington."

[133]Indeed, there was some suggestion that repairs undertaken in New Zealand were by this stage of a higher quality than in Australia. ANZ, N1, Box 94, 6/10/10A, "HM Ships—Repairs to ships—Refits of HM and US and allied Ships—Correspondence." As late as December 1944, however, shortages still persisted in New Zealand, and Wellington sent a mission to Australia to learn from the Australian repair establishment, showing how long it took for these developments to actually come about. ANZ, EA1, Box 764, 95/4/11 Part 1, "Transport—Supply of Shipping—Shipping construction and repairs in NZ."

[134]Such destructive losses were not avoided everywhere. While Churchill stated that there was "no reason why all parts of the British Empire should not feel the pinch in the same way as the Mother Country has done," the truth was that the UK and dominions were spared the worst. Conditions in colonies such as India and Mauritius were significantly worse due to London's decisions. As Collingham has pointed out, "although Britain did not set out with the explicit intention of exporting wartime hunger to their empire, this is in fact what happened." Howard, *Grand Strategy*, Vol. IV, p. 293; Collingham, *Taste of War*, pp. 125–26.

[135]Comparing the period January to June 1939 with April 1944 to March 1945, a commission in New Zealand found that improvements in ports meant that foreign vessels were spending on average eighteen less days in New Zealand. Improvements were undoubtedly made during the war, but this figure is significantly (and deliberately) exaggerated by the dates chosen—for instance, the earlier period was one of dock worker unrest and deliberate "go-slow" protests. Butlin, *War Economy*, pp. 402–03.

Conclusion

Mahan prefaced his seminal work on sea power by stating that "it is easy to say . . . the use . . . of the sea is and has been a great factor in the history of the world; it is more troublesome to seek out and show its exact bearing at a particular juncture."[136] The maintenance of the shipping lanes during the Second World War and the eventual overwhelming dominance of the Allies at sea was one of these junctures. The hard fought war of attrition on the oceans was a testament to the importance placed on the sea as a decisive factor in the war by all belligerents.

Shipping was the foundation upon which the Commonwealth war effort was based. The UK was an island at the heart of a global empire—drawing on large-scale imperial support necessitated the safe passage of a tremendous number of vessels. To disperse its own strength of armaments and manpower to theaters distant from home required that same safety. Unfortunately, war exerted mounting pressures on the finite shipping resources of the Commonwealth: the loss of the seaborne support and dock facilities of European allies; the closure of the Mediterranean; and the advancement of the Axis' submarine potential, tactics, and technology. Even when the USA entered the war and fought beside the British Empire, the support it could provide depended upon shipments traveling across vast oceans.

The Commonwealth's role proved to be essential in preventing defeat in the shipping war. Ironically, the island dominions of Australia and New Zealand had the smallest part to play. Largely dependent upon London for shipping, often reluctantly in Australia's case, the Antipodes worked to ensure the efficiency of their ports, the safety of their waters, and furthermore offered navies which proved a useful adjunct to British and American naval power. Antipodean vessels and naval personnel gave these dominions a global presence in the war at sea; however, wartime developments meant that it was the other two dominions which had the more crucial roles with merchant shipping.

The importance of South Africa's shipping contribution was a somewhat surprising development. Its geographical location on the Cape of Good Hope had always been the strategic basis for RN interest in the area,[137] but the Suez Canal had long since superseded the Cape as the primary passageway to the east. The closure of the Mediterranean subjected South Africa's resources to a level of shipping which it could not readily cope with and that, until the Mediterranean was once

---

[136]Alfred Thayer Mahan, *The Influence of Sea Power Upon History, 1660–1805* (London: Sampson Low, Marston, 1890), p. iii.

[137]In 1890, Mahan had written of the naval basis for such a presence in South Africa, describing "the demand for stations along the road, like the Cape of Good Hope . . . for defence and war; . . . posts whose value was chiefly strategic." Mahan, *Sea Power*, p. 28. Almost a century before, in 1797, the Governor of the Cape had called it the "master link and connexion between the Eastern and Western world." Turner et al., *Southern Oceans*, p. 1.

more secured, it failed to deal with comprehensively. Yet if the Union could not find the answers to the crisis levels of British tonnage seeking both a dominion haven on the Cape Route and the facilities of its ports, neither did the imperial supply system break down at this pressure point.[138] Repairs proceeded as fast as the Union could handle despite the lack of any considerable Allied support in materials or advisors for long periods; the British Empire maintained its link from east to west; and the necessary provisions despatched from the USA, Canada, and the UK to the forces in North Africa and the Middle East continued to flow.

If South Africa was central to maintaining imperial unity and keeping the British ground forces supported in their primary land theaters, Canada was an essential part of keeping the UK in the war. North American imports and military support proved increasingly essential to British victory and Canada's support in the North Atlantic ensured this flow of men and materiel was maintained. From the first convoy that departed Halifax in 1939, the RCN defended merchant vessels as they made the perilous journey across the Atlantic to such an extent that it became a fundamental element of first survival, and ultimately victory. Given the RCN's narrow responsibilities, Roberts' recent claim that the "Royal Canadian Navy . . . contributed almost as much to victory as the Royal Navy" greatly overstates the case, but Canada undeniably played an irreplaceable part in sustaining the UK's capacity to wage war.[139]

The dominions were reliant on the UK for merchant shipping and lacked notable naval forces on the outbreak of war. By the war's conclusion, they had all developed substantial power at sea and, rather than carving out individual roles, had taken on the tasks that Allied strategy and Axis successes forced upon their relatively limited resources. Perhaps it is unsurprising then that these naval roles have often been easily incorporated into the all-encompassing "British effort." Closer examination shows, however, that in the war at sea British strategy, survival, and victory would have been impossible without the junior partners in the Commonwealth alliance.

---

[138] Indeed, Leighton and Coakley have stressed how close the empire was to breaking into two pieces and falling back upon "its two centres of gravity," the UK and the Antipodes. Richard M. Leighton and Robert W. Coakley, *Global Logistics and Strategy, 1940–1943* (Washington, DC: Office of the Chief of Military History, U.S. Army, 1955), p. 49. South Africa proved crucial in keeping the two poles connected.

[139] Andrew Roberts, *The Storm of War: A New History of the Second World War* (London: Penguin, 2010), p. 367.

5

Schnellboote, Strategy, and the Defense of *Festung Europa*, 1943–1944

G. H. Bennett

The slaughter on Omaha Beach on 6 June 1944 has had a lasting effect on the historiography of Operation Overlord. The struggle to get off the beach and break through the crust of coastal defenses has dominated the attention of historians from Stephen Ambrose to Joseph Balkoski and beyond.[1] The Battle for Caen, Allied generalship, and the eventual Allied breakout from the bridgehead have been the subjects of considerable debate amongst historians.[2] By contrast, the seaborne phase of the invasion (Operation Neptune) remains a neglected area: an overlooked precursor to the battle for the beaches. The rather dated official World War II histories of the Royal Navy and United States Navy by Roskill and Morison still provide the best discussion of the process of assembling, convoying, and landing the invasion force.[3] The recent scholarship on Neptune has concerned gunfire support for the assault troops and the nature of Anglo-American amphibious

[1]Stephen E. Ambrose, *D-Day, June 6, 1944, The Climactic Battle of World War II* (New York: Simon & Schuster, 1994); Joseph Balkoski, *Omaha Beach: D-Day, June 6, 1944* (Mechanicsburg: Stackpole, 2004). See also Stephen Badsey and Tim Bean, *Omaha Beach* (Stroud: Sutton, 2004); Antony Beevor, *D-Day: The Battle for Normandy* (London: Viking, 2009). The research for this chapter was supported by a grant from the Wheatcroft Collection and I would like to record my thanks to Kevin Wheatcroft. This chapter was originally published in *Global War Studies* 11:1 (2014).

[2]See, for example, Simon Trew and Stephen Badsey, *Battle for Caen* (Stroud: Sutton, 2004); Christopher Pugsley, *Operation Cobra* (Stroud: Sutton, 2004); Steven J. Zaloga, *Operation Cobra 1944: Breakout from Normandy* (Oxford: Osprey, 2001); Stephen E. Ambrose, "Eisenhower's Generalship," *Parameters* (June 1990), pp. 90–98.

[3]Samuel Eliot Morison, *History of United States Naval Operations in World War II*, Vol. XI, *The Invasion of France and Germany, 1944–1945* (New York: Castle Books, 2001); Captain S. W. Roskill, *The War at Sea, 1939–45*, Vol. III, Pt. 2 (London: HMSO, 1961).

doctrine.[4] The success of the Allied navies in bringing the invasion fleet to the shores of Normandy is treated as the almost inevitable outcome of superiority of numbers over the German opposition.

Likewise, the brief battles off the coast of Normandy in 1944 have little place in the historiography of the *Kriegsmarine* dominated by the U-boat campaign and the destruction of the major surface units such as *Bismarck, Scharnhorst,* and *Tirpitz.* In part, this is a reflection of the fact that the German Naval Archives remained closed to most researchers until the 1970s, which amplified the concentration on famous naval actions, together with the on-going influence of Admiral Dönitz on the post-war historiography of the war at sea. Publication of his memoirs in 1958,[5] and his willingness to give interviews, meant that the hand of Dönitz is detectable in works as late as the 1969 publication of Edward Von der Porten's *The German Navy in World War II.*[6] Much of the post-war historiography was shaped by Dönitz's concerns and thoughts: particularly his involvement with the U-boat campaign. As Theodore Ropp noted in 1960, Dönitz's preoccupations meant that his memoirs were "chiefly a document . . . on the U-boat war."[7] The U-boats played a minor and undistinguished role in the brief battles off Normandy in June 1944. It was the minor surface units of the *Kriegsmarine* which had to contest the approaches to the invasion beaches on 6 June. Their role in the war at sea has been almost completely ignored even in more recent work by historians. For example, Vincent O'Hara's *The German Fleet at War, 1939–1945* examines only those surface actions fought by ships of over 500 tons.[8]

The omission is significant if it is not altogether surprising. Failure to stop the Allied invasion fleet in 1944 was a decisive moment in World War II, and the *Kriegsmarine's* role in it was neither large nor glorious. In war the *Kriegsmarine* was geared to the offensive and that spirit has dominated the post-war historiography carefully shepherded by Dönitz until his death in 1980. Increasingly, since the

---

[4]See, for example, William B. Kirkland, Jr., *Destroyers at Normandy: Naval Gunfire Support at Omaha Beach* (Washington, DC; Naval Historical Foundation, 1994); Adrian R. Lewis, "The Navy Falls Short at Normandy," *Naval History* 12:6 (1998); Adrian R. Lewis, *Omaha Beach: A Flawed Victory* (Chapel Hill: University of North Carolina Press, 2001).

[5]Karl Dönitz, *Memoirs: Ten Years and Twenty Days* (London: Weidenfeld and Nicolson, 1958).

[6]Edward P. Von der Porten, *The German Navy in World War II* (London: Arthur Baker, 1969). In his review of the book published in 1970, Roskill commented, "some parts of the book are strongly tinged with . . . [Dönitz's] outlook and prejudices," *International Affairs* 46:4 (1970), p. 781.

[7]Theodore Ropp, book review, *The Journal of Modern History* 33:1 (1961), p. 98.

[8]Vincent P. O'Hara, *The German Fleet at War, 1939–1945* (Annapolis: Naval Institute Press, 2004).

1990s historians have questioned Dönitz's strategy, but their focus has mostly been on the conduct of the U-boat war.[9] A more wide-ranging reassessment of his strategy as head of the *Kriegsmarine* from 1943 to 1945 has yet to be developed. Through a micro-historical methodology focusing on the role of the *Schnellboote* in German preparations to meet the Allied invasion, this chapter seeks to provide answers to a number of strategic questions that have not been fully examined by historians. Why wasn't the German Navy better prepared to defend the invasion beaches? Could the pursuit of alternative policies have allowed the *Kriegsmarine* more of a chance against the Allied invasion force?

During 1943, the *Kriegsmarine* suffered a series of blows which forced it away from the strategy which it had pursued since the Fall of France in June 1940. The *guerre de course* waged against Britain by the submarines and surface vessels of the *Kriegsmarine,* almost invariably without the effective cooperation of the *Luftwaffe,* could not be maintained in the face of a series of challenges. With German armies under pressure on all fronts, the attention of the German High Command increasingly turned to the moment when Allied armies would attempt an amphibious landing on the coast of Western Europe. This chapter argues that by default Admiral Dönitz and the *Kriegsmarine* were reluctantly and hesitatingly, and they hoped temporarily, forced to adopt a defensive posture as the danger of an Allied invasion of Western Europe mounted. Yet Dönitz continued to regard the U-boat campaign, and the development of a next generation submarine force to fight it with, as the priority for the *Kriegsmarine.* There was a consequent failure to accept, articulate, and imaginatively manage a strategy of coastal defense. For Germany, the defense of occupied Europe would begin at the water's edge instead of in the sea approaches to the invasion beaches. With the U-boat campaign receiving priority, German industry was ill-equipped to turn out significant numbers of vessels suitable for coastal defense. The chapter will further argue that the *Schnellboote* as a design was ill-suited to a program of mass production to meet the threat of invasion, and that consequently the German Navy had to contest the initial landings in Normandy with an inadequate number of conventional vessels of high quality. Nevertheless, the performance of the *Schnellboote* against the Allied invasion flotilla in June 1944 was more significant than is acknowledged in subsequent accounts, serving as an indicator of what might have been if a strategy of littoral defense had been properly thought through and pursued. Failure to stop, or seriously hamper, the launching of the second front meant that the war would end before Dönitz could realize his dream of re-starting the war on commerce with Type XXI and XXIII U-boats.

---

[9]See, for example, David Syrett, *The Defeat of the German U-Boats: The Battle of the Atlantic* (Columbia: University of South Carolina Press, 1994).

The Development of German Naval Strategy

Guntram Schulze-Wegener has demonstrated that naval thinking at the time of the formation of the German state continued to influence the Navy until at least 1943.[10] For most German naval historians, such as Holger Herwig, the German Navy's pursuit after 1919 of Admiral Tirpitz's goal of an all powerful High Seas Fleet represents a collective and continuing failure of imagination and strategy.[11] Despite her small coastline, the development of the Kaiser's High Seas Fleet in the late nineteenth century signaled that the newly unified German state would pursue naval development and strategy based on the offensive. While the performance of the High Seas Fleet between 1914 and 1918 left much to be desired, the offensive intent of the Navy was underlined by the development of the German submarine arm. Even before the outbreak of war in 1914, the German Navy had conceived the submarine "not as a static element of coastal defence but as a flexible reconnaissance and torpedo unit."[12] Following the Battle of Jutland in 1916, the German Navy was forced into an ostensible strategy of coastal defense. The peacemakers at Paris envisioned a German Navy capable of fulfilling only this role. Covertly, however, the Navy continued to pursue offensive oceanic warfare. Preparations were made for the day that Germany would be able to break free of the shackles of Versailles and begin laying down a fresh generation of U-boats and battleships. The development of the pocket battleship class of heavy cruisers in the late 1920s saw the German Navy once again in a position to project its power around the globe. With the advent of Hitler in 1933, the new *Kriegsmarine* under Admiral Erich Raeder commenced rapid expansion. The pattern of development for the German Navy in the nineteenth and early twentieth century would cast long shadows on the service, and indeed the writing of the history of the *Kriegsmarine* after 1945. The Navy of the German Reich after 1870 conceived itself to be an offensive force capable of global power projection: a conscious rejection of the idea of a smaller force capable of defending Germany's coasts and her interests in the Baltic.

The development of a large German Navy would take considerable time. Raeder was reassured by Hitler that "he would not need the fleet as an instrument of his purposes until 1946."[13] Raeder's assessment of the strategic position of the *Kriegsmarine* on the outbreak of war in 1939 was gloomy: "No one was in any

---

[10]Guntram Schulze-Wegener, *Die deutsche Kriegsmarine-Rüstung 1942–1945* (Hamburg: Mittler & Sohn, 1997).

[11]Holger H. Herwig, "The Failure of German Sea Power, 1914–1945: Mahan, Tirpitz, and Raeder Reconsidered," *The International History Review* 10:1 (February 1988), pp. 68–105; See also Keith W. Bird, "The Origins and Role of German Naval History in the Interwar Period 1918–1939," *Naval War College Review* 32:2 (March-April 1979), p. 52.

[12]Philip K. Lundeberg, "The German Naval Critique of the U-Boat Campaign, 1915–18," *Military Affairs* 27:3 (Autumn 1963), pp. 105–18 [p. 106].

[13]Erich Raeder, *My Life* (Annapolis: Naval Institute Press, 1960), p. 125.

doubt that as things stood there was little chance of winning any decisive success in a naval war with Britain. . . . The present Fleet would be able to do little more than fight and go down to an honourable defeat."[14] Despite Raeder's misgivings, the opening phases of the war were marked by impressive German naval successes, especially in the unrelenting destruction of Allied merchant shipping. Dependence on the importation of food, oil, and munitions by sea made Britain particularly vulnerable to this type of warfare, a weakness well understood by both Raeder and Dönitz. They were frustrated by the inadequacy of the naval force at their disposal, but they still saw a "tonnage war" on merchant shipping as their best hope for enfeebling Britain, restricting her military options, and forcing her to make peace. By the end of 1942, the worldwide loss of British, Allied, and neutral merchant shipping amounted to almost seventeen million gross registered tons, of which about eleven million tons had been sunk by Axis submarines.[15] The emphasis on commerce warfare as the primary strategy against the British Isles marked a throwback to the earliest days of German naval thought.[16]

The Impact of Hitler on Naval Strategy

Following the poor showing of German naval units in the Battle of the Barents Sea in December 1942, Hitler was determined to force a fundamental shift in the strategy of the *Kriegsmarine*. The *Führer* Naval Conference on 6 January 1943 opened with a monologue by Hitler on the development of the German Navy from the nineteenth century onwards.[17] Hitler declared that henceforth the big ships of the *Kriegsmarine* were to be decommissioned. While the small ships, especially the submarines, had more than proved their worth in Germany's fight for survival, the bigger ships were costly liabilities. They spent most of their time hiding from the enemy and were too vulnerable to Allied air attack. In 1942, most of the surviving large ships had been sent to the relative safety of Norway. Now Hitler wanted them decommissioned. Their guns could be deployed as coastal batteries: their crews and other resources recycled into the submarine program. Without specifying it, Hitler was pushing the *Kriegsmarine* towards a strategy of coastal defense. Grand Admiral Raeder, the head of the German Navy, preferred to step down rather than decommission the big ships. Admiral Dönitz (the head of the submarine arm) was asked to take over. Hitler's insistence on decommissioning the big ships also meant that Dönitz would be required to carry out a reassessment of the strategy of the German Navy.

---

[14] Ibid., p. 135.

[15] Roskill, *The War at Sea*, p. 479.

[16] David H. Olivier, *German Naval Strategy, 1856–1888: Forerunners to Tirpitz* (London: Frank Cass, 2004), p. 1ff.

[17] *Führer* Naval Conference, 6 January 1943, *Fuehrer Conferences on Naval Affairs, 1939–1945* (London: Greenhill Books, 1990), pp. 318–19.

The new head of the *Kriegsmarine,* who had served in U-boats during World War I and who had survived the sinking of *UB-68* in the Mediterranean in 1918, viewed naval strategy through the narrow confines of the periscope. While still in a British prisoner-of-war camp in Yorkshire in 1918, he had formulated the wolf pack tactics which would become one of the hallmarks of the German submarine campaign during World War II.[18] Single-minded to a fault, Peter Padfield has pictured Dönitz as "rigid in thought and slow in comprehension."[19] His intellectual and personal rigidity were one of his strengths and his principal failing. At U-boat command, Dönitz had surrounded himself with men prepared to quietly agree with him, and he had taken an instinctive dislike to officers who voiced their opinions too strongly. Dönitz had also demonstrated an ability for self-delusion and the capacity to believe that through sheer willpower and determination seemingly impossible difficulties could be overcome. His belief in the ultimate victory of his submarines was as unshakeable as Hitler's belief in the eventual outcome of the war. The two men shared similar character traits which would bring them increasingly close after January 1943. In many respects Dönitz, and his approach to command, conforms with Eric C. Rust's analysis of the German Navy as an institution where intellectual thought was discouraged and conformity highly prized.[20]

An early sign of the harmonious relationship between the two men was that Dönitz managed to talk Hitler around from decommissioning the big ships. For the new head of the German Navy, the big ships had value in tying down escorts which might otherwise be released for convoy work in the Atlantic. Likewise, Dönitz saw value in the small ships, the *Schnellboote,* torpedo boats, and destroyers deployed along the coast of occupied Europe. Since the fall of France in 1940, the *Schnellboote* arm in particular had taken the offensive in the English Channel and North Sea. Meanwhile, the destroyers and torpedo boats of the German Navy had been engaged in more defensive operations protecting coastal convoys, port facilities, and movements of some of the big ships of the German Navy. Achieving considerable successes against the coastal convoys traveling up and down the English coast, the *Schnellboote* were deadly effective from 1940 to 1941. Capable of speeds in excess of forty knots, equipped with two torpedo tubes and two reloads, together with 20mm/40mm cannon and heavy machine guns, the *Schnellboot* was a formidable weapon of war. They were more than 110 feet in length and powered by three Mercedes-Benz diesel engines (2,500 to 3,000 bhp). The British struggled to deal with the *Schnellboote* in the Channel battles of 1940 and 1941. The successes grew steadily fewer in 1942 and 1943 as a result of increasing deployments of escort

[18] Peter Kemp, "Grand Admiral Karl Dönitz," in Field Marshal Sir Michael Carver, ed., *The War Lords* (London: Weidenfeld and Nicolson, 1976), p. 473.

[19] Peter Padfield, *Dönitz: The Last Führer* (London: Gollancz, 1993), p. 495.

[20] Eric C. Rust, *Naval Officers under Hitler: The Story of Crew 34* (New York: Praeger, 1991), pp. 56–57.

vessels along the English coast, the growing dominance of Allied Air Forces, and the effectiveness of Allied radar and signals intelligence. However, the *Schnellboote* found new opportunities in the Mediterranean (3rd and 7th Flotillas, 1942–45) and the Black Sea (1st Flotilla, 1942–45).[21] Despite the problems in the English Channel, Dönitz was convinced that the offensive had to be maintained. A destroyer based at Plymouth or Portsmouth to counter the 5th and 9th *Schnellboote* Flotillas at Cherbourg was one less destroyer facing the U-boat crews in the Atlantic.

The Faltering U-boat Campaign

As analyzed by Werner Rahn and others, within weeks of taking over as head of the German Navy, Dönitz's instinctive policy of *U-boat uber alles* as the basis for the strategy of the *Kriegsmarine* was dealt a series of profound blows in the Atlantic.[22] The hotly contested convoy battles of March 1943 gave way to Black May as the German submarine arm suffered unacceptably high losses in the Atlantic. Dönitz was forced to withdraw his submarines from the Atlantic, although limited modernization programs would allow the Type VII and Type IX submarines to return later in the year. Dönitz was, however, a realist. The generation of submarines with which Germany had entered the war would be sent back into the Atlantic because the *Kriegsmarine* lacked other means to take the offensive. Dönitz conceived the move not in terms of renewing the full *guerre de course* of 1940–43, but as a very temporary measure. He could not hope to defeat the Allies in a war on tonnage, but he could fight a delaying action to allow new weapons of war to emerge. The *Kriegsmarine* would prepare to renew the campaign against trans-Atlantic commerce with the new Type XXI and XXIII electro-boats, which had far superior performance to the current generation of submarines then in service. Dönitz hoped that by early 1945 he would be ready to take the offensive once more.

What is interesting and significant is the extent to which Dönitz was prepared to concede the switch to the defensive after May 1943 to the German Navy and Hitler. From his memoirs and interviews, Dönitz appears to have been unwilling to give recognition to the fact that in the period from Black May 1943 to the introduction of the electro-boats, the *Kriegsmarine* would have to pursue a defensive strategy: to return to the policy of *kleinkrieg* which had marked German naval operations in the Baltic and on the coast of Flanders in World War I. The German Navy could not defeat the Allies in the Atlantic, but it could play a role in maintaining the

---

[21]22nd and 24th S-boat Flotillas operated in the Adriatic, Aegean, and Mediterranean in 1943–44.

[22]See Rahn's chapter on the climax and collapse of the U-boat war in Horst Boog, Werner Rahn, Reinhard Stumpf, Bernd Wegner, *Germany and the Second World War*, Vol. VI (Oxford: Oxford University Press, 2001), pp. 379–404.

borders of the German Empire. Dönitz could take some steps to prepare for the invasion, and in his memoirs he describes the submarine campaign in late 1943 to early 1944 in defensive terms, but in 1943–44 he seemingly could not bring himself to fully develop and articulate a properly thought-through strategy of littoral defense. To concede that, for the moment, the offensive had been paused seems to have been as much as he could bring himself to do. With so many lives being lost in the submarine campaign—the second that Dönitz had participated in—the idea of acknowledging that his submariners in 1943–44 were now dying in what amounted to a holding action was perhaps more than he could stomach. Amongst the casualties was Peter Dönitz, the Admiral's youngest son, lost in the destruction of *U-954* on 19 May 1943.

The Admiral was unshakeable in his belief that the commerce war against Britain had to be continued. He explained in his memoirs: "After my assumption of the post of Commander-in-Chief no fundamental change occurred in the duties which devolved upon the Navy. The existence of the British people depended on sea communications. . . . The war against merchant shipping remained as before the principal task confronting the German Naval High Command."[23] And yet by 1943, with the German Empire increasingly prey to Anglo-American amphibious operations (North Africa, Sicily, Italy), the situation perhaps merited a fundamental rethink and a move from oceanic offensive war to defensive coastal war.

The surviving documentary record does not allow us to determine whether there was some kind of debate on strategy at the highest levels in the *Kriegsmarine* in 1943. As Peter Padfield has argued: "We are unlikely ever to know the precise course of the discussions among the staff at U-boat command. . . . Undoubtedly there was much soul-searching. Undoubtedly Dönitz took part. . . . The fundamental tactics remained unchanged for Dönitz's response was unchanged."[24] Moreover, with Dönitz seemingly molding his commands in his own image, there is every chance that any debate about pursuing a defensive strategy was very limited in scope. Dönitz's continued pursuit of the *guerre de course* had been depicted as the product of emotion, "inertia or blind determination."[25] S. M. Ritchie has put it even more forcefully: "Throwing more boats and crews away instead of husbanding them carefully while seeking new types and tactics, and a new strategy, was anachronistic and prodigal folly."[26]

While there may have been a lack of debate at U-boat command, within the wider service arguments certainly raged as Clay Blair has noted: "Two of the most highly decorated skippers, Reinhard Suhren and Erich Topp, argued against . . .

[23]Dönitz, *Memoirs,* p. 342.

[24]Padfield, *Dönitz,* p. 284.

[25]Ibid., pp. 288 and 495.

[26]Lieutenant Commander S. M. Ritchie, RAN, "The effectiveness of the leadership of Admiral Karl Dönitz," Geddes Papers, 2003, pp. 36–42 [p. 40].

[continuing the tonnage war] because of the anticipated 'immense losses' and the lack of even [the] 'slightest prospect of success' as Topp put it in his later memoir. On the other hand, the even more highly decorated skippers Albrecht Brandi and Wolfgang Lüth thought the campaign should continue with the greatest intensity possible."[27] The German Navy would remain on the offensive, publicly and rhetorically, and there would be no thorough-going debate within the *Kriegsmarine* about a defensive strategy of littoral protection. The defensive duties of the *Kriegsmarine* came second to the offensive in Dönitz's mind. While he was willing to concede the need for more small ships, and greater use of coastal artillery units, to defend Germany's conquests in Europe, the U-boat campaign was Dönitz's principal concern. In his mind, "offensive U-boat warfare" remained the *Kriegsmarine's* "foremost task" even as "the defensive duties of the Navy . . . rose in importance" in 1943.[28] Soon after taking over from Raeder, Dönitz decreed that the German Navy would require more U-boats, and it would also require more small ships. This was not part of a carefully thought through shift in strategy, but because "the need for these coastal forces would increase enormously with the enemy's approach to the German frontiers."[29]

Resourcing the *Kriegsmarine*

Even as he began to frame the 1943 building program, Dönitz recognized that certain structural problems within the German war economy would pose a grave challenge in increased resourcing of the *Kriegsmarine.* When Dönitz took over as head of the German Navy there was no authority in the *Reich* charged with overall coordination and oversight of armaments production programs.[30] Albert Speer, the Minister of Armaments, was responsible for the production of armaments for the Army, but the Navy had certain shipbuilding yards and other facilities at its disposal, while key aspects of military production ranging from labor supply to steel allocation were in the hands of the Central Planning Committee, headed by Speer. The *Kriegsmarine* had no representative on the Committee, unlike the Army and Air Force. In framing the 1943 building program, Dönitz was forced to appeal to Hitler for a greater allocation of steel and for more skilled labor. Hitler in turn backed Dönitz who went on to reach an accommodation with Speer by which the Ministry of Armaments would take over responsibility for providing the Navy with the weapons it required. To coordinate relations between the two departments,

[27]Clay Blair, *Hitler's U-Boat War: The Hunted, 1942–1945* (London: Random House, 1998), p. 353.

[28]The National Archives, London (hereinafter TNA), ADM 223/688, Essay by Grand Admiral Dönitz on the war at sea, 24 September 1945.

[29]Ibid.

[30]Dönitz, *Memoirs,* p. 347ff.

a new Shipbuilding Commission was set up under Rear Admiral Topp to advise on the nature, number, and specifications of vessels required by the *Kriegsmarine*. Representatives from the shipbuilding industry also sat on the Commission in the hopes that they could use their expertise to remove bottlenecks and other difficulties from the process of providing the *Kriegsmarine* with the weapons it required. Structural changes to the program of weapons procurement in 1943 made considerable sense, but such was the pressure on resources that the practical results were limited.

To a certain extent, the increased construction of U-boats and new building of coastal vessels impinged on each other such was the shortage of steel and dockyard resources. Increasing submarine construction to twenty-seven per month in late 1943, and to thirty per month by late 1944, meant that there was no scope for the construction of new destroyers, and torpedo boat construction would be increased to just eighteen vessels *per annum*.[31] Sufficient steel allocation was, however, found for ambitious plans to turn out *per annum* seventy-four *Minensuchboot* (a general-purpose light naval vessel). An old design built to a new specification, they would be built in prefabricated sections to ease the demand on dockyard space and labor. In practice, difficulties were immediately encountered and in the event only seventeen vessels of the class would ever be built. Privileging U-boat construction in the allocation of steel to the Navy meant that, by default, Dönitz turned to the idea of using the German boat building industry to deliver the means to defend the coast of Europe. Working in wood rather than steel, the boat builders of the North German coast would be asked to provide seventy-two *Schnellboote* and seventy-two *Räumboote* (a large, but lightly-armed general-purpose patrol boat). The primary responsibility for meeting and defeating an invasion force off the coast of France in late 1943 and into 1944 would fall on the more heavily-armed and faster *Schnellboote*.

The 1943 Production Program for *Schnellboote*

Little of an academic nature has been written on German coastal forces in general (as Gordon Williamson has observed)[32] and *Schnellboote* in particular.[33] James Foster Tent's *E-Boat Alert: Defending the Normandy Invasion Fleet* is a notable exception, although in the circumstances of this chapter his primary concern is to analyze Allied efforts to deal with the *Schnellboote* threat rather than to explore German

[31] Dönitz memorandum to Hitler, presented at the *Führer* Naval Conference, 11 April 1943, *Fuehrer Conferences*, pp. 318–19.

[32] Gordon Williamson, *Kriegsmarine Coastal Forces* (Oxford: Osprey, 2009), p. 5.

[33] On the history of the *Schnellboote*, see Jean-Phillippe Dallies-Labourdette, *Deutsche Schnellboote 1939–1945* (Stuttgart: Motorbuch, 2006); Hans Frank, *Die Deutschen Schnellboote im Einsatz* (Hamburg: Mittler & Sohn, 2006).

naval strategy.[34] Overall there has been a failure to understand the *Schnellboote* arm and the intricacies of the construction priorities of the *Kriegsmarine*. That failure is significant: the interconnections are important and in commanding the *Kriegsmarine*, Dönitz had to balance a series of competing demands and capabilities. As early as August, the 1943 building program was being reexamined as it became clear that German industry would struggle to produce the suggested number of *U-boote*, *Räumboote*, and *Minensuchboote*.[35] Construction of auxiliary vessels including landing ferries could not be entirely dispensed with no matter how bad the need for other types.[36] A construction target of seventy-two *Schnellboote per annum* was an ambitious target. Contrary to popular perceptions of the German war economy, the *Schnellboot* was not a mass-produced product resulting from modern and efficient methods of design and production. The *Schnellboot* was a bespoke product resulting from a 1927 design for a super motor yacht by Lürssens of Bremen.[37] In terms of naval strategy, the need for such a design originated from the 1923 Ruhr Crisis when the *Reichsmarine* identified fast torpedo boats as a counter to a potential French naval threat in the Baltic. The *Schnellboot* design was a combination of modern engineering (especially in the form of three Mercedes-Benz marine diesel engines and an aluminium reinforcing endoskeleton) and old-fashioned wooden boat building techniques. The hull of the *Schnellboot* was a composite design made up of two layers of wood (including tropical hardwoods) and a canvas layer between the two (proofed with palm oil). While building wooden boats did not place an obviously significant demand on the *Kriegsmarine's* allocation of steel, engines, engine beds (steel), armor (including bridge), endoskeleton (aluminium), armament and fastenings (including brass screws, bronze keel bolts, and propeller shafts) were not inconsiderable requirements.

Building the *Schnellboot* to perform to the potential of the design was a considerable challenge. To get the required level of performance from an already complex design required hundreds of hours of highly skilled labor. Corners could not be cut. For example, in 1935 the performance of the first *Schnellboote* flotilla in maneuvers had been "dogged by bad luck—especially by bad weather and by constructional defects."[38] Vice Admiral Carls, Commander-in-Chief Fleet, had concluded that the

---

[34]James Foster Tent, *E-Boat Alert: Defending the Normandy Invasion Fleet* (Annapolis: Naval Institute Press, 1996), p. 208.

[35]Entry 5 October 1943, *Kriegstagebuch der Seekriegsleitung* (hereinafter *KTB der SKL*), Teil A, Band 50, October 1943 (Berlin: Mittler & Sohn, 1994), pp. 82–83.

[36]Entry 25 October 1943, *KTB der SKL*, Teil A, Band 50, October 1943, p. 508ff.

[37]The design of the *Schnellboot* was an evolution of that for *Oheka II*, a private motor yacht built for German investment banker Otto Hermann Kahn.

[38]ADM 223/28, Tactical and Staff Duties Division (Foreign Documents Section), German E-Boat Operations and Policy 1939–1945, War Diary Schnellboote Command (hereinafter E-Boat Operations and Policy).

design did not constitute a "reliable weapon" and had asked the High Command to cancel further investment in the program.[39]

The High Command had not agreed to the request, but with the outbreak of war in 1939 there had been no rush to build a large fleet of *Schnellboote:* a fleet of 40–50 operational boats (with a yearly replacement of sixteen boats) would suffice. With the victory in Scandinavia and the Low Countries in 1940, and the need to defend a coastline that stretched from the North Cape to the Spanish border, the Naval High Command had thought in terms of a fleet of 160 *Schnellboote,* but this never came close to realization. The production of engines remained a critical bottleneck. Production of Mercedes-Benz 500 series marine diesels simply could not keep pace with demand. Priority was given to overhauling old engines (to maintain a ratio of 50% spare engines to the number in operational service) rather than turning out new units for new *Schnellboote.* Shortage of copper in 1941 meant that some of the new engines were built without oil coolers. It had not been until the latter stages of 1942 that a monthly production target of twenty new engines per month had been attained, but this was affected by the transfer of the Marienfelde engine works to *Luftwaffe* control leaving the Daimler-Benz works at Untertürkheim as the main producer of *Schnellboote* engines.[40]

Even with an improvement in the supply of engines, hull construction remained a critical problem. Lürssens remained the principal builder of *Schnellboote* with the Schlichting Werft at Travemünde turning out a small number from the launch of *S-101* in 1940.[41] Lürssens and the Schlichting Werft simply did not have the capacity to meet the need for increased *Schnellboote* construction, and by 1942 the need for increased construction, simply to maintain the threat to British coastal convoys let alone to meet the threat of invasion, was becoming acute.

The steady strengthening of anti-*Schnellboote* forces in the English Channel and improvements in British radar meant that by mid-1942 the head of the *Schnellboote* service [*Führer der Schnellboote,* or FdS] was ready to conclude that his boats were "no longer master of the situation in the Western area."[42] With his boats reliant on mass attacks against convoys for their effectiveness, the FdS considered it "vital to increase the number of boats in the West, and to develop new types of boats, mines, torpedoes, gunnery and signal equipment."[43] His report was a clear signal that the qualitative superiority enjoyed by the *Schnellboote* in 1940–41 was

---

[39]Ibid.

[40]Schulze-Wegener, *Die deutsche Kriegsmarine-Rüstung,* pp. 173–74, p. 167.

[41]The yard was founded in 1898 by Johannes Schlichting (1872–1946). See Eike Lehmann, *100 Jahre Schiffbautechnische Gesellschaft: Biografien zur Geschichte des Schiffbaus* (Berlin: Springer, 1999), pp. 432–33.

[42]ADM 223/28, Report by FdS to Group West, 25 May 1942 in E-Boat Operations and Policy.

[43]Ibid.

now being eclipsed thanks to the productivity of Allied industry. A production target of thirty-six *Schnellboote per annum* was under consideration in October 1942 by the *Hauptamt Kriegsschiffbau.*[44] Increased numbers of *Schnellboote* and weapons enhancement programs were now necessary to redress the balance.

It was against this background, and the growing threat of an Allied invasion of Western Europe, that the 1943 *Schnellboote* building program was developed. To increase the number of hulls in production, the *Kriegsmarine* turned to Danziger Waggonfabrik, a Danzig-based coach works. Its skilled craftsmen were more used to turning out railway carriages and trams than high-performance motor torpedo boats. With oversight from Lürssens and the Schlichting Werft, it was hoped that Danziger Waggonfabrik would turn out a lot of the new *Schnellboote.* The response of the FdS to the projected construction program was very positive: "Increased . . . [*Schnellboote*] production, and further development of the speed, stability and armament of the existing type, should bring about a great improvement; and should justify the building of more boats in the future."[45]

Despite the hopes of the FdS, and the fact that the invasion was expected to take place in the summer of 1944, it took some time to put in place the agreements and understanding necessary to deliver on the new *Schnellboote* building program. Discussions took place in Berlin in the late summer of 1943 between representatives of the shipyards and the *Konstruktionamt* of the *Oberkommando der Marine.* On 16 December, representatives of the Danzigger Waggonfabrick and Schlichting Werft arrived in Vegesack for a three-day meeting with the staff at Lürssens.[46] In early 1944, additional labor began to arrive at the Schlichting Werft from Holland and Denmark. Rudolf Schlichting rented an old hotel, the "Kurgarten Haus," to cope with the extra workers.[47] Later they would be joined by Russian laborers, which necessitated the building of a work camp.

There was no lack of enthusiasm amongst shipyard workers for the task at hand. The monthly works magazine of the Schlichting Werft, which has survived in family hands, provides evidence of the professionalism and determination of the workers involved in *Schnellboote* manufacture.[48] They saw themselves as helping to provide the weapons by which the German empire would be defended. Some of their former coworkers were serving in the *Schnellboote* arm and their

---

[44]Entry 18 October 1942, *KTB der SKL,* Teil A, Band 38 (Berlin: Mittler & Sohn, 1993), p. 359ff.

[45]ADM 223/28, Report by FdS to Group West, 31 December 1943 in E-Boat Operations and Policy.

[46]Rudolf Schlichting Diary, 16–18 December 1943, Archiv Schlichting. My thanks to Marcus Schlichting for this.

[47]Information from Marcus Schlichting, 12 February 2012.

[48]The newsletter started out as a relatively poorly produced newssheet, but by 1943 it had grown into a professionally printed mini-magazine, Archiv Schlichting.

letters would be reproduced in the works magazine. The successes of every boat turned out by the Schlichting Werft were closely followed and senior *Schnellboote* officers would visit the yard to decorate workers for their efforts and to exhort everyone on to higher production levels.[49] The handover of each new *Schnellboot* would be celebrated by a football match between the crew and workers from the Schlichting Werft.[50] Each commander would sign a special friendship book to celebrate the relationship between the crew of the boat and the Schlichting Werft.

The enthusiasm of the workforce and the extra labor and resources allowed an increase in *Schnellboote* production in 1943–44. However, the results were disappointing and patchy: scarcely surprising given the lack of urgency in laying vital ground work for the new *Schnellboote* building program. Lürssens was able to increase its production of *Schnellboote* considerably while that at the Schlichting Werft remained stable. Production by Danziger Waggonfabrik took until the end of 1944 to realize, although highly ambitious plans for production in 1945 were in place. Drilling down farther into the figures, the trends in *Schnellboote* production become even more apparent. Despite the 1943 building program, and despite Dönitz's accommodation with Speer, the number of *Schnellboote* entering service was effectively static until the end of 1943/start of 1944. During the first months of 1944, the increase in production from 1943 levels represented only a 20–40% improvement (the equivalent of one or two boats). Significant increase in *Schnellboote* construction (in excess of 100% improvement on the 1943 figures) took place in July and August 1944 (ie. after the Normandy invasion had taken place).[51] In other words, the attempt to provide the *Kriegsmarine* with sufficient numbers of conventional weapons suitable for coastal defense really began to bear fruit only after it was too late.

Inevitably, the number of *Schnellboote* at a state of operational readiness in June 1944 was decreased by routine maintenance and breakdowns. The three Mercedes-Benz marine diesels of a *Schnellboot* required overhaul every 300–350 hours use. Engines were exchanged after 500–1000 hours of use, and in late 1944 they were again in short supply as a result of Allied air attack. Air attacks against Stuttgart and the vital factory at Untertürkheim hit engine production badly (in October 1944 only one new engine was produced).[52] Further complications were caused by the complex hull design. The woodwork was easily damaged and required considerable labor and time to repair. Structural repairs would usually take two to three weeks to effect. As a result, in June 1944 FdS Petersen would face a paucity of resources. This is laid bare in the war diary for his command:

---

[49]For example, visit by Werner Toniges, 20 February 1942, text of speech and 16mm footage in Archiv Schlichting.

[50]Schlichting Werft friendship books 1–3, Archiv Schlichting.

[51]Schulze-Wegener, *Die deutsche Kriegsmarine-Rüstung*, pp. 173–74.

[52]ADM 223/28, Report by FdS to Group West, 25 May 1942 in E-Boat Operations and Policy.

Table 5.1: 1 June 1944 Disposition of Units[1]

1st Flotilla	8 boats operational	Black Sea
2nd Flotilla	5 boats operational	Ostend
3rd Flotilla	2 boats operational	Cattaro
4th Flotilla	8 boats operational	Boulogne
5th Flotilla	5 boats operational	Cherbourg
6th Flotilla	8 boats operational	Gulf of Finland
7th Flotilla	5 boats operational	Cattaro
8th Flotilla	4 boats operational	Ijmuiden
9th Flotilla	7 boats operational	Cherbourg
10th Flotilla	2 boats operational	Swinemünde

[1]ADM 223/28, War Diary Schnellboote Command, 1 June 1944, E-Boat Operations and Policy.

At least at Ostend, Cherbourg, Rotterdam, Ijmuiden, and Boulogne Petersen's flotillas were protected by purpose-built S-boat bunkers. However, the program had not gone far enough. At Cherbourg, only one of the two flotillas could be accommodated inside the bunker. Plans for additional facilities had foundered on a lack of materials and labor. Moreover, the bunkers were built to a lower standard of protection than the U-boat bunkers at places such as Lorient. This left them vulnerable to the Barnes Wallis-designed Tallboy bombs.

The Rise of the Small Battle Units (*Kleinkampfverbände*)

The slow acceleration in the *Schnellboote* building program, and the unsuitability of the design for mass production, seems to have been recognized in late 1943 following the British attack on *Tirpitz* using midget submarines. The combined efforts of the attack by *X5, X6,* and *X7* were dramatic as charges exploded underneath the German battleship. Eight hundred cubic meters of water had entered the ship within 2.5 hours of the attack and serious damage was being reported to electrical and propulsion systems.[53] *Tirpitz* would be non-operational for months, and even

[53]The Attack on the Tirpitz by Midget Submarines on 22 September 1943, Despatch submitted to the Admiralty, 8 November 1943, by Rear Admiral C. B. Barry, published Supplement to *The London Gazette,* 11 February 1948, pp. 993–1008. See also Ultra decrypts relating to Operation Source in TNA, HW1/2036, 2039 and 2041.

then questions would remain over whether she was truly fit to go to sea.[54] Within the *Kriegsmarine* there was surprise at the attack and intense interest in the method. The previous year coastal defense specialist *Kapitänleutnant* Hans Bartels had submitted a memorandum in which he advocated the use of midget submarines.[55] Fearing that his suggestions would not be given a hearing, Bartels arranged for the memorandum to be smuggled into the briefcase belonging to Dönitz's adjutant. With Dönitz's submarines seemingly bringing victory in the Atlantic, it was felt unnecessary to proceed with new weapons of war. The attack on *Tirpitz,* however, and the changed strategic position meant renewed interest in the idea. Midget craft offered the possibilities of mass production away from heavily bombed shipyards. Their production would not face the limitations facing the *Schnellboote* building program or the U-boat building programs. Bartels was promoted to *Korvettenkapitän* and Vice Admiral Hellmuth Heye was entrusted with the task of forming a new command for *Kleinkampfverbände* (small battle units) at Timmendorfer Strand. Heye would oversee the development of a range of midget submarines and motor boats. Development and training would take time that the German Navy was increasingly short of in late 1943 and early 1944, and the *K-Verband* program was not wholly without consequence for the *Schnellboote* service. Amongst the first volunteers to the service were two junior officers, Pettke and Potthast, who had served with 3rd *Schnellboote* Flotilla in the Mediterranean.[56] Ideally, such combat-experienced officers would have been the commanders of the wave of *Schnellboote* being turned out to meet the threat of invasion in the West. Looking forward to the production of large numbers of *Schnellboote* as a result of the 1943 program, FdS had warned: "The bottleneck to the building up of the . . . arm is the supply of trained personnel, especially of . . . commanders."[57] In the event, inadequate production of *Schnellboote* meant that the bottleneck would not materialize.

[54]"Tirpitz hit in Fjord Raid—Gallant Midget Submarines," *The Times,* 12 October 1943, p. 4.

[55]C. D. Bekker, *Swastika at Sea: The Struggle and Destruction of the German Navy, 1939–1945* (London: William Kimber, 1953), p. 137. Bartels (5 July 1910–31 July 1945) was awarded the Knight's Cross of the Iron Cross on 16 May for his outstanding bravery and initiative in the battle for the fjords while commanding minesweeper *M-1.* After the occupation of Norway, he was given responsibility for developing local coast defense flotillas using only local resources, particularly trawlers. In doing this, he established a reputation for resourcefulness and imagination.

[56]C. D. Bekker, *K-Men: The Story of the German Frogmen and Midget Submarines* (Maidstone: George Mann, 1968), pp. 21–22.

[57]ADM 223/28, Report by FdS to Group West, November 1943 in E-Boat Operations and Policy.

Waiting for the Invasion

The small number of naval units at the disposal of Navy Group West, and the increasingly vigorous Allied patrols from early 1944, placed significant limits on the disposition of German surface vessels along the Channel coast. German forces were engaged routinely as they sortied from their bases along the Channel coast and found great difficulty in engaging the coastal convoys running along the South and East coasts of England. In the context of the Allied invasion, those convoys in early 1944 were carrying a significant proportion of the supplies called for under Operation Bolero: the logistical build up to D-Day. At the end of 1943, the Royal Navy had rather nervously instituted more frequent convoying and the movement of ocean-going ships along the coastal routes. With the build up "lagging behind schedule" (coal stocks[58] were estimated to be three weeks below minimum Bolero requirements for the Spring of 1944), on 25 November 1943 the Naval Staff instructed the commands at Nore, Rosyth, and Portsmouth to begin seven-day convoying.[59] Successful attacks on these coastal convoys could have inflicted serious damage on the Bolero schedule for D-Day.

With regard to the period in the late Spring of 1944, as the prospect of invasion loomed, the head of the *Kriegsmarine* noted in his memoirs: "The meagre forces at the disposal of the Commander-in-Chief, Navy Group West, could not be kept permanently at sea as patrols against possible enemy landings. . . . The losses and damage we sustained were so severe that even the maintenance of a permanent patrol line, let alone any reconnaissance sortie into enemy coastal waters, was out of the question."[60] These are significant comments and they point towards what might have been possible with a greater number of *Schnellboote* at the disposal of the *Kriegsmarine* in the English Channel in June 1944: a standing patrol line to detect the approach of the invasion fleet, reconnaissance sorties along the enemy coast, and the possibility of spoiling attacks on concentrations of Allied shipping. Those spoiling attacks would have involved torpedo attacks by *Schnellboote*, but also their use as fast minelayers. *Schnellboote* were equipped with deck rails to allow them to lay mines (the type 38 could carry six conventional LMB mines in addition to two torpedoes in the tubes, but at the expense of the two reloads).[61] Shortage of ships and aircraft meant that in May and June 1944 there were only limited mining operations against the harbors and anchorages along the southern coast of Britain from which the invasion would be launched. The concentration of shipping at those points presented a target-rich environment of considerable

[58]ADM 1/15815, Director Trade Division, 28 November 1943.

[59]ADM 1/15815, Admiralty to Commanders-in-Chief, Nore, Rosyth and Portsmouth, 25 November 1943.

[60]Dönitz, *Memoirs*, p. 395.

[61]Frank, *German S-Boats*, p. 150.

strategic significance. The British Naval Staff later noted: "had the [limited mining] attacks been on a more massive scale they might well have caused the Allies serious embarrassment."[62]

Difficulties also hampered other potential preparations by the *Kriegsmarine* to combat the Allied invasion fleet. Considerable problems were encountered with the development of coastal batteries for the *Kriegsmarine* because the construction of fixed defenses for the Army, and launching sites for V-1 flying bombs, occupied most of the available construction resources. Only a handful of additional coastal batteries were constructed in France for the Navy between 1943 and 1944, and Dönitz profoundly disagreed with the Army's policy of siting batteries behind the beaches where they were incapable of direct fire.

What made these failures particularly galling was the fact that the high command of the *Kriegsmarine* correctly guessed that the beaches of Normandy would be selected by the Allies as the point for the invasion: "The naval command considered the area east of the Cotentin Peninsula as far as Boulogne to be the most probable place for a landing. The coast of the Pas de Calais was regarded as the second possibility."[63] The steady build up of shipping in the ports of South Western Britain was noted in the *Kriegstagebuch der Seekriegsleitung* and it pointed towards an invasion in Western France in May/June 1944.[64] Taking no chances, in early 1944 a barrage of mines was laid on the ocean bed parallel to the coast as far north as Holland. They were, however, to play a minimal part in the German defense in June 1944. Concerned about restricting their own freedom of movement, the mines were set to flood and drop to the ocean bed on 31 May.[65] The enormous investment represented by the mine barrage had been an almost complete waste of time, effort, and resources.

Properly equipped for littoral warfare, with effective reconnaissance to provide some warning of the assault, more effective use of mine barrages and close collaboration with the Army in the construction of fixed positions and the conduct of the defense, the light coastal forces of the *Kriegsmarine* should have inflicted substantial casualties on the Allied invasion force. The 5th and 9th *Schnellboote* Flotillas had demonstrated the potential effectiveness of the light forces against the invasion fleet when on 28 April they had torpedoed three LSTs (Landing Ship,

---

[62]ADM 239/415, Naval Staff History, The Defeat of the Enemy Attack on Shipping 1939–1945: A Study of Policy and Operations, Volume 1A, Text and Appendices, CB3304(IA), Historical Section Admiralty, 1957, p. 200.

[63]ADM 223/688, Essay by Grand Admiral Dönitz on the war at sea, 24 September 1945. Dönitz's recollections were also confirmed by Vice Admiral Hellmuth Heye, see ADM 223/690, essay by Heye on the war at sea, 15 October 1945.

[64]See, for example, entries 5, 10 May 1944, *KTB der SKL*, Teil A, Band 57 (Berlin: Mittler & Sohn, 1995), pp. 69 and 163.

[65]John Frayn Turner, *Service Most Silent: The Navy's Fight against Enemy Mines* (Barnsley: Pen & Sword, 2008), p. 165.

Tank) engaged on a training exercise. Two ships had been sunk and 749 U.S. Army and U.S. Navy fatalities sustained. The episode demonstrated the danger that a small number of sinkings could generate heavy losses of invasion troops. Preventing the invasion in 1944 was undoubtedly beyond the light forces, but they were potentially in a position to inflict sufficient casualties to give the German Army a chance of victory at the water's edge and behind the invasion beaches.

As James Foster Tent has conclusively demonstrated, senior British and American officers in 1944 were deeply troubled by the *Schnellboote* threat.[66] A pre-invasion analysis by the Royal Navy revealed that the principal problem for the *Schnellboote* between 1940 and 1944 was in the location of coastal convoys.[67] When they were located, the *Schnellboote* were deadly effective, sinking an average of 2.4 ships per engagement. With the invasion, *Schnellboote* would have no problem in locating targets. Moreover, the same analysis demonstrated that in 1944 Allied forces still faced massive problems in locating *Schnellboote* once at sea, and an even bigger problem in sinking them once engaged. Such an analysis amply demonstrates the potential outcomes if there had been a conscious shift towards a strategy of littoral defense in mid-1943.

D-Day for the *Kriegsmarine*

Those possibilities are perhaps underplayed in most accounts of the D-Day landings. In the historiography of D-Day, there is only limited discussion of ship-to-ship engagements off the Normandy coast in June 1944. Samuel Eliot Morison, for example, devotes just five pages of Volume XI of his *History of United States Naval Operations in World War II* to the *Kriegsmarine's* responses to the invasion fleet.[68] There is general agreement that on D-Day the *Kriegsmarine* reacted slowly to the invasion threat. At 01.45 hours the telephone lines between Paris and Cherbourg were cut by the *Maquis*. Navy Group West was forced to rely on radios for communication as they received the first reports suggesting that the invasion was underway. Senior officers hurriedly reported for duty to headquarters with Chief of Staff Admiral Hoffmann arriving in his dressing gown. It was not until 03.48 hours on the morning of 6 June that Krancke (Commander-in-Chief Navy West) ordered 8th Destroyer Flotilla to move from the Gironde to Brest, and for inshore patrolling by the Cherbourg-based 5th and 9th *Schnellboote* Flotillas. The former had barely left harbor when it came under heavy attack by fighter bombers and was compelled to seek the safety of its bunkers. The 9th Flotilla followed suit after making a torpedo attack at long range on the ships of the invasion force. In the east, it took until 05.30 for the ships of the 5th Torpedo Boat Flotilla (*T28*, *Falke*,

---

[66]Tent, *E-Boat Alert*, pp. 63–74.

[67]ADM 223/172, Note on E-boat Operations in the English Channel, January–May 1944.

[68]Morison, *History of United States Naval Operations*, Vol. XI, pp. 170–75.

Jaguar, and *Möwe*) to sortie from Le Havre. They fired their torpedoes at extreme range and then promptly returned to port. The Norwegian destroyer *Svenner* was the only casualty of the attack.

The opening engagement seemed to set the tone for naval engagements over the next few days. Available German naval forces rushed to the invasion area with varying levels of success. The 4th *Schnellboote* Flotilla was able to move from Boulogne to Le Havre, but heavier German units were not successful. Faced with overwhelming Allied naval power, German surface vessels appeared highly reluctant to close with the enemy. Admiral Dönitz later commented: "The German light naval forces found themselves confronted with a vastly superior enemy. We had thirty . . . [*Schnellboote*], four destroyers and nine torpedo boats. The British and the Americans, on the other hand, had concentrated between seven and eight hundred warships in the sea area of the invasion. . . . The German naval forces did what they could and managed even to achieve a few successes."[69]

Academic opinion has concurred with Dönitz's assessment. As Peter Smith has commented: "The limited German forces found themselves overwhelmed by the huge size of the Allied fleet. They could never hope to make more than a few sorties before destruction, but, in the event, made even less of an impression than had been expected."[70] The apparent pattern of long-range, and largely unsuccessful, torpedo attacks appeared to mark *Schnellboote* operations over the next few days.

Then on 14 June Royal Air Force (RAF) Bomber Command staged a large raid on Le Havre that severely depleted the *Schnellboote* force in the Channel. Le Havre was not a regular *Schnellboote* base. It lacked sufficient protected space, dedicated dockyard and repair facilities (in near constant operational use since 6 June, maintenance and repair was a growing issue), and torpedo storage and handling capacity (torpedo shortage was placing a growing restriction on operations).[71] The dock area of Le Havre was specifically targeted with considerable success by three waves of RAF bombers. One Royal Canadian Air Force pilot wrote in his diary: "not a bad trip" and that Air Chief Marshal Harris, the Head of Bomber Command, had congratulated his crews on "the greatest ever air to sea victory."[72] The number of vessels

[69] Dönitz, *Memoirs*, p. 396.

[70] Peter C. Smith, *Hold the Narrow Sea: Naval Warfare in the English Channel, 1939–1945* (Annapolis: Naval Institute Press, 1984), p. 227.

[71] Within five days of daily operations, almost every boat in the 9th *Schnellboote* Flotilla required significant levels of repair and maintenance ranging from engine overhaul and replacement through to deck damage. In the midst of the purpose-built Cherbourg *Schnellboote* bunker, the maintenance and repair issues were dealt with speedily. See U.S. National Archives and Records Administration (hereinafter NARA), Record Group (hereinafter RG) 242, Records of the German Naval High Command, 1935–1945, Microfilm Publications, T1022, roll 3150, Kriegstagbuch Abschnitt Machine, 1 June–10 June 1944.

[72] Bruce Johnston, 115 Squadron RAF, diary entry for 14 June 1944, http://lancasterdiary.net/June%201944/june_14_1944.php.

lost was staggering: "twenty minesweepers and patrol-boats, nineteen tugs, three torpedo boats . . . and fourteen S-boats [were destroyed and others damaged]. Only one S-boat (S167) was left operational in Le Havre."[73] The attack was quite literally devastating, removing the remaining surface threat to the invasion fleet.

The conventional reading of the Normandy invasion would seem to go against the idea that German forces might have been able to achieve a significantly better performance with the greater availability of weapons that should have resulted from a properly developed strategy of littoral defense. However, there is scope for a reevaluation of the effectiveness of the short-lived *Schnellboote* campaign against the invasion force in June 1944. While Dönitz and the historians have been content to write off the operations of the 4th, 5th, and 9th *Schnellboote* Flotillas in June 1944 as an inevitable poor performance in the face of the overwhelming power of the Allied navies, FdS Petersen formed a rather more optimistic assessment of the situation. While recognizing the disparity between the invasion force and German coastal forces, and other disadvantages such as the shortness of the summer nights for the conduct of operations, Petersen also realized that some factors were in favor of his *Schnellboote* flotillas:

Operational Advantages:

1) Contrary to expectation, the enemy did not attempt to take . . . the two harbours of CHERBOURG or LE HAVRE immediately, so that the E-boats had bases close to either flank from which they could operate against the SEINE ESTUARY.
2) The enemy was forced, at least in the early days, to operate with large surface forces and transport units, in a narrow area, without any regular convoy rhythm, so that there were always worthwhile targets in the area.
3) Enemy defence systems were not fully worked up, so that in the early days they were reasonably easy to break through.
4) The enemy had many new units—especially American units—which were neither used to the area nor to E-boat warfare, and which were therefore easier targets for . . . [*Schnellboote*] than the experienced British Channel destroyers.[74]

Petersen's recognition that his flotillas in the Channel did enjoy some advantages perhaps suggests that *Schnellboote* operations against the invasion fleet

[73] V. E. Tarrant, *The Last Year of the Kriegsmarine: May 1944–May 1945* (London: Arms and Armour, 1994), p. 69. See also Gerhard Hümmelchen, *Die Deutschen Schnellboote im Zweiten Weltkrieg* (Hamburg: Mittler, 1996), pp. 175–77.

[74] ADM 223/28, War Diary Schnellboote Command, 1 June 1944, E-Boat Operations and Policy.

deserve reassessment. Despite Dönitz's rather negative comments, torpedo attacks by *Schnellboote* in June 1944 were not entirely without result. In the four nights between 7 and 13 June, as V. E. Tarrant has calculated: "the S-boats sank three tank landing ships, one tank landing craft, one infantry landing craft, one motor-torpedo boat, one motor gunboat, the US tug *Partridge*, three small merchant ships grossing 1,812 tons, a component of the artificial Mulberry harbour, and the frigate *Halstead* . . . written off as a constructional loss."[75] They also damaged one destroyer, a tank landing ship, and a landing craft. By contrast, in the two months following D-Day, Dönitz's submarines managed to sink just five escort vessels, two naval landing ships, and four merchant ships.[76] Clay Blair comments tersely that "submarines were useless against alert invasion forces."[77] John Terraine writes of "a real U-boat fiasco." He asserts that boats not fitted with *Schnörkel* "achieved precisely nothing" and their prudent withdrawal was "an abject confession of powerlessness."[78] In Peter Padfield's judgement, "the U-boats had been literally overwhelmed by the forces massed to prevent them reaching the transports, as indeed had the German light surface forces."[79] Correlli Barnett sums up the intervention of the U-boats as "a pitiful performance for a weapons-system and a service which had come so near to winning both world wars; the last limp punch of a beaten prizefighter whose legs were already buckling."[80]

Meanwhile, FdS Petersen was so pleased with the reports reaching him of *Schnellboote* attacks that on 10 June he signalled: "S-boat successes against the invasion fleet first class! . . . Keep at it! Now you have plenty of targets—at them!"[81] Hitler was sufficiently impressed with the performance of the *Schnellboote* arm as to award Petersen the Oak Leaves to his Knight's Cross on 13 June.[82] *Kapitänleutnant* von Mirbach, head of the 9th *Schnellboote* Flotilla, received the same honor the following day and the German press trumpeted the invasion as "The

[75]Tarrant, *Last Year*, p. 67.

[76]G. H. Bennett and R. Bennett, *Hitler's Admirals* (Annapolis: Naval Institute Press, 2004), p. 200.

[77]Blair, *Hitler's U-Boat War*, Vol. II, p. 601.

[78]John Terraine, *Business in Great Waters: The U-Boat Wars, 1916–1945* (London: Leo Cooper, 1989), pp. 646–47.

[79]Peter Padfield, *War Beneath the Seas: Submarine Conflict, 1939–45* (London: John Murray, 1995), p. 429.

[80]Correlli Barnett, *Engage the Enemy More Closely: The Royal Navy in the Second World War* (London: Penguin, 2000), p. 832.

[81]Führer Der Schnellboote War Diary, 10 June 1944, translated and reproduced in Hans Frank, *German S-Boats in Action in the Second World War* (Barnsley: Seaforth, 2007), p. 112.

[82]"Sie erhielten das Eichenlaub zum Ritterkreuz des Eisernen Kreuzes" and "Der Erste Eichenlaubtrager der Invasionsfront," *Die Kriegsmarine* (August 1944), pp. 2–3.

great hour of the *Schnellboote*."[83] Although the reports reaching the public of success by the *Schnellboote* were exaggerated, there was real substance behind the claims.[84] Had a larger number of *Schnellboote* been available in the Channel in June 1944, the *Kriegsmarine* would have been in a position to mount localized swamping attacks, and to have taken greater risks to get through the screening forces to the troop-carrying landing ships and craft heavily laden with ammunition and other flammable materials. The casualties could have been considerable and that might have made a difference to the struggle on Omaha Beach, and to the subsequent fight to link the different beachheads before German armored units could move up from the interior.

The potential effectiveness of a larger *Schnellboote* force is reinforced by the fact that a new type of mine was available to the *Kriegsmarine* and *Luftwaffe* in June 1944. The use of the three new types of pressure mine (usually referred to as oyster mines by the Allies) had been limited so as not to compromise the new weapon before the invasion.[85] Some 4,500 of them were held in reserve in depots in Germany with preparations to rush supplies to the critical area once the invasion was underway.[86] However, even with preparations for the deployment of the oyster mine, the failure to develop a coherent defensive strategy was evident. Early in 1944, 2,000 of the new types of mine were being held at underground bunkers at Le Mans. In May, concerned that they might be captured before they could be used, Göring ordered the stocks withdrawn to German depots with the last mines arriving in Germany on 4 June.

Attempts to reinforce the defenses of the Normandy coast with conventional mines proved similarly problematic. An attempt to lay mines in the Bay of the Seine estuary on 23 May was signaled by Ultra decrypts leading to interdiction by RAF aircraft.[87] The loss of two minelayers meant that it was not until 5 June that

---

[83]"Die Grosse Stunde Der Schnellboote," *Die Kriegsmarine* (August 1944), pp. 4–5. See also "Schnellboote im Kampf mit der Invasionflotte" and "Schnellboot Manner," *Die Kriegsmarine* (August 1944), pp. 8–15.

[84]The Schlichting family maintained a diary of significant mentions of *Schnellboote* activity which allows some understanding of the reports reaching the German public. On 8 June 1944 they had heard reports that thirteen landing craft had been sunk during the invasion or on the nights following. On 9 June they heard that *Schnellboote* had the previous night accounted for one cruiser, two destroyers, and five landing ships, entries for 8 and 9 June 1944, Press and Scrapbook, Archiv Schlicting, Travemünde. In the monthly newsletter of the Schlichting work, Rudolf Schlichting referred to the "blood drenched soil of Normandy" and "thousands drowned" in the English Channel when their landing craft/ships were sunk, "Arbeitskameraden an der Front under in der Heimat!," June–July 1944, *Der Werftbote*, Archiv Schlichting.

[85]ADM 223/688, Essay by Grand Admiral Dönitz on the war at sea, 24 September 1945.

[86]ADM 239/415, Defeat of the Enemy Attack on Shipping 1939–1945, p. 201.

[87]Steven J. Zaloga, *D-Day Fortifications in Normandy* (Oxford: Osprey, 2005), p. 14.

operations began to lay new minefields in the Bay of the Seine estuary.[88] Use of *Schnellboote* and *Räumboote* to lay the mines was an option, but it was not taken by a Navy Group West anxious to preserve its fleet-in-being, which meant that the chance was squandered to lay invasion-imminent "*Blitzsperren*" (lightning barrages of mines). As Vice Admiral Ruge (the naval expert on the staff of Commander-in-Chief West) later commented: "Even a few hundred mines in the Seine Bight could have caused losses and considerable delays."[89] This would have been especially true if the minefields had been defended by coastal artillery and additional *Schnellboote* operations. The new oyster mines (once they had arrived from Germany), laid by minelayer, *Schnellboote,* and aircraft proved as effective as Ruge had expected with useful cooperation between the *Kriegsmarine* and *Luftwaffe.*[90] As the commander of U.S. naval forces wrote in 1946, the first problems with mines were encountered on 7 June near the Cardonnet shoal off Utah Beach: "The field claimed seven casualties including two U.S. destroyers . . . and the U.S. Fleet minesweeper *Tide.* In the eastern area . . . the air mining effort mounted steadily. . . . Fifteen casualties, three vessels sunk and twelve damaged, were suffered in the ETF area by D-plus 16, while some ninety mines had been swept."[91]

The mine laying campaign had more than nuisance value even if it was compromised to a certain extent in the early phases by the use of unreliable fuses.[92] It was further compromised by the fact that the Germans lacked sufficient numbers of aircraft and ships to deliver the new mines. As the British Naval Staff later concluded: "[The Germans] found themselves in the exasperating position of possessing weapons of a most formidable nature while deprived of the ability to lay them offensively except on a relatively small scale."[93] By 21 June, some forty-four Allied ships had fallen victim to German mines (oyster and conventional), although it is impossible to determine how many of these can be attributed to *Schnellboote* oper-

[88]ADM 223/689, Essay by Admiral Krancke on the war at sea, 2 October 1945, pp. 197–98.

[89]Vice Admiral Friedrich Ruge, "The invasion of Normandy," in Hans-Adolf Jacobsen and Jürgen Rohwer, *Decisive Battles of World War II: The German View* (London: Andre Deutsch, 1965), p. 327.

[90]Sönke Neitzel, *Der Einsatz der deutschen Luftwaffe über dem Atlantik und der Nordsee 1939–1945* (Bonn: Bernard & Graefe, 1995), pp. 221–26. On page 225, Neitzel cites a figure of 4,407 mines laid by the *Luftwaffe* in the Seine Bight between 6 June and 26 July 1944. See also Tony Le Tissier and Sönke Neitzel, *Die Zusammenarbeit zwischen Schnellbooten und Luftwaffe* (n.p.: Mittler, 1995), pp. 55–63.

[91]Commander, U.S. Naval Forces in Europe, *Administrative History of U.S. Naval Forces in Europe, 1940–1946,* Vol. 5 (London: U.S. Navy, 1946), p. 443.

[92]ADM 219/392, Review of the German Mining Campaign in the British Assault Area, 16 October 1944. See also Sönke Neitzel, "Kriegsmarine and Luftwaffe Co-operation in the War against Britain, 1939–1945," *War in History* 10:4 (2003), pp. 448–63.

[93]ADM 239/415, Defeat of the Enemy Attack on Shipping 1939–1945, p. 200.

ations.[94] The effectiveness of the new pressure mine forced the Allies into desperate innovation and large-scale countermeasures. Royal Navy minesweepers responded by taking on board pipes and hand grenades.[95] Sent down the pipe to explode underwater, a hand grenade could initiate a sympathetic explosion from an oyster mine. The RAF targeted the airfields used by minelaying aircraft and night fighters tried to put combat air patrols over the invasion fleet by night. Restrictions were placed on the movement, speed, and channels used by Allied shipping to maximize the sweeping effort and to decrease the chances of setting off pressure and acoustic mines.[96] The number of Allied vessels sunk and damaged by mines during June 1944 did not pose a threat to the success of the Allied landings, but it was certainly respectable in the circumstances.[97] The logistical problems caused by the mining effort were felt on the battlefield where restrictions sometimes had to be imposed on such things as the expenditure of artillery ammunition.

Contrary to some post-war perceptions, the *Schnellboote* gave a reasonable account of themselves in the days between 6 and 14 June when the RAF's raid on Le Havre ended that threat to the invasion force. Some twenty operational *Schnellboote* had enjoyed some success against a well-defended invasion force of over 1,000 Allied vessels. Following the raid on Le Havre, surviving German coastal forces, with limited reinforcement from elsewhere, were left to stage a fighting withdrawal up the Channel coast. From ever more distant ports such as Le Havre and St Malo, they continued to undertake operations against the fleet anchored off Normandy.[98] The burden of contesting control of the Normandy coast had fallen to the forces of the *K-Verband* long before the last vessel left Le Havre at the end of August.[99] Rushed into use, the *K-Verband* proved highly ineffective.[100] As Hellmuth Heye later commented: "Trials were, virtually speaking, made against the enemy."[101]

---

[94]Tarrant, *Last Year,* p. 68.

[95]Dunstable Town Centre, "Life Aboard the 6th Minesweeping Flotilla," BBC WW2 People's War Project.

[96]Commander, U.S. Naval Forces in Europe, *Administrative,* pp. 443–44.

[97]On the performance of the *Schnellboote* in the English Channel 6–14 June, see: Hümmelchen, *Deutschen Schnellboote,* pp. 168–75; Mike Whitney, *Deutsche Seestreitskräfte 1939–1945* (Stuttgart: Motorbuch, 1995), pp. 72–74. See also RG 242, Microfilm Publications, T1022, roll 3146, Fuehrer Der Schnellboote Kriegstagbuch und Anlagen, June 1944.

[98]For example, *S-130* became the last *Schnellboot* to vacate the Cherbourg bunker at 23.30 hours on 23 June. Its destination was St Malo. See RG 242, Microfilm Publications, T1022, roll 3871, 9th Schnellbootsflotille, Kriegstagbuch und Anlagen Entry for 2330 hours 23 June 1944. See also entry 23 June 1944, *KTB der SKL,* Teil A, Band 58 (Berlin: Mittler & Sohn, 1995), p. 319ff.

[99]Tent, *E-Boat Alert,* p. 208.

[100]On the *K-Verband,* see Paul Kemp, *Underwater Warriors* (Annapolis: Naval Institute Press, 1996).

[101]ADM 223/690, Essay by Heye on the war at sea, 15 October 1945.

The successes of the *K-Verband* in the closing months of 1944 were minimal although Dönitz seems to have had trouble in coming to terms with their failure. In his memoirs, he remarks that the forces of the *K-Verband* were handled with "courageous determination" and that they "succeeded in sinking a number of enemy ships."[102] More tellingly, Vice Admiral Heye commented after the war that "whatever their nature, small battle weapons can never be a substitute for regular weapons of war."[103] Nevertheless, even as the war was drawing to a close, the development of new weapons for coastal warfare proceeded. For example, on August 1944 a prototype was commissioned of a new four-man motor torpedo boat. Named the *Hydraboot,* the prototype was capable of thirty-six knots.[104] The first batch of fifty was commissioned on 4 December 1944 and a further batch of 115 was ordered on 8 February 1945. Approximately thirty-eight boats were ready for service at the war's end. It was a case, however, of too little, too late. The critical moment for the development and realization of these programs to supplement the *Schnellboote* program had been in 1943. The *Kriegsmarine's* obsession with the U-boat offensive had meant that critical programs for coastal weapons had not borne fruit until after the point at which the invasion had become an established reality.

Conclusion

This study suggests a need to reappraise the historiography of the D-Day landings, and further suggests that Dönitz's failure as Commander-in-Chief of the German Navy after 1943 was even more profound than Peter Padfield and others have identified. During and after the Second World War, Dönitz was unwilling to accept that the *Kriegsmarine* had been forced into a coastal war. As late as the *Führer* Naval Conference of 3 December 1944, Dönitz still added the caveat "with the exception of submarine warfare" to the admission to the *Führer* that the *Kriegsmarine* was waging "a purely coastal war."[105] In the last months of the war, Dönitz continued to give optimistic assessments of the offensive performance of the *Kriegsmarine.* On 3 January 1945, Dönitz told Hitler that he expected the *Seehund* midget submarine to sink 100,000 tons of Allied shipping.[106] While in Allied hands in 1945, Dönitz continued to maintain that German coastal forces had enjoyed "considerable suc-

[102] Dönitz, *Memoirs,* p. 396.

[103] Foreword by Admiral Hellmuth Heye to Bekker, *K-Men,* p. 6.

[104] The hydra boat was to be built by a number of different yards: Krögerwerft, Warnemünde and Stralsund; Gebrüder Engelbrecht, Köpenick; Lürssen, Vegesack; Danziger Waggonfabrik; Karl Mathan, Berlin; Robert Franz; Karl Vertens, Winning; H. Heidtmann, Hamburg; Hamburger Werft, Hamburg; Hinr. von Cölln, Hamburg; Schlichtingwerft, Travemünde.

[105] *Führer* Naval Conference, 3 December 1944, *Fuehrer Conferences,* p. 420.

[106] *Führer* Naval Conference, 3 January 1945, *Fuehrer Conferences,* p. 425.

cess" even if they had not been able to "bring about a change in the situation" as a result of Allied material superiority.[107] In the decade between entering Allied captivity and completing his memoirs, Dönitz seems to have shifted from self-delusion to denial. The tone of Dönitz's memoirs is striking at the point at which he discusses the invasion. Matter of factly he raises central questions about German strategy in the face of the invasion. The failures of the *Heer* and *Luftwaffe* are to the fore. He raises no question about his own strategy in privileging the U-boat war at the expense of a strategy of littoral defense.[108] Most strikingly of all, he refuses to consider the failure of the high command to create a command structure and defensive plan in the West which could have unified and made best use of the efforts of the German Army, Navy, and Air Force. As a result, on D-Day there would be "no uniform defence plan. . . . The three services did not cooperate enough in many details. . . . Almost every commander had his own ideas, not always very clear, on how to defend his area."[109] Dönitz appears to have been one of those commanders guilty of what Vice Admiral Ruge described as a "fatalistic acceptance of the deteriorating situation, and a lack of alertness in looking for possible improvements."[110] The head of the *Kriegsmarine* appears to have accepted Germany's defeat with the equivalent of a shrug of the shoulders. Historians have not subjected Dönitz's memoirs to an appropriate level of analysis.

Germany's failure to pursue policies to adequately secure the coast of *Festung Europa* between 1943 and 1944 provides confirmation of a range of failings within the Nazi state: a failure of leadership; a failure of production and organization; a failure to respond to the realities of the changing tide of war. That failure extended even to the electro-boats upon which Dönitz was prepared to gamble all. Although post-war German writers, such as *Kriegsmarine* veteran Cajus Bekker, would suggest that in the spring of 1945 the U-boat war was about to be renewed with a vengeance, in reality the rot was evident even in this most cherished of programs.[111] As Grier has noted, delays resulted from "simple poor planning," poor coordination, and a failure to meet construction specifications.[112] Even if the *Schnellboote* had been able to buy Germany more time in June 1944, it is unlikely that the electro-boats would have revolutionized the strategic position at sea in 1945.

Finally, the failure to discuss, articulate, and pursue a strategy of littoral war highlights significant failings on the part of Karl Dönitz as head of the *Kriegsmarine*. As Peter Kemp concluded in his 1976 study: "Brilliant as had been his direction

---

[107]ADM 223/688, Essay by Grand Admiral Dönitz on the war at sea, 24 September 1945.

[108]Dönitz, *Memoirs,* p. 397.

[109]Ruge, "Invasion," p. 323.

[110]Ibid., p. 321.

[111]Bekker, *Swastika at Sea,* p. 207.

[112]Howard D. Grier, *Hitler, Dönitz, and the Baltic Sea: The Third Reich's Last Hope, 1944–1945* (Annapolis: Naval Institute Press, 2007), p. 181.

of the U-boat campaign, Dönitz was out of his depth as Commander-in-Chief. . . . As Admiral Commanding U-boats Dönitz was, without any doubt, the most brilliant of the war. As Commander-in-Chief Navy he was a sorry failure. He had neither the grasp of strategy nor the breadth of vision necessary to survive in the rarified atmosphere of top naval command."[113] Lothar-Günther Buchheim put it more personally and more bluntly: "When Dönitz proved unable to deliver the promised weapons to his men, his exhortations became stereotyped appeals to their 'spiritual conduct,' their 'spiritual force,' and their 'spiritual tenacity of purpose.' This constant escape into vagueness at a time when all the evidence damned his strategy is typical of the man and the period."[114]

[113]Kemp, "Dönitz," pp. 482–83.

[114]Lothar-Günther Buchheim, *U-Boat War* (New York: Knopf, 1978), pages unnumbered.

6

The Development of Combined Operations Headquarters and the Admiralty during the Second World War

Personalities and Administration

C. I. Hamilton

Introduction: Keyes and Mountbatten

A large organization is necessarily mysterious, and it can be helpful to study it through another that is smaller and linked to the first. But how does one address the smaller one in the first place? There the personalities of some of its leading staff can serve as a useful introduction, an approach to what one could regard as the more fundamental matters. These considerations lead students of the Admiralty in the Second World War to Combined Operations Headquarters (COHQ), and initially to the colorful and determined men who headed it from 1940–43, Sir Roger Keyes and Lord Louis Mountbatten.

It is surely true that the differences in their characters and abilities explain a great deal about why the first failed and the second succeeded. Much is shown by contrasting their tactics in Whitehall battles: Keyes relied on the frontal assault, Mountbatten on the indirect approach—sometimes too indirect for the good of his reputation. There is, of course, a great deal in favor of the Keyes' method, but not as a constant recourse, as one sees as the months rolled by after Keyes' appointment as Director of Combined Operations (DCO) in July 1940. His regular appeals to the Prime Minister, and complaints about those who allegedly were trying to fetter him—notably the Chiefs of Staff (COS) and the planning committees under them—led in the end to irritation and curtailment of the broad powers he had initially been granted: the culmination was reached in mid-1941 when his title was altered to Adviser on Combined Operations (ACO). The loss of his executive pretensions was too much for Keyes; he balked, and the chance was taken to force his resignation (October 1941), Captain Lord Louis Mountbatten being appointed instead.[1] By contrast, the new man stated that he had no objection at all to being an Adviser; but he, too, was a naval officer, and like Keyes, he wanted a title with connotations of command. His methods, though, were those suitable to a bureaucracy. When, a little later, he was promoted Commodore, he began to call himself Commodore Combined Operations (CCO), tolerated on the understanding that when he was dealing with the COS, he would remain ACO. But more and more it was as CCO that he appeared in correspondence, and once he was promoted to Vice Admiral in March 1942 it was accepted that—rather than cause confusion with a new acronym—his substantive title should be altered to Chief of Combined Operations. Moreover, Mountbatten became one of the COS. Here as elsewhere there was much substance to Keyes' later complaint that what he had sought in vain, Mountbatten was given.[2]

The differences between the two men are shown also in the organizations they created. The starkest contrast was in simple size. When Keyes resigned, COHQ had a total staff of only twenty-three, including typists. Within six months, under Mountbatten, this had expanded to 400. In the process, the organization also altered remarkably in nature and composition, and with only a little exaggeration one may say that it came up to date by about two decades in the process. Keyes' staff was rather more suitable to the end of the Great War than the beginning of the Second. The naval officers at its heart were a somewhat elderly band of brothers, veterans of the earlier conflict, and all on the retired list.[3] The approach to planning was also somewhat backward-looking. Keyes had been Director of Plans (D of P) at the Admiralty in 1917, and had been notable for his great enthusiasm, but that included a tendency to underestimate logistical problems. In January 1918, he went on to an operational command, but it must have been of him as well as his successor that the Deputy Chief of Naval Staff, Sir Sydney Fremantle, was thinking when in May he proposed the Plans and Operations Divisions be combined, precisely to stop the articulation of plans without due reference to available means. Keyes had not changed by the later war, and caused similar exasperation to other senior officers—for instance, Sir Andrew Cunningham.[4]

[1]See Keyes' correspondence in The National Archives, London (hereinafter TNA), CAB 121/169. Unless otherwise noted, all archival references are to TNA. All quotations in this chapter from archives are Crown Copyright, bar the phrase from a letter of Commander Robin Todhunter, where I have not been able to trace the copyright.

[2]CAB 121/169, fos. 105, 129, 130–33A, 138F-137. Also, Directives and drafts, 30 October 1940 and 14 March 1941, fos. 13, 44. See also Jeremy Langdon, "Too old or too bold? The removal of Sir Roger Keyes as Churchill's first Director of Combined Operations," *Imperial War Museum Review* 8 (1993), pp. 72–84.

[3]University of Southampton, Hartley Library (hereinafter HL), Mountbatten Papers, MB1 B74, Mountbatten to Vice Admiral H. T. Baillie-Grohman, 31 August 1953.

[4]TNA, ADM 1/8524/135, May 1918. Bernard Fergusson, *The Watery Maze: The Story of Combined Operations* (London: Collins, 1961), p. 74. See also Stephen Roskill, *Churchill and the Admirals* (London: Collins, 1977), pp. 171–77.

Furthermore, it appears that to Keyes the model for combined operations remained the attack he had organized back in 1918, against Zeebrugge. Captain John Hughes-Hallett—who became Mountbatten's "Naval Adviser" in December 1941, and was the naval force commander during Operation Jubilee, the raid on Dieppe—wrote in his memoirs that as far as he could see, under Keyes, COHQ did not think in terms beyond those of "a bigger and better Zeebrugge-type raid."[5]

The arrival of the much younger Mountbatten brought an unequivocal advance in method and modernity. Among his very first acts were the appointment of a Chief of Staff and creation of a signals department. Even more indicative of change, he determinedly introduced the inter-service principle. Mountbatten later said that when he arrived, he found just one RAF officer, supposedly there only because he was an MP. Hughes-Hallett's memoir is also again pertinent, arguing that Keyes, like many of his older officers, did not really understand the need for the three Services to cooperate closely, and came to concentrate on operations in Norway since he then had to work only with the C-in-C, Home Fleet.[6] Mountbatten, by contrast, strove to get a balance between the Services in numbers, and also to ensure the various colors of uniform worked together as one.[7] The intelligence organizations can be contrasted, too. Here Keyes had showed ambition, at one stage looking to create a unit of his own to gather topographical information—of obvious importance when it came to planning raids—though was persuaded by Vice Admiral John Godfrey, the Director of Naval Intelligence (DNI) from 1939–43, instead to send a liaison officer to the Inter-Service Topographical Department (ISTD) he had set up, largely at Oxford.[8] Mountbatten, on the other hand, whilst encouraging copious use of the ISTD, "flatly refused to have anything to do with" another of Godfrey's enthusiasms, the Intelligence Section (Operations) (IS(O)), created in December 1941 and intended as a central intelligence collation agency. He preferred instead to build up his own existing collation unit under his Senior Intelligence Officer, and later praised the latter for creating "the finest intelligence organisation in the Empire."[9]

His attachment to modernity, though, is best shown in another example. It is a cliché that the Second World War was far more a conflict of science than 1914–18

[5]Philip Ziegler, *Mountbatten: The Official Biography* (London: Collins, 1985), pp. 160f.

[6]Imperial War Museum, J. Hughes-Hallett Papers, JHH 67/70/1, J. Hughes-Hallett, memoirs, "Before I Forget," p. 111.

[7]Mountbatten Papers, MB1 B74, Mountbatten to Baillie-Grohman, 31 August 1953; MB1 B51, 15 February 1943 COHQ Directory.

[8]The Inter-Service Topographical Department, not to be confused with the ISTDC (Inter-Service Training and Development Centre) set up in 1938, the beginnings of the later Combined Operations organization. See ADM 223/90, Godfrey note, para 29, "Inter-Service Topographical Department" by Bassett and Wells.

[9]ADM 223/465, Lord Dunwich, "The Development of I.S.(O)," 2 November 1943. Mountbatten Papers, MB1 B26, Mountbatten to the Marchioness of Casa Maury, 28 February 1943.

had been, and the new COHQ was not left behind. Two highly capable scientists—Solly Zuckerman and J. D. Bernal—were brought in to advise on the numerous technical problems that arose in planning combined operations, and they proved invaluable. Bernal set the crucial precedent. He was asked by an officer to devise an echo sounder, but pressed to be told why. The officer was reluctant to share such secret information with a mere civilian, but his eventual answer showed Bernal that the initial inquiry had been quite misconceived, and the basic problem could be solved in a more simple fashion. From this grew the doctrine that scientists should not be asked for solutions, but had to be brought in to help formulate the problems. Mountbatten gave his support not only to that, but was willing to break down conventional technical responses through using the untrained, wild talent of Geoffrey Pyke, now principally known as the planner of a vast sea-going aerodrome of ice, codenamed Habbakuk. Pyke had approached Keyes, but had been turned away unseen; Mountbatten recruited him, which suggests not just an attachment to the modern, but perhaps even a mischievous delight in the bizarre.[10]

It is, of course, possible to be too modern and risk being thought "unsound." One source wrote that the Admiralty as a whole viewed Habbakuk as a "frivolous scheme," and thus was suspicious of COHQ subsequently.[11] However, complaints about one exotic idea should not damage a reputation badly. More dangerous was the broad swell of objections that one finds during Mountbatten's time in command and afterwards. Various of these, one notes, were essentially objections to Mountbatten himself, and specifically that he chose too many of the wrong sort of people, and anyway was an empire-builder who recruited far too many in total.

There was a particularly vicious attack on COHQ personnel in the memoirs of Lord Lovat, one of the few commanding officers to gain much credit at Dieppe, and deserves repeating even though it has been quoted before—or perhaps especially because of that. According to him, the place "swarmed with red-tabbed gentlemen," its busy passages "honeycombed with rooms filled with every branch of the Services, including the powder puff variety, who looked elegant in silk stockings. There was said to be a fair proportion of drones among the inmates." On "drones," we might also recall what M. R. D. Foot said about his introduction to COHQ, an interview with Mountbatten's Senior Intelligence Officer, the Marquis

[10]Solly Zuckerman, *From Apes to Warlords: The Autobiography, 1904–1946* (London: Hamish Hamilton, 1978), p. 150. David Lampe, *Pyke: The Unknown Genius* (London: Evan Brothers, 1959), pp. 88f. J. R. M. Butler, *Grand Strategy,* Vol. 2, *September 1939–June 1941* (London: HMSO, 1971 impression), p. 355. Mountbatten Papers, MB1 B52, Interview with Maurice Goldsmith, 17 November 1977, p. 5. Strictly, Habbakuk was Habbakuk I, since there was a number II, a prefabricated floating aerodrome, see Mountbatten Papers, MB1 C24, J. D. Bernal to Mountbatten, 6 September 1943.

[11]Mountbatten Papers, MB1 B50, Major General Harold Wernher, "C. M. & S. F. Activities 1942–1944 at COHQ and SHAEF," (post 17 May 1945) p. 9. Wernher had been promoted from Brigadier during the war. He succeeded his brother as Baronet in 1948.

of Casa Maury, allegedly given simply for finding out whether Foot was socially acceptable. Or there is what is argued in another memoir by a COHQ veteran, the novelist Robert Henriques, who praised Mountbatten for his "quick, logical mind and an even swifter intuition," commanding "an uncritical loyalty from almost everyone;" but also alleged that he had no eye for good men, and appointed many lightweights simply because they were old friends.[12]

For broader assertions that COHQ staff was inflated, we can look confidently to high Treasury officials, always quick to perceive "superfluity." A. J. D. Winnifrith, the Establishments officer for the War Cabinet and Ministry of Defence, believed there were "lots of frills" at COHQ, for instance, finding the intelligence section entirely unnecessary; Edward Bridges, the Cabinet Secretary, agreed with him. Another very influential voice was the Chief of the Imperial General Staff, the formidable Sir Alan Brooke. He also thought Mountbatten had far too large a staff, and seems to have been one of the first to warn against it, actually using the word "inflated" in a discussion.[13] However, what proved the most dangerous challenge to COHQ came from Mountbatten's own Service ministry.

We must not forget that Mountbatten had some great allies in the Admiralty, for instance, Captain Richard Symonds-Tayler, Director of the Tactical and Staff Duties Division (DTSD), or Commander Robin Todhunter, Deputy Director in the section of the Department of Naval Equipment dealing with combined operations.[14] Various members of the Director of Naval Construction's department were also of great help, and offered interesting and invaluable designs for specialized landing craft; although admittedly there was an understandable reluctance on the part of what was a critically short-staffed department to be overly distracted from dealing with planning for construction, reconstruction, and repair for the whole fleet, and Mountbatten was fortunate to be able to rely on the support of Bruce Fraser, Third Sea Lord and Controller between 1939 and 1942, and thus in charge of matériel.

On the other hand, we have to note what Todhunter described in a letter to Mountbatten long after the war as resentment by part of the Admiralty establishment, above all some senior "salt-encrusted" officers. The latter presumably partook of varied prejudices, whether against Mountbatten on grounds of his birth, or against combined operations as stealing resources from the "real" war at sea, or fully supporting combined operations, but thinking that if anyone should control them, it ought to be the Admiralty itself.

---

[12]Lord Lovat, *March Past: A Memoir* (London: Weidenfeld & Nicolson, 1978), p. 238. M. R. D. Foot, *Memories of an S. O. E. Historian* (Barnsley: Pen and Sword, 2008), pp. 69f. Robert Henriques, *From a Biography of Myself: A Posthumous Selection of the Autobiographical Writings of Robert Henriques* (London: Secker and Warburg, 1969), p. 53.

[13]CAB 121/169, Winnifrith minute of 21 September 1943, f. 201; Bridges' minute same date, f. 204; Bridges to Hollis, 23 September 1941, f. 96A. See also fos. 140, 143, 171, 173, especially COS meetings, 17 and 30 March, 19 October 1942; also L. Hollis to A. Nye, 25 October 1942.

[14]Mountbatten Papers, MB1 B49 file 2, T. Hussey to Mountbatten, 17 February 1978.

In his book *Watery Maze,* Bernard Fergusson alludes to two naval officers who though seconded to COHQ, clearly fell into the third of those categories.[15] He does not name them, but they must be C. S. Daniel and R. M. Ellis. Daniel had become Admiralty Director of Plans in 1940, but had irritated Churchill by pointing out the logistical objections to overly-ambitious schemes, earning himself the title of "that Maaaster of negation."[16] Ellis, on the other hand, had a successful initial wartime career at sea, being Captain of HMS *Suffolk* during the *Bismarck* chase. However, both officers moved to COHQ, Ellis at the start of 1942, Daniel right at the end. By January 1943, the latter was Assistant Chief of Combined Operations (ACCO), that is, one of the three joint Service chiefs immediately under Mountbatten, with Ellis as his deputy (DACCO). Here, their loyalties to their new chief were put under great stress, and in at least Ellis' case, one finds a definitely querulous note entered in his minutes. For instance, in April 1943, he tried to brush off what he described as another attempt to allow COHQ staff to liaise between two Admiralty Departments or Divisions, ending with the clear implication he thought the CCO was both over-extending himself and also "meddling." In a minute of 21 September, he reflected that any success on the naval side of amphibious operations had been in spite of COHQ, not because of it—though in his memoirs admitted that was a little unfair.[17]

Querulous minuting aside, there was the substantial question of the Combined Operations (CO) naval assault fleet. By mid-1942, that had attained a very respectable size, in matériel and personnel, and it was argued by Admiralty heavyweights that there would be significant gains in efficiency if it were absorbed within the Royal Navy. Here, both Ellis and Daniel supported the Admiralty side. It is perhaps not surprising that Mountbatten retained sour memories of naval opposition to his plans.[18] The sourness would have been all the greater since the opposition was

---

[15]Fergusson, *The Watery Maze,* p. 121.

[16]Ibid., p. 223.

[17]ADM 116/4731, R. M. Ellis minute, 25 April 1943. Churchill Archives Centre (herinafter CAC), Ellis Papers, ELLS 2/3B, 21 September 1943. CAC, ELLS 3/2, Ellis Memoirs, "When the Rain's Before the Wind," ch. 29, pp. 9f.

[18]ADM 116/4731, Ellis minute of 21 April 1943, COHQ docket (dkt), 29 March. Mountbatten Papers, MB1 B49 file 2, Mountbatten to Robin Todhunter, 6 February 1978, reply of 10 February; Hussey to Mountbatten, 17 February 1978. See the comment by Major General Leslie Hollis, a Royal Marine, and a Military Secretary in the Cabinet Office, in a letter to Sir Hastings Ismay, his chief, 20 August 1944, CAB 121/169, f. 254, that, once back in the Admiralty, Ellis acted as an *éminence grise* to influence the First Sea Lord (Cunningham) to try to take over complete control of COHQ. Hollis showed bias, but was substantially correct: see Ellis' minutes of 19 March and 13 July 1944, and Cunningham's submission to the COS of 22 July; however, Ellis was not alone—see the Charles Hughes-Hallett paper, also of 13 July, though comments in blue pencil, presumably by Ellis, seem to imply Hughes-Hallett was copying Ellis, Ellis Papers, ELLS 2/4.

successful. The Admiralty case was presented to the COS, who agreed, and in the course of 1943, the CO assault fleet and linked COHQ naval functions were taken over by the Admiralty. That in turn suggested COHQ could well lose some personnel, as it did after a hurried inquiry in October under Air Marshal N. H. Bottomley, the Deputy Chief of Air Staff. By that time, Mountbatten had already accepted a new post, that of the Supreme Allied Commander in South-East Asia (SACSEA), leaving his reduced inheritance to his successor as CCO, Major General Robert Laycock.[19]

From Personalities to Administration

The personal factor seems to account for much, both for the way COHQ developed and also certain attacks on the organization itself, not least by what might be thought of as a naval cabal intent on the narrow interests of its own Service. To see whether that is accurate, we have to test the accusation about decorative staffing at COHQ, and also assess the Admiralty takeover. For the first, we can look at the COHQ section where the accusation of "drones" might have credibility, and one that was also attacked viciously by at least one flag officer.

Wing Commander the Marquis of Casa Maury, head of COHQ intelligence, was an old friend of Mountbatten, dating back twenty years or so, and very much part of the inter-war smart set. His Cuban birth and foreign title gave him an exotic rank. His activities as racing driver, savior of Bentley motorcars, and manager of two first-class London cinemas displayed an enterprising style; and marriage to Freda Dudley Ward—a former intimate friend of the Prince of Wales—enhanced his place in fashionable society, incidentally making him Robert Laycock's step-father-in-law. He was also a man of taste, a sympathetic listener as well as engaging talker, and all in all exactly the sort you would expect Mountbatten to like and value. For exactly the same reasons, more traditional Englishmen could find him anathema, despite his service in the Royal Flying Corps during the First World War.

We have seen some implied criticism of him, or at least his section, from M. R. D. Foot and Mr. Winnifrith, and more can be found, actually suggesting incompetence, most sharply from Vice Admiral H. T. Baillie-Grohman, who had been naval force commander during Operation Rutter, the prelude to Jubilee. In a letter published in 1953, he referred to "the no doubt too hastily formed C. O. H. Q. Intelligence staff," adding that though he was expected to rely on it, in practice it "simply did not function so far as the Naval Force was concerned, though it improved right at the last minute." This would have been "a most serious handicap" had Baillie-Grohman not gone directly to the Naval Intelligence Division. Later commenting

[19]TNA, DEFE 2/1773, *History of the Combined Operations Organisation, 1940–1945* (London: Amphibious Warfare Headquarters, 1956), p. 67. CAB 121/169, Bottomley report, f. 208.

privately, he emphasized the frustrations and even worse of working at COHQ, and said he had had to restrain himself in his public letter. Here, therefore, was something of a general attack, and Mountbatten's reaction was both dismissive and ruthless, telling Major General V. D. Thomas, R.M., the Chief of Amphibious Warfare (1950–54), and under whom the official history of wartime combined operations was then being prepared, that Baillie-Grohman had proved "a complete failure," was "a very sick man at the time," and had to be removed at the request of all of his colleagues.[20]

If Mountbatten's fierce protectiveness of his own reputation is less than pleasing, mitigation comes from his willingness to extend his shield over his friends—including Casa Maury. We have seen already his phrase "the finest intelligence organisation in the Empire," which occurs in a letter to Casa Maury's wife. That was written after her husband insisted on resigning in February 1943, when it proved impossible to get him promotion from the Air Ministry, though he was also reacting to Laycock criticizing some of his intelligence sources, albeit not his treatment of them. It is telling to add that in his letter Mountbatten also drew a parallel between the treatment of Casa Maury as a foreigner and what had happened to his own father, Prince Louis of Battenberg, when hounded out as First Sea Lord in 1914. The sense of identification was strong, and it is not surprising to find that he tried to persuade Casa Maury to stay on, and was certainly sorry when he insisted on going. But had friendship blinded him? Perhaps Mountbatten always tended to see his geese as swans; it is still difficult to believe he would have retained someone who had shown he was not very good at such an essential task; would he really have fought to keep a drone?

We have to look at what others thought, and they tend to cast doubt on Baillie-Grohman's more sweeping judgements. First, there is his successor, John Hughes-Hallett. Here was someone with a reputation for strictness and uncertain temper: famously, his nickname was "Hughes-Hitler." Yet he had only praise for Casa Maury's intelligence reports. And Vice Admiral Rushbrooke, DNI from late 1942, wrote to Casa Maury after he resigned to pass on praise from both himself and Godfrey, his predecessor, on the way the "extremely exacting duties" in COHQ intelligence had been carried on, and that they had heard nothing but good about Casa Maury's "constant zeal and invaluable co-operation." Again, we are not citing easily-

---

[20]Mountbatten Papers, MB1 B74, transcript of letter to *Journal of the Royal United Services Institute;* Baillie-Grohman to P. Horan, 8 August 1953; Mountbatten to Major General Thomas, 31 August 1953. Baillie-Grohman retired at the end of 1943, though was fit enough to serve later in the war. Note the following comment on "Bee-Gee" from earlier in the Mediterranean: "He had inherited a game leg from World War I which gave him hell." Bill Strutton and Michael Pearson, *The Secret Invaders* (London: Hodder & Stoughton, 1958), p. 22. In his autobiography, Baillie-Grohman shows a highly personal prejudice against Casa Maury, National Maritime Museum, Greenwich, GRO 33, "Flashlights on the Past," Vol. 2, p. 157.

pleased men. There are other similar testimonies, including from Colonel Sam Bassett, the head of ISTD, who noted in passing in one letter to Casa Maury that they both knew how to run "a truly inter-service team," and, more tellingly, later offered him an intelligence job.[21]

Some letters between Laycock and Casa Maury, especially one exchange of 9 and 10 December 1942, are more helpful on the details. Laycock offered heavy praise of COHQ intelligence dockets as "clearer, better and more thoroughly set out, and more easily assimilated by O[perational] Staff Officers" than any he had seen, and it was obvious "an unlimited amount of work and inspiration had been put into them." But he also thought Casa Maury did not get all the intelligence he should have done, notably from the Special Intelligence Service (SIS, or MI6), or from Military Intelligence, either because of lack of proper contacts or due to suspicion of him as a "sinister Grandee." Here Laycock elaborated on the criticism he had already given, which the other had taken personally, but this time elicited a reasoned answer. First, Casa Maury emphasized his section was not supposed to gather raw intelligence, merely "To choose and mount collated Intelligence suitable for planners and Force Commanders." It depended on others for the basic data, even for photographic interpretation, so could not be blamed if they were inadequate. Second, he argued he had very good relations with the intelligence departments, specifying SIS as well as Military and Naval Intelligence.

Here is a defense against any who might now blame him for the gross intelligence failures in Operation Jubilee, in particular the unpleasant surprises about enemy gun positions in the cliffs, and the way the stony beach threw tank tracks. The information had not been gathered, so could not be passed down to COHQ. But one notes that neither Baillie-Grohman nor Laycock were talking about those particular lacunae; still, both agreed there was some relevant material that was known and yet not passed down the chain. And Laycock, we should add, is not open to the kind of *ad hominem* attack Mountbatten made on the other; what with his experience and army rank, his allegations about inadequate military intelligence seem weighty. Thus it is highly relevant to ask whether COHQ's means of collation were suitable to the times. Casa Maury put great stress in his defense on the personal element in his work: he noted he had dined with the DNI recently; that the Assistant Chief of Air Staff (Intelligence) was a personal friend of his; and though COHQ had no official access to SIS, he had "a personal relationship" with the organization. But reliance on personal links was fallible if there was personal prejudice against Casa Maury, perhaps as a socialite or foreigner, and it is quite apparent that that was to be found: for instance, Mountbatten told the Marchioness that one or two officers in the Air Ministry actually loathed her husband. Moreover, in that December 1942 letter to Laycock, Casa Maury admitted to a

[21] J. Hughes-Hallett, "Before I Forget," p. 120. Mountbatten Papers, MB1 B26, Rushbrooke to Casa Maury, 4 March 1943; Bassett ditto, same date.

good cooperation only with "most" Military Intelligence sections, heavily implying that the personal element compromised links with the other one.[22]

There is also a more fundamental reason for distrusting personal links: the growing weight of information and the thickening web of intelligence organizations as the war progressed. A continuing reliance on personal linkages could restrict the growth of more stable and dependable connections. One recalls that Mountbatten had refused to deal with IS(O), Godfrey's inter-service collation unit. One cannot but feel that that was unwise, and it would have been an advantage for Casa Maury to have an alternative organization to depend on, especially since IS(O) grew more quickly than his own section. Significantly, after he had gone, inspired by inconsistencies between enemy strength estimates from the different Service ministries, two COHQ intelligence staff officers pushed for the development of a new inter-service agency. The war showed the importance of personal liaison, but even more so of effective administration.[23]

Let us now begin to consider the transfer of the CO assault fleet to the Admiralty, looking first at one of the individuals concerned in that, Captain Ellis. He was an impressive officer, and only in part because of his height (his nickname was "Lofty"). His administrative ability was also admired by John Hughes-Hallett for one, and also by Mountbatten, who had plucked him from *Suffolk* to serve at COHQ, recalling their effective work together in the Admiralty Naval Air Division during the 1930s. Someone else who had known him before the war—Ellis had served as his staff officer in HMS *Hood* in 1934—was Vice Admiral F. T. Tower (Vice Controller 1939–45), who described him in one minute in 1943 as "very capable and knows what he wants." In that case, of course, the praise is a little coded, implying that Ellis could raise difficulties, something substantiated elsewhere. His once admitting to dealing with disagreeable dockets by locking them away for six months perhaps proves no more than unwise frankness about a vital tactic of the effective administrator, but one cannot ignore the sometimes minatory tone of his minutes, nor his occasional quarrels. Here one can set aside his unpleasant dispute with Lord Reith, the former head of the BBC, later Chamberlain's Minister of Information, but by 1943 reduced to command of the Combined Operations Material Department at the Admiralty: both men were authoritative and apt to think they were correct; furthermore, both were very tall, a factor not to be ignored in a personal conflict. It would have been almost shocking had they not

[22]Mountbatten Papers, MB1 B26, Laycock to Casa Maury, 9 December 1942; reply 10 December; Laycock to Casa Maury, 23 February 1943.

[23]Foot, *Memories of an S. O. E. Historian*, p. 81. For previous overlapping between COHQ intelligence and IS(O), see ADM 223/465, Lord Dunwich's notes of conversation with the DNI, November 1942. An anonymous piece in the same volume, "The Birth of I. S. O. Charters Etc.," states IS(O) was first conceived at a meeting between Mountbatten and the Joint Intelligence sub-committee and the Joint Planning Staff.

clashed. It is much more telling about Ellis that after he had moved to the Admiralty to head the CO Naval Staff Division, he also got into an argument with "M" Branch officials, to the degree it drew in the Permanent Secretary himself, an unusual breach of the harmony with which the Naval Staff and "M" Branch generally worked together.[24]

However, should one gather that Ellis was so much the brisk executive officer as to qualify as one of the "salt-encrusted" that Todhunter referred to? Here we should look first to Todhunter himself. He retired from the Royal Navy in 1927 and sat at a desk at ICI for twelve years, returning to the Service only with the war. Initially he served at sea, but ill health forced him to an Admiralty desk, where he proved invaluable as an administrator, especially when it came to landing craft, to the extent it looks as though his return to fitness to serve again at sea was covertly ignored. This career gives a certain context to his suggesting some naval officers could be too naval. More important, there is an interesting parallel with Ellis, who retired from the Navy in 1947 and made a successful career as a management consultant in the USA. If not a model bureaucrat, neither was he the stereotype of the sailor stranded at a desk; we should also note he had only good words about a man he worked with closely at the Admiralty, R. C. Donovan, another officer who had retired before the war and gained much industrial and commercial experience, only to return with the hostilities, proving himself to be a very effective administrator. That quality—effectiveness—was the key to Ellis' good opinion: he even praised Reith's work, if not all his methods. And not only could Ellis recognize good administrators whatever their backgrounds, his rationale for absorbing the CO assault fleet within the Admiralty lay in good administration, as we shall see.[25] So the Ellis case, like the Casa Maury one, leads us away from personalities towards administration. We have to assess whether it was quite right for the Admiralty to take over various COHQ functions, and that first means we should look in more detail at the subject of Second World War administration, in particular at the Admiralty.

[24]J. Hughes-Hallet, "Before I Forget," p. 133. ADM 1/14054, Tower minute of 30 June 1943. Mountbatten Papers, MB1 B49 file 1, T. Hussey to Mountbatten, 22 September 1970. Ellis, "When the Rain's Before the Wind," ch. 28, pp. 7f. Yellow sheet, pencil comments, Ellis Papers, ELLS 2/4. For unkind comments by Reith about Ellis, with allegations he was generally unpopular, including at COHQ, see the diary for 1 January—31 December 1944, pp. 207B, 212A, 254A, 255A, BBC Archives, Caversham, Reith Papers. ADM 1/15630, Ellis minute, 14 October 1944; comment by C. G. Jarrett, head of "M" Branch, 7 November 1944.

[25]On Todhunter, see ADM 1/14056. For Ellis on Donovan, see his letter to the Hon. G. H. E. Russell, 23 December 1944, Ellis Papers, ELLS 2/4. On Reith (DCOM), Ellis, "When the Rain's Before the Wind," ch. 28, p. 9.

The Second World War, Central Administration, and the Admiralty

Come the Second World War, there was an unprecedented degree of mobilization of the British government as of the economy, entailing the development of a bureaucratic-military state on a scale not known before, even in the First War; office returns tell us that the April 1945 Admiralty was double what we know it had been in November 1918, yet they do not compare like with like, and in reality the department had more than tripled.[26] Increase of size (and, as Lovat said, of rank) also marks the change from Keyes' COHQ to Mountbatten's. Doubtless, some of this was due to a certain extravagance by the CCO; though here some blame should attach to the initial lack at COHQ of a civil service Establishments Officer, so the Treasury watchdog could not exercise its usual restraint. There was also—and here Mountbatten cannot be blamed—the dangerous "Trinity" principle, where—for reasons of status and authority—officers from the three Services who carried out the same function together had to be of the same rank, and that meant all moved up to the highest common denominator. Any inter-service organization was thus inherently inflationary. But it was administration that was central to the growth, and that had a firm base on need, not empire-building. Mountbatten's first reforms were to extend the administration, going well beyond the appointment of a Chief of Staff, notably the structural division of COHQ between administration and operations, swelling the size of both. The idea was to prevent rote matters from distracting the higher military planners, a time-honored principle, evinced back in 1917 when the First Lord divided the Board of Admiralty into what were called the Operations and Maintenance Committees. COHQ saw a further important administrative development in early 1942. The USA was now in the war and there were serious prospects of invading the Continent; under what was called Operation Bolero, Whitehall arranged for the reception and maintenance of the large U. S. forces due to arrive in the UK. That had its echo within COHQ, as Mountbatten sought to make the country into a proper operational "jump-off" base with the appointment of a "transport Dictator," Brigadier Harold Wernher, albeit disguising him under the unthreateningly vague title of Coordinator of Ministry and Services Facilities (CMSF). Wernher had only a small staff at COHQ, though became involved in a large network of committee and liaison work covering an impressive variety of activities, including studying the production of artificial rubber, providing vast numbers of "Jerricans" for petrol, building "hards" along the south coast for landing craft, and managing the Mulberry artificial harbor project. On a less operationally significant level, he also helped set up and ran the popular, if

---

[26] 10,637 on 11 November 1918, ADM 116/2577, "Staff of the Admiralty Office (Vote 12) in 1914 and 1927." 20,748 on 1 April 1945, ADM 116/5523.

rather smart, club that provided meals and recreation for COHQ members and visitors.[27]

With respect to wartime government administration beyond COHQ, again one finds very significant development: furthermore, we should notice there were commonly some causal connections with combined operations (whether the latter were organized by or through COHQ or not). We scarcely have time to observe the whole subject in detail, but should note four basic subdivisions: the administration of operational planning, administration of distribution and supply, administration of production, and the administration of administration. For the first two, we must look to the heart of the beast—the committees and bureaucracy in and around the Cabinet. They had been greatly augmented after Churchill took over as Prime Minister in May 1940, notably through the creation of a dual Defence Committee to advise the War Cabinet, one part of which was devoted to various aspects of administration. The civil and military secretariats within the Cabinet Office also swelled, so—*inter alia*—they could support not merely the COS, but also Churchill in his new role as Minister of Defence. Furthermore, we should note a complex of bodies created later in the year, served by the secretariats, which supervised as well as undertook operational planning. Under the existing COS subcommittee consisting of the Directors of Plans from the three Services, a Joint Planning Staff was articulated, composed of the Strategical Planning Section (STRATS), Future Operational Planning Section (FOPS), and the Executive Planning Sections (EPS). The first two were inter-service, and were based in the Cabinet Office; the EPS, though, were in three parts, one for each Service, and based in the relevant Service ministries.[28]

The Joint Planning Staff was divided rationally, to deal respectively with the long-, the medium-, and the short-term. Keyes, for one, did not approve. He thought the COS obstructive, and regarded the Directors of Plans with deep suspicion—especially C. S. Daniel—but had a special animus against the "everlasting committees," above all the EPS, which had the nerve to alter his arrangements, or to point out where they depended on inadequate resources. "Inter-Service Committees," he grudgingly admitted, "can, of course, give invaluable help," but added "joint authority and responsibility is bound to reproduce the miscarriages and confusion which have characterised almost all Combined Operations in the past."[29] He yearned instead for the divine simplicity of the single directing mind, and at

[27]Mountbatten Papers, MB1 B50, Mountbatten to Wernher, 24 March 1942; (Wernher), "C. M. & S. F. Activities 1942–1944."

[28]STRATS was based on the previous JP sub-committee; EPS on the former Inter-Service Planning Staff; FOPS was a new creation. STRATS was to be more forward-looking than FOPS. CAB 121/169, Directors of Plans proposal, 6 September 1940, approved by COS 9 September.

[29]CAB 121/169, Keyes to Ismay, 27 August 1940; Keyes to the COS, 22 October 1940; 24 April 1941.

one stage proposed he be made superior to the COS, and presumably would then have swept away much of the planning organization under them. Mountbatten, on the contrary, realized very well that the path to power was now through using the existing system. Rather than trying to be made superior to the COS, he succeeded in becoming one of their number. And instead of fighting the Directors of Plans, he got his own Director added to their committee.

Those examples where combined operations came to swell the central planning organization were mere promises of what was to happen once the USA had joined the war. Under the aegis of Operation Bolero, various central committees and secretariats came into being, and necessarily these were primarily concerned not so much with the administration of operational planning as the administration of supply, though it will be understood that the frontier between the two cannot always be distinguished. There was a ministerial Bolero committee advising the Defence Committee, and under that was created the Principal Administrative Officers Committee (PAOC), consisting of the heads of matériel of the Services. According to its terms of reference of 17 June 1942, it advised the COS on the administrative aspects of operational and strategic questions, supervised some of the Bolero work on somewhat lower levels, and could discuss administrative questions of common interest to "the four defence services." (The number is significant: Mountbatten's organization came to be counted as a separate Service, at least before it was emasculated in 1943.)[30]

Although the PAOC worked well enough, it slowly became obvious there were faults in following through on decisions, suggesting there was a need for an interservice agency slightly lower down in Whitehall, so the supply administrators could be fully apprised of the operations being planned, and the planners would know of the administrative implications of the operations. Lacking that, as the Directors of Plans committee reported in January 1944, "there is a tendency for plans to reach the Chiefs of Staff before the administrative implications have been fully considered." As had already been argued in a report from the Air Ministry, previous attempts to bridge the gap had been through a loose liaison network, one heavily dependent upon the abilities—and personalities—of certain individuals, implying the lack of more formal bureaucratic links had been felt. The EPS had not suffered much, given its staff were rooted in the Service ministries, so had ready and immediate contact with officers deeply involved with the nuts and bolts. But STRATS and FOPS were in a worse case, it was said, since they were somewhat cut-off in the Cabinet Office from the service "coal face." The COS approved the January report, hence in Spring an additional two-level organization was set up: first, a committee composed of the three heads of the Services' "Q" (or supply) organizations, called the Directors of Administrative Planning (DAP); second, a staff to the DAP, called the Joint Administrative Staff—the DAP and their staff

[30]See papers in CAB 21/1502 and 1507, especially the PAOC terms of reference in the latter, as approved by the Prime Minister.

together comprised the Joint Administrative Planning Staff (JAP). The Joint Administrative Staff was to be in continual dialogue with the Joint Planners about planning and supply, and had to keep the DAP up to date accordingly—the DAP were thus to be able properly to advise the PAOC and hence the COS. Here, rather late in the war, you had the same kind of relationship that had existed for some time between intelligence and operational planning. Another similarity was the JAP was served by the Cabinet Office secretariat (specially reinforced so it could take on its new administrative responsibilities), just as the planning staff was.[31]

The need for such better linkages with supply, though, had already been understood some time before with regard to what was going on farther down the chain, and where the part played by combined operations was more intimate. In passing, we can refer again to the administrative development under Wernher at COHQ, but otherwise should observe the greater one over at the Admiralty, which is where we can now focus. Charles Hughes-Hallett, the Director of Plans(Q) (DofP(Q)), commented in a memorandum of September 1944 on the poor state of the supply organization in the Admiralty in 1939, blaming the attitude that the constant replenishment necessary to an army was neither wanted nor even proper in the case of a fleet. Famously, sailing warships, once stored, could be self-sufficient for months on end; steam might have breached that independence, but ships retained an enviable ability to operate and fight without constant resupply—or at least were thought to have done. Change came rapidly during the war, signified by the Admiralty's acceptance—sometimes highly reluctant, admittedly—of a new primary meaning for the word "logistics." For a long time, the primary sense had been mathematical, and the synonym for supply was usually "administration;" one exception came from the occasional borrowing from Army practice where—since Quartermasters dealt with supply—the letter "Q" would represent that. In the 1940s, though, the Royal Navy was faced with the fact that the modern usage of logistics caught on in the USA and crossed the Atlantic, propelled by Washington's growing dominance in matters of production and supply.[32] What also contributed was the enormous increase in demand together with the new organization needed to cater to it, since new organizations in themselves call for new words.

That greater demand in part originated in Combined Operations. Where Keyes had complained regularly about the exiguous resources he could call on during his

---

[31]TNA, AIR 9/188, draft Air Ministry paper, 16 December 1943; 19 January 1944 report to COS; Report by the PAOC, 19 January 1944; Air Ministry report to COS, 25 January 1944; COS recommendation, 26 January 1944; also draft paper of 16 December 1943 to the Directors of Plans, arguing that an inter-service equivalent was needed to the U. S. logistical planners.

[32]ADM 1/17058, 9 September 1944. Charles Hughes-Hallett was John's brother. See also ADM 1/13169, dkt 7 July 1943. John Sweetman, *War and Administration: The Significance of the Crimean War for the British Army* (Edinburgh: Scottish Academic Press, 1984), pp. 2f. For a fine survey, see Peter V. Nash, *The Development of Mobile Logistical Support in Anglo-American Naval Policy, 1900–1953* (Gainesville: University Press of Florida, 2009).

time as DCO, Mountbatten was able to draw on the profusion of production available from 1942 onwards. The Admiralty was his main supplier of craft (including those produced in the USA), as well as of personnel, and it was there that much of the supporting "logistical" organization spread. However, the Admiralty had to satisfy far more than COHQ requisitions. Mountbatten's force was itself only part of a broader Allied development of amphibious warfare, as combined operations came to be called, a form the war had increasingly to adopt once the western powers could take the offensive against Germany and Japan, and from 1942—as we shall note—the organizing of the larger Allied operations fell outside COHQ's control. The Admiralty was also faced with two other serious changes in priority. There was a shift towards coastal-water operations, due to the German occupation of France, which necessitated the construction of the necessary swarms of "mosquito" vessels. Furthermore, there was the vast enhancement of maritime air power. These three great developments were all the more important because they cannot be fully distinguished operationally since they shared one characteristic, nothing less than maritime heresy—a move away from Blue Water. That might seem a paradoxical assertion, given the strategic achievements of naval air power, above all in the Pacific. But as Charles Hughes-Hallett put it in that memorandum, "aircraft carriers and assault shipping are essentially shore-based." The two might strike across great distances, but demanded regular and copious resupply, and that entailed very large and thoroughly efficient logistic organizations (or what Hughes-Hallett called "good 'Q' organisation"). Resupply on such a scale was best done in close proximity to established bases; and if, as in the Pacific, resupply had to be improvised, that simply implied a logistics organization on an even greater scale: the logistics—or the "administration"—still asserted themselves. Moreover—to return to linkages—even better organizational means had to be found to keep operations and logistics in constant dialogue both within the Admiralty and at the sharp end.[33]

The subject of supply ineluctably leads one to production, and there, too, an ever greater bureaucratic engagement was required. Again, in 1939 there were grave weaknesses. That applied in central government, where the machinery that monitored production lacked key elements, ones provided only at the last moment before the war.[34] The Admiralty's problem was even greater, since there was no equivalent machinery at all; as was said in one of the preparatory official histories written just after the war by D. C. M. Bolster, "the smallness of the Admiralty Vote

[33]For a graph showing the dialectics of "administration" and operations, one whose crudity illustrates the freshness of the concept, see "Organisation for Joint Planning. AIR 9/188, Note by the Air Ministry," Appendix "A," 11–15 November, 1943, circulated 28 November. For some similar comments to Hughes-Hallett's, see the paper "Some Thoughts on Naval Administration in 1945," by Captain Godfrey French, CAC, French Papers, GDFR 2/5.

[34]In April–August 1939. John Ehrman, *Cabinet Government and War, 1890–1940* (Cambridge: Cambridge University Press, 1958), pp. 114–20.

did not permit the engagement of staff whose precise function could not be tabulated."[35] Here was a neat little blow at the rigorous attempts by the Treasury pre-war to ensure that all posts were justifiable! Again the war wrought miracles, and by its end most Admiralty matériel departments had what were called "progress sections," and it is interesting to see how they might be set up at first with very limited powers, but take on more as circumstances demanded. Take the case of the one under the Engineer-in-Chief from early 1943, and the comment—again in one of the preparatory histories—that "it was not long before it began to undertake the other essential feature of successful production, namely planning."[36] It appears, though, that the planning in question was largely procedural: there is a better instance of the war forcing delegation of function and initiative in the statistical section created in May 1942 in the British Admiralty Delegation (BAD) in Washington, and where one can perhaps again see the U.S. example, the high regard for statistical returns this time, encouraging change on the British side. This section quickly developed into being the secretariat for the supply side of the delegation, which meant keeping production as well as supply records, programming of requirements, and general coordination. In short, from merely keeping records it matured into an organization responsible for monitoring and even assessing what had become the crucial matter of priority of production.[37]

The main model for such enlargement of function, though, is the central priority organization set up within the Admiralty Secretariat in Britain on 9 September 1940: the Production and Priority Branch, ("P" Branch), under a new officer, the Principal Assistant Secretary (Priority), (PAS(Pr)).[38] This straddled the original Admiralty in London and the matériel branches (now largely moved out to Bath), but also rapidly got associated with a whole regional production organization, linked progressively to private industry—going well beyond the circle of firms with which the Admiralty had been associated in the 1930s—and also trade unions, local worthies, local government, and other central government departments, notably the Ministries of Supply and Production. Through this network, "P" Branch was able to monitor and encourage production, and also to seek for new capacity; perhaps more importantly, it was able to rejig priorities at the local level, though here, we should note, it had essential technical and detailed monitoring support from the newly developing progress sections in the Controller's departments, especially the one under the Engineer-in-Chief, mentioned above. In 1940, production priorities

---

[35]CAB 102/524, D. C. M. Bolster, "Execution of the Naval New Construction Programmes 1939–45," pp. 121, 126. The work was based on interviews with staff in Armament Supply, the Engineer-in-Chief's department, and "P" Branch.

[36]CAB 102/861, "Progress and Production, Planning Organisation," anon., u/d, 19A, pp. 1–6.

[37]ADM 199/1236, H. A. Cooper, "Civil Administration: Statistical and Procurement Functions," para. 2–8. See also CAB 102/862, A. R. W. Kent, "Admiralty Organisation," para. 24f.

[38]Cooper, "Civil Administration," para. 3.

had been set very crudely, through dictation from above, or what H. V. Markham, then PAS(Pr) and soon to be the Admiralty Permanent Secretary, called in one October minute "the present rotten form of administering priorities by comprehensive Directions."[39] He had given more details of what should replace that in a long letter a few days after "P" Branch had been set up, where he made clear it was to coordinate all the relevant departments of the Admiralty on priority matters, and to ensure any details and statistics called for were supplied. Whether it was realized at the time, that was eventually to entail "P" Branch drawing information about local production from the Area Organisations, and intervening not just about the efficiency of production, but also what was being produced. Here, the branch's work was given powerful support when a Board of Admiralty Priority sub-Committee was constituted in December 1942. It was of course the War Cabinet that decided on what the main priorities were, but an intermediate body was needed to interpret the big decisions, to decide on policy over the lesser clashes of interest, and enable "P" Branch to act with decision. The result, as noted at the time, was that "P" Branch became equivalent to the longer established and omni-competent Secretariat organ, "M" Branch (or branches), which interpreted the Naval Staff to Admiralty departments and Whitehall generally, and vice-versa, showing both authority and—one should underline this—initiative.[40]

There had been no real equivalent to "P" Branch even in the First World War Admiralty, no similar combination of higher-level analysis connected with a dedicated statistical unit within the branch itself. Nor were there the same formal progress units in other departments that one finds in 1939–45. The Second War made far greater administrative demands than the First had done, particularly given it was more mutable, militarily, as with the swings in the Battle of the Atlantic, and in the bombing campaign against Germany. For combined operations, mutability consisted essentially in changing operational possibilities, first with Lend-Lease production, even more so with the USA formally joining the war; COHQ could thus become more demanding, leading to complaints in one paper for the 4 December 1942 Board meeting about "the considerable but uncertain incidence of C.C.O's requirements," and "a constant menace to ordered priority."[41] However irritating, the changing war meant numerous new choices and altered priorities, as one sees most obviously in some major War Cabinet decisions. The 1942 crisis in the Battle of the Atlantic led in October to approval of a program to build 200 escort vessels, one given the highest national priority the next January. But a swing in the fortunes at sea—and pressure from Bomber Command—meant

[39] ADM 116/4383, 28 October 1940, dkt 24 September 1940.

[40] CAB 102/861, Markham to D. P. Walsh, 12 September 1940; N. MacLeod, "Notes on Statistics in P Branch," p. 10, copy sent to R. Hughes, 27 December 1951.

[41] ADM 116/4862, "Board Priority Committee" paper, December 1942, approved 4 December.

the Ministry of Aircraft Production could be awarded "Super-preference" in June 1943. Come February 1944, though, a high priority was given to production for Operation Overlord.

Behind the changing high policy, the real problem remained of implementation. In 1917, there were still the resources to allow expansion in shipbuilding when the U-boats threatened starvation. But for the later war, mobilization began earlier, and though it did not reach its full extent until the end of 1943, already a year before that there was not much spare capacity in the British economy. Thus new demands could be catered to largely only by replacing existing production, and that promised clashes and bottlenecks. Particular problems could be envisaged substituting landing craft for escorts in 1943–44, and here—in line with what Markham had said—it was not enough to hand down *diktats*. Instead, better organization had to be provided. For one thing, the CCO could be brought on to the new Board Priority sub-Committee, so compromises could be struck and useful targets set. It was essential to get the targets met, too, and thanks to "P" Branch and its network out in the regions, together with the Controller's departments and their networks, and of course cooperation with other Ministries, the difficult task could be managed of altering production at the workplace according to the latest priorities, preferably without the delay of referring back to higher authority. Admittedly, part of the administrative achievement—though this in itself needed meticulous work—was finding new capacity through ordering some prefabricated ship construction by structural engineer firms, hitherto underexploited, and arranging for the final putting-together of the main pieces to be done at once-abandoned shipyards hastily brought back into use. But much of the job was even trickier: reallocating existing resources in midstream from one type of construction to another, meaning planned escort production was reduced and landing craft substituted. All in all, given near-full mobilization, this was a remarkable administrative job.[42]

At last, one turns to the administration of administration. In part, this was straightforward: dealing with procedure, or anomalies, or the proper arrangement of work, or fitting rank to function, or just ensuring proper representation on all relevant bodies. One notes the appointment of Admiral Sir Charles Kennedy-Purvis in 1942 as a Deputy to the First Sea Lord, to take such matters off his shoulders. That was another sign of the growing weight of administration, since his predecessor in 1917–18 had largely dealt with operations: it might also be mentioned that Kennedy-Purvis was one of a number of senior officers at the Second World War Admiralty who worked too hard and died early—something that in the previous war had been found rather among senior civil servants.[43]

---

[42]Ibid., Markham to CCO, 17 December 1942. CAB 102/539, "Naval Production 1939–45 with special reference to Landing Craft." J. D. Scott and Richard Hughes, *The Administration of War Production* (London: HMSO, 1955), see generally the sections written by Hughes.

[43]See ADM 211/4, Office Acquaint, 29 July 1942. Sir Henry Markham, we should note, threw himself into the war effort with great energy, and died short of fifty, of cancer, in 1946.

However, the administration of administration covers also a much more ambiguous area. The formulation might appear a little inexact, except it pales beside one suggested for the purpose in November 1943 by a member of the Civil Establishments branch of the Secretariat, Mr. T. Lyon: "the planning of the co-ordination of the administrative etc."[44] A certain vagueness was perhaps inescapable, since Lyon was describing liaison work, a very plastic activity; though the unit in question that was engaged in it gained more substance over time, and exemplifies a formalizing tendency of the war, one noted with reference to the JAP, where under the pressure of events, casual and personal linkages eventually crystallize into bureaucratic structures. What developed in this case was a whole new Naval Staff division, Plans(Q)—or PD(Q)—where "Q" did not imply the direct administration of logistics, but rather the relevant liaison and coordination, with the prime object of linking operational planners with those responsible for providing the necessary resources.

In the development of PD(Q), one again sees the influence of COHQ and combined operations. By early 1942, spurred by the increase of business but also to help control Mountbatten's proclivity for taking unorthodox bureaucratic routes,[45] it was found necessary to have a high-powered section created within the Admiralty to liaise with COHQ. Hence an Administrative Planning Section was set up in the Naval Staff Plans Division, through absorbing the (naval) EPS already there, and adding the Combined Operations Section taken from the Tactical and Staff Duties Division. It had the advantage of being sited next door to the CO section in the Department of Naval Equipment, and otherwise liaised with the Second, Third, and Fourth Sea Lords and the numerous departments under them, so was able to link COHQ with the main Admiralty planning, matériel, and doctrinal units. This unit accumulated more and more business as time went on, because it had taken up an intermediary role for which there was a crying need. In December 1943, it emerged as a full division—that is, as PD(Q)—and by then the Admiralty had taken over CCO's assault fleet, so liaison duties with COHQ could be cut back; but there was all the more coordination work to be done within and without the Admiralty, particularly as the Pacific theater began to occupy more attention. D. C. M. Bolster's list of some of the new tasks gives us an idea of the sheer variety of divisional work: planning and progressing the Fleet Train for the Pacific Fleet; indoctrinating officers going to join staffs of commanders-in-chief; putting flag officers in command in touch with relevant officials in the Admiralty; advising the French Navy on strategic and administrative planning; and maintaining liaison with civil affairs and "Q" offices in the War Office and Air Ministry. As he commented: "There

[44]ADM 1/14205, T. Lyon minute, for Head of Civil Establishments branch, 29 November 1943.

[45]See ADM 1/15894, dkt 17 August 1942, concerning orders for "Alligator" craft, and plastic armor.

appeared, as time went on, to be practically no limit to the functions which the new Division was asked to perform—so great had been the need for a centralised Division."[46]

COHQ and Admiralty, 1940–45

Personalities explain something; administration, it has been implied, can explain more. What we can do now is ask whether we are better able to understand the development of COHQ, the rationale for the partial coalescence with the Admiralty in 1943, and finally something of the nature of the wartime Admiralty itself. To begin with, it is surely now clear that the pretensions of the Keyes' COHQ were quite impossible, given the size. It might appear ludicrous that at one stage, on top of everything else it had to do, his tiny organization was supposed to supervise the production of landing craft, though that task was dropped early in 1941.[47] And if not quite ludicrous, Keyes' aim of superseding the COS, and of dismissing or at least greatly diminishing many of the central committees and planners—especially the EPS, whose contribution as part of PD(Q) we have now seen—at least seems unrealistic. In short, one confirms the impression of Keyes given earlier as personally over-ambitious, careless of "logistics," and, overall, retrograde. Furthermore, a better idea of what was entailed by CO responsibilities helps put the tales of over-manning and over-ranking under Mountbatten into context. If twenty-three were absurd for Keyes' total HQ staff, is it fair to describe his successor's 400 as patently luxurious? For instance, the problem about Casa Maury's intelligence unit was scarcely that it was inflated: it contained only ten people—not much for the finest in the Empire. The comparison lacks rigor, but Mountbatten as CCO was certainly very restrained in comparison with his later efforts as SACSEA, when his headquarters at Kandy quickly raced past 7,000. Here we might note that one of the initial tasks Churchill gave Mountbatten on his CO appointment was to plan for the invasion of Europe, thus we might compare COHQ with General Eisenhower's eventual Supreme Headquarters, Allied Expeditionary Force, which numbered over 20,000.

As to the transfer of part of COHQ responsibility, above all the assault fleet, to the Admiralty, it would be a travesty to ascribe that even in part to a treacherous naval cabal. It was circumstances that were decisive. Here we should first recall the evidence in the previous section about how effective organization, in particular with regard to supply and logistics, depended on effective administrative integration. But the separate COHQ and Admiralty were not so integrated, even if their

---

[46]ADM 1/14205, Office Memorandum no. 23, 25 August 1942. ADM 1/15894, F. W. Mottershead, Head of MII, minute of 7 March 1944. Bolster, "Execution of the Naval New Construction Programmes," pp. 121, 125, 218f.

[47]CAB 169/121, Instructions 30 October 1940 and 14 March 1941, fos. 13, 44.

supply and production were increasingly interwoven. When originally set up as a raiding organization, Combined Operations' needs were hard enough to satisfy, but grew alarmingly as duties grew, too. That authorization to Mountbatten to plan for the reentry to Europe, which opened up almost infinite liabilities, was terminated with the creation in 1942 and early 1943 of separate naval and inter-service allied commands, and they were the ones to organize the larger amphibious operations;[48] however, COHQ retained widespread if ill-defined responsibilities for training, infrastructure, and organization, plus, as yet, the maintenance of its own, expanding amphibious assault fleet—thus it called for more and more resources, largely from the Admiralty. Already in May 1942 one finds a frantic note to the DTSD from the Naval Assistant to the Second Sea Lord—and largely responsible for naval manning and appointments—noting his "gravest concern" over Combined Operations' expansion, which "was not forecast, or allowed for in our future statistic. . . ."[49] The situation got only worse, what with Bolero and the coming of Overlord, for instance in the outports in Southern England, where the naval commanders-in-chief found themselves under increasing pressure to cope with the shortages faced by Mountbatten's organization in trying not just to maintain its fleet, but open up numerous new CO operational and training bases. Since the CO local staff were scarcely sufficient to cope with even their immediate duties, it fell to the naval commands to find many of the administrators and repair crews, and that demanded experienced petty officers and leading seamen, exactly the categories in the shortest supply of all, bar certain types of executive officer.[50]

Already by the second half of 1942 the Navy was being sucked dry—or so various Admiralty officials thought. It was also thought there was a wasteful overlap of COHQ and Admiralty staff. Ellis bore the brunt of much of this, though Daniel came in at the end of the year to aid him. They wanted to draw on an increasing proportion of the resources of a naval ministry concerned about its own priorities. That raised crucial questions about decisions and the power to enforce them, particularly difficult of solution when two separate organizations were in question. First, who was to apportion the resources being competed for? Second, if COHQ was to be more than a supplicant to the Admiralty, it needed some kind of executive authority the latter would recognize, but how could that be the case when the Navy looked to the Board and the Lords Commissioners—or "Superintending Lords"—as the source of power? COHQ could hope to act through the COS, but

[48]In particular, the appointment of Lieutenant General Sir Frederick Morgan as Chief of Staff Supreme Allied Commander in early 1943. In any case, the further one got from British Home Waters, the less the writ of COHQ ran: thus, to send a part of the CO assault fleet even to the Mediterranean meant losing control of it.

[49]ADM 199/1777, Alec Madden to R. V. Symonds-Tayler, 8 May 1942.

[50]For instance, ADM 199/1777, "Remarks on Schemes of Complement approved for Combined Operational Bases," for D of P, 17 June 1943; Ellis to CCO, 29 April 1943; Commodore Commanding, CO Operational Bases, Portsmouth Command, to CCO, 28 April 1943.

that was better as a last resort, and instead depended on requests and appeals, in part dealt with through a liaison network rather than formal channels. The method usually worked, but was irritating to officers like Ellis who were accustomed to more immediate procedures: more importantly, it was time-consuming and difficult to justify when it began to be clear an almighty rush would be needed to get such a vast operation as Overlord underway according to the timetable.

Various attempts were made to give COHQ more weight in its dealings with the Admiralty; here Ellis cooperated with the Deputy First Sea Lord, though he was never able to get COHQ any real clout, failing, for instance, to get Mountbatten made an associate member of the Board of Admiralty.[51] The resulting administrative rejigging was outrun by what were increasing CO necessities. Within the Admiralty, the opinion grew that the only real solution was the complete takeover of the CO assault fleet and its administration, so decisions about resource priorities could be taken centrally and then enforced effectively, all while reducing overall staff. Ellis came to agree, but was constrained by his loyalty to and admiration of a Chief, who had chosen him for his job in the first place, and who wished to retain the integrity of COHQ's responsibilities. It is no wonder a tetchy note entered his minutes. In the end, he worked towards the Admiralty takeover, partly by helping to convince Mountbatten there was no choice. The latter, perhaps disillusioned by others taking over direction of the serious amphibious operations, and anyway with hopes of a grander appointment for himself, fell back to fighting merely for the continuation of COHQ in a reduced form, rather than having it completely swallowed by the three Services, and he succeeded.[52]

The takeover should also be tested by what happened subsequently. One ought at least to glance at the local level, where, as we saw, the naval commands had been called on to support CO functions, exacerbating personnel shortages. The takeover meant scarce resources could be more easily pooled and then promptly directed where needed, as with the mobile repair and maintenance units that were created.[53] Similar advantages of resource allocation and direction were also felt at the Admiralty. Furthermore, a rational structure evolved there to deal with the administration of combined operations, though it took some months. After much discussion and some false starts, a tripartite administrative structure was developed. One part dealt with personnel, and worked with the Director of Personnel Services, the Second Sea Lord, and his staff. Another dealt with matériel, and was linked with the

[51]See Ellis' 1943 papers of 20 February, p. 6, and 25 June, Ellis Papers, ELLS 2/3A. Also, "When the Rain's Before the Wind," ch. 26, p. 11.

[52]See the alternative schemes submitted by Mountbatten to the COS, 20 September 1943, CAB 121/169, f. 202. For an important naval viewpoint by Kennedy-Purvis, see his minute of 14 July, ADM 116/4979.

[53]See, for instance, ADM 199/1777, R. C. Donovan (PD(Q)) to office of the C-in-C, Portsmouth, 5 October 1943, reply to inquiry of 18 September; F. W. Mottershead circular, 26 October 1943.

numerous departments under the Controller. A third—the Combined Operations Division (COD), under Ellis as Director, or DCOD—looked essentially to operational administration, in particular ensuring the right craft and equipment got to the right place at the right time: this was the only Naval Staff Division of the three, and thus maintained close contact with the rest of the Naval Staff. Yet all these departments were linked, and with the rest of the Admiralty, through formal as well as more casual administrative structures, made it easier to do two important things: to get information around and to facilitate informed discussion. Geography helped here, at least for the Staff Division. The main Admiralty was very crowded, so the CO matériel department had to be found accommodation outside, though still in Whitehall. The heads of CO Personnel remained at their old lodgings in Richmond Terrace along with COHQ, though the main clerical staff were in a stone frigate in Chelsea with many of the Admiralty personnel administrators. But room was found for COD in the main Admiralty buildings, squeezed next to the Plans Division (PD) and PD(Q); it was also close to the Citadel, where the operational staff and the War Registry were located.[54]

Admittedly, not all the previous difficulties were removed. Tempers did not necessarily improve. Charles Hughes-Hallett said some of the extended discussion that led to the new structure was "rampant."[55] And there was disappointment over the expected economy of staffing. Certain Admiralty/COHQ overlaps of function were indeed found, which implied reductions of staff, but the growth of responsibilities that had caused difficulties in the first place only continued, absorbing any excess. There was a similar result due to the discovery in the Admiralty that COHQ had been doing certain things—notably planning landing craft movements—whose complexity had not been realized, and had to be maintained. One other problem that remained was that of undue rank. There were some demotions. Ellis himself ceased being a Commodore, reverting to the substantive rank of Captain. But though the new departments were all within the Admiralty, some of their officers still had to deal with colleagues in the other Services, and it was thought they would be at a disadvantage if they were of too junior a rank: thus the baleful influence of "Trinity" still exerted itself.[56]

However, the takeover was scarcely confounded. Even if, say, discussion was "rampant," at least it was played out within the same organization instead of encouraging rifts between separate government departments. Nor did the emergence of

---

[54]Ian McIntyre, *The Expense of Glory: A Life of John Reith* (London: HarperCollins, 1993), p. 271. See ADM 1/12833; also ADM 116/4979, Lyon minute, 2 November 1943. ADM 1/15630, Confidential Admiralty Fleet Order 88, 18 January 1945.

[55]Ellis Papers, ELLS 2/4, paper by Charles Hughes-Hallett of 16 March 1944. He was specifically referring to the creation of Ellis' division, thus implies colleagues could find Ellis difficult.

[56]See minute (presumably by Lyon), ADM 116/4979, "Proposed Complement for D.C.O.D.," u/d (late 1943).

some new difficulties show an error had been made—indeed, quite the reverse. In July 1944, the First Lord, A. V. Alexander, minuted as follows: "This inflation of Admiralty headquarters has got to stop, not only because we cannot afford the extra man-power, but because the multiplication of Divisions and Departments can reach a point where the circulation of papers is slowed down to such an extent that the speedy and efficient conduct of business is hindered, and not helped."[57]

Alexander had been prompted to write as he did because of the further development of the war. By early 1944, with victory over Germany now foreseeable, prospective operations against Japan had to be considered, with all their fearsome logistical problems. There were crucial discussions about how to organize the large fleet train and advanced bases needed to support a Pacific Fleet. And again, unsurprisingly, combined operations were deeply involved, with questions such as how to transfer landing craft to the East or the apportioning of the new amphibian vessels whose value in seaborne assaults was being shown by the Americans. Given the paucity of the previous Admiralty logistical organization, it seemed rational to create a brand new department for the purpose. But that ran directly into the objection we find the First Lord making. The desired end had to be achieved more economically, so thoughts turned to adapting an existing office for the new purpose, with various candidates being suggested from the Controller's departments and from the Naval Staff. There were no volunteers, the buck being tossed from one side to the other; eventually, in August, PD(Q) was chosen, since it was already linked to almost all the organizations that would have to be brought in to the new effort in the Pacific. But it needed new powers and capacity. Back in January, Charles Hughes-Hallett, still DofP(Q), wrote that it was no part of his division's duties to set technical requirements: that was the task of other departments. Nevertheless, a few weeks later we find one of his officers outlining what the proposed new organization would have to do, including "Giving guidance on policy," "Co-ordinating and stating detailed requirements," and "Organising technical development." These, he said, would be done through coordinating various departments, though—part of the buck-passing, perhaps—he envisaged another Staff division than his own, the Local Defence Division, doing most of the work, and suggested greatly strengthening its staff so it could cope, including adding three sorts of engineer. This was objected to by one of the senior men from the Secretariat, H. N. Morrison, the Principal Assistant Secretary (Staff), since it would create "a miniature Admiralty." But he was fighting the trend.

We saw what that trend was in the previous section: growing staffs, more linkages, and more local initiative. The first had long been an Admiralty problem: Alexander—and the Permanent Secretary, too—had been fighting against it for years, but the numbers continued to creep up until April 1945. As to linkages and initiative, we need only think of the statistical department in BAD, and "P" Branch. They also characterized PD(Q), right from its origins in 1942. Perhaps the

[57]ADM 1/16559, 23 July 1944.

Ur PD(Q) was at first just a means of sending faster communications; a bureaucratic alternative to the "yellow jacket" priority docket used in the Admiralty. But the growing functions gave the Division more authority. That was above all the case after the further expansion thrust on it in 1944; it emerged as just the kind of "miniature Admiralty" Morrison had feared. It started to recruit some engineers—albeit slowly, given the great constraints on personnel. It took on policy-making attributes—or what the Controller (by now Vice Admiral Sir Frederic Wake-Walker), approved of in a meeting in June 1944 as "thinking." And it took on executive powers to push through that thinking: that was not what was supposed to happen, strictly speaking, but—again the Controller said this—if PD(Q) saw work had to be expedited, it would follow "the normal practice" and report as much to the relevant Superintending Lord, that is, send orders at just one remove. And of course in reality any "suggestions" it sent in its own right would be understood as being rather more than the expression of an opinion.[58]

It might well seem that the rationale for the 1943 Admiralty takeover has been substantiated. It is clear the linkages between COHQ and the Admiralty were inadequate before then. The two organizations had functions and purposes that were at many points the same, but there was dangerous administrative separation. Reliance on mere liaison and general understandings was not sufficient; that remained true even after PD(Q) was first formulated to overcome the gap. There had to be more formal bureaucratic links; proper discussion and circulation of the vital streams of information meant administrative union. Furthermore, there had to be a single recognized source of executive authority so orders would be carried through expeditiously, implying common subordination to the Board of Admiralty: again, that meant administrative union. Finally, once together, it would be possible to have a really effective centralized means of coordination, both economical of personnel and expertise and able to understand and tackle problems as they arose. Local initiative, or delegation of power, was here as important as linkages. But that had anyway been developing apace as the wartime administration had grown, and gives us what is surely a clinching argument in favor of the Admiralty takeover. How, one wonders, could two separate government departments have been maintained to deal with the same business, where not only did the larger recognize only its own source of executive authority, but where there was an increasing delegation of decision-making within it?

Conclusion

We saw that back in 1918, Sydney Fremantle had attacked the Plans Division for articulating plans without considering resources. His attempt to strangle the new division was certainly unwise, but he referred to a real difficulty, one found in

[58] ADM 1/16559, 29 June 1944.

Whitehall more generally, where policy-making was being improved, but administration rather less so, and burgeoning departments could still be divided into watertight compartments.[59] In the Admiralty case, the inadequate links between operational plans and "logistics" were typical of the rest of the offices. The structure was very much a vertical one, and communicated principally from top downward to bottom. In short, the Admiralty—even the large civilian element—was strongly hierarchical, still too much a mechanism for passing orders down. That position was improving at the end of the First War; one should point in particular to the development from 1917 of the Secretariat "M" Branch as a means of communication and coordination, and also the new Plans Division, which already in the 1920s was helping to coordinate the whole Naval Staff.[60] However, by the early 1930s, due in part to economies, further advance elsewhere was limited, or could even be choked-back, at the very time when technological advance was notably complicating the whole process of decision-making, and calling for ever greater interconnectedness, sometimes of the most surprising kinds (as, say, with scruffy scientists). Then came rearmament and rapid expansion: the departments were more numerous, bigger, and contained more newcomers. Also, in 1939 there was a great splitting of the Admiralty, as most of the matériel departments went out to Bath, followed by the hectic and highly dispersed growth of wartime. What now threatened was actual sclerosis.

In the emergency, the first response was to try to hold everything together through *ad hoc* measures, or what Admiral Godfrey described as "the invisible liaison, conferences, telephonic conversations and so on. . . ."[61] These helped greatly, though had their limits: certainly, after a given size of administration, it is not enough to have a few liaison officers—however bright and persuasive—telephoning or galloping madly between the offices. Formal bureaucracies are better fitted for the task of regularly passing streams of data. In any case, to be effective, good liaison officers need talent, but they are in short supply: one benefit of formal bureaucracies, one might suggest, is that they can be adequately staffed by those who are competent, but not exceptional. Really good people should be kept in the offices and given their head, vital in keeping monstrous bureaucracies moving.

The development of maritime war helped propel these developments, and here we might point in particular to combined operations. In 1939, it still seemed one could fight according to Blue Water principles, and one senior officer in the Admiralty said he could envisage only one possible combined operation in the war at

---

[59]For central government at the end of the First War, see Ehrman, *Cabinet Government and War*, pp. 96–99.

[60]C. I. Hamilton, *The Making of the Modern Admiralty: British Naval Policy-Making, 1805–1927* (Cambridge: Cambridge University Press, 2011), pp. 259ff.

[61]ADM 1/9531, John Godfrey, "Local Defence and Auxiliary Patrol Questions in War Time," 7 October 1935.

hand.[62] That position soon became untenable. Back in 1911, Second Sea Lord, Prince Louis of Battenberg, had written that "In an island Kingdom like ours all warfare becomes sooner or later 'amphibious.'"[63] If a little astray for the First War, his judgement was right on target for the Second. The consequent need for close Service cooperation put severe strains on the Whitehall machinery. By 1942, the amphibious element was emerging strongly, and COHQ and Admiralty were forced to be more interdependent; but if they were one animal, the tail was energetically wagging the dog, or so the dog asserted. Outsiders, on the other hand, could well conclude the real problem was insufficient integration. There had to be either a retreat or an advance: but the first was impossible, and the second all the more advisable since the relationship was shifting. The Admiralty was anyway becoming more "amphibious" in its work and attitudes: it was becoming a COHQ on a greater scale. At least that meant by this stage of the war that there was no danger, as there had been in 1939, that the cause of combined operations would be lost within an unsympathetic Admiralty devoted to Blue Water operations.

The takeover of 1943 suggests that total war means total administration. Even if integration of British war-making machinery was by now going far, it continued to go further in 1944–45. Then, of course, came the peace, when everything changes and you can get back to dealing with your serious enemies. As Captain Roskill wrote, all operations of war are to an extent combined operations, while in peacetime the Services tend to go their own ways.[64] But the experience after 1945 was to hold a certain novelty. The financial shortages of peace can enforce rationalization just as the personnel shortages of wartime can. And however much of a relief peace may be, it does not destroy memories of what had worked during war. Better linkages and a wider common organization had produced a higher efficiency. That in turn suggested one might even think of a single, unified Defence department. We can scarcely turn to a new subject now, but ought to point to the curious irony that it was Battenberg's younger son, by now Earl Mountbatten of Burma, who came back to create the Ministry of Defence of 1964, something which also serves as a reminder that the personality factor can suddenly crash back in to take the lead over more orderly and apparently fundamental developments.

[62]Fergusson, *The Watery Maze*, p. 41.

[63]CAB 1/31, Part 2, "Naval War Staff," c. December 1911.

[64]Stephen Roskill, *War at Sea*, 3 vols. in 4 (London: HMSO, 1954–61), Vol. III, part II, p. 362. Whilst referring to the official naval historian, one should mention that in his copy of "When the Rain's Before the Wind," where Ellis states he is convinced COHQ's naval administrative and logistical functions had to be assumed by the Admiralty, Roskill made the marginal comment that they should never have been given to CCO in the first place. See CAC, Roskill Papers, ROSK 6/2, 26/15.

7

To the Jungle Shore

Australia and Amphibious Warfare in the South West Pacific Area, 1942–1945

Peter J. Dean

Australia is the sixth largest country in the world and the only one of the top six to be completely surrounded by water. Covering 7,618,493 square kilometers (km), it consists of 59,736 km of coastline and is not only an island, but also a country and a continent, a nation heavily dependent upon the sea for its security and trade.

The first European settlers set sail from England in 1787 and reached Australia's shores in January 1788. Despite its maritime birth, and its continued dependence upon the sea for its lines of trade and communication, Australians have spent their time mythologizing their vast, and largely empty, interior. As a number of senior naval officers have often noted, Australia is not the "girt by sea" that the national anthem talks of, but rather "girt by beach." Despite the fact that since European settlement nine out of ten Australians live in large coastal cities and towns, Australia's identity has been formed around its unique "outback" environment. This focus on the bush and the land means that culturally Australia sees itself "first and foremost as a continent and not as an island."[1] This has been reinforced in Australia's military traditions.[2] The infamous Anzac spirit emanates from one of the most famous amphibious operations in history, the landing at Gallipoli in 1915, but this fighting spirit did not derive from the amphibious aspects of this operation, but rather the battles that occurred once the troops were ashore.[3] As

[1]Michael Evans, *The Tyranny of Dissonance: Australia's Strategic Culture and Way of War, 1901–2004* (Canberra: Land Warfare Studies Centre, 2005), pp. ix, 35. This chapter was originally published in *Global War Studies* 11:2 (2014).

[2]John Grey, "Opening address by the Chief of the General Staff," in Glenn Walhert, ed., *Australian Army and Amphibious Operations in the South West Pacific, 1942–45* (Canberra: Army Doctrine Centre, 1995), p. 2.

[3]"ANZAC" has multiple meanings. It can mean: a soldier—an ANZAC, originally a member of the AIF that had served at Gallipoli, and later any Australian or New Zealand soldier; a place, such as ANZAC Cove, as the site of the landing near Ari Burnu became known; a day of commemoration to remember those who have both died and served the nation at war; a battle and/or campaign; a fighting spirit; a folklore tradition amongst the soldiers; and a

Jeffrey Grey has pointed out, Australia's approach to war fighting has always been distinguished by the quality of its expeditionary infantry, who are usually sent overseas as part of a wider coalition and depend on a larger ally for logistical and other support.[4]

Strategy and the South West Pacific Area (SWPA), 1942–1945

The maritime environment has and will always be exceptionally important for the defense of Australia and its strategic posture. Geography dictates that Australia needs to embrace a maritime strategy for both the defense of the continent and for force projection in the region.[5] However, in the inter-war period, 1919–1939, Australia had relied on the Royal Navy to provide the first bastion of its defense via the deeply flawed, one dimensional, Singapore naval strategy.[6]

The war in the Pacific in 1941–42 exposed the fallacy of this strategy and caused a drastic change in Australia's military position. With the collapse of resistance in Malaya, Java, and Rabaul in early 1942, Australia's strategic circumstances had radically changed. With the bulk of Australia's trained forces fighting overseas in the Middle East or lost in the islands of the southern Pacific, the country remained largely defenseless. Australia in 1942, while never threated with invasion by the Japanese,[7] largely relied upon the U.S. Navy to secure its sea lines of com-

---

legend/myth/spirit that is endorsed by institutions and governments that has come to personify Australian "values." For a more detailed discussion of its influence on the contemporary writing of Australian military history, see Peter J. Dean, "Recent Scholarship in Military History and the ANZAC Legend: Down Under 2010," *Global War Studies* 7:2 (2010), pp. 230–37.

[4]Jeffrey Grey, *A Military History of Australia* (Melbourne: Cambridge University Press, 2000), p. 5.

[5]Despite this requirement, it was not until 2009 that Australia adopted a maritime strategy. Prior to this, Australia relied on Great Britain to provide its security in the pre-World War II period via the Singapore Strategy, then it adopted a forward defense posture in combination with Great Britain, and then the United States in the Asia-Pacific region during the 1950s and 1960s, before adopting a defense of Australia construct from 1976–2000. For Australia's current strategic posture, see Commonwealth of Australia, *Defence White Paper 2013: Defending Australia and its National Interests* (Canberra: Commonwealths of Australia, 2013). Section 3.42 outlines a maritime strategy for the ADF.

[6]Australia was well aware of the failing of this strategy, but chose to ignore the reality of its strategic situation in the inter-war period. See Augustine Meaher IV, *The Road to Singapore: The Myth of the British Betrayal* (Melbourne: Australian Scholarly Publishing, 2010). See also David Stevens, ed., *In Search of a Maritime Strategy: The Maritime Element in Australian Defence Planning since 1901,* Working Paper No. 119 (Canberra: Strategic and Defence Studies Centre, ANU, 1997).

[7]The supposed potential for a Japanese invasion has been the cause for much public debate in Australia. This sentiment was heightened by the decision of the Australian government

munication. The victory at the Battle of the Coral Sea provided this security and also safeguarded Australia's last defensive bastion in Papua, Port Moresby.[8]

Soon after the establishment of the South West Pacific Area, it became apparent to the U.S. C-in-C General Douglas MacArthur that a maritime strategy,[9] formed around land-based air power and amphibious operations that embraced the concept of maneuver, was needed to defeat the Japanese in the region. MacArthur's concept of operations evolved throughout 1942–43 and moved from the re-capture of the Japanese stronghold at Rabaul to its isolation. The strategy involved the use of maneuver warfare aimed at shattering the enemy's morale and physical cohesion. This was to be achieved by physically dislocating the enemy through superior operational mobility to achieve superiority in mass and firepower at critical points. MacArthur's principle strike weapon was his air force. His strategy rested on air superiority enabling sea control to allow his ground force to leapfrog forward using amphibious warfare, isolating large numbers of Japanese, and establishing airfields, ports, and logistic bases to allow the advance to continue. This was to be achieved through a joint operational approach contained within a maritime strategy that used the ocean as the space for maneuver.[10]

In order to execute this maritime doctrine, MacArthur had to develop a number of capabilities: a fleet to be able to support combat operations including amphibious transportation and assault; an air force capable of gaining air superiority and undertaking strike and reconnaissance operations in support of the fleet;

---

to introduce a commemorative Battle for Australia Day, held on the first Wednesday in September each year. The commemoration is in remembrance and recognition of those "who served in defence of Australia in 1942 and 1943 when we faced the gravest threats to our nation . . . [when] there were direct attacks on the Australian mainland, particularly in Darwin, and battles in the Coral Sea and Papua and New Guinea, including Milne Bay and the Kokoda Track." The vast majority of academic historians reject the notion that Australia was under direct threat from the Japanese. See Peter Stanley, *Invading Australia: Japan and the Battle for Australia, 1942* (Camberwell: Viking, 2008) and Peter J. Dean, ed., *Australia 1942: In the Shadow of War* (Melbourne: Cambridge University Press, 2012).

[8]See David Stevens, "Australia's Thermopylae? The Kokoda Trail," in Craig Stockings, ed., *Zombie Myths of Australian Military History* (Sydney: University of New South Wales Press, 2010), pp. 172–74.

[9]MacArthur never used the phrase "maritime strategy" to describe his approach to operations in the SWPA, but it contained all the fundamental elements of this type of strategy.

[10]This strategy is articulated in a GHQ SWPA report entitled "Defensive and offensive possibilities." The report discusses strategy for the offensive against Japan and outlines the use of island hopping of no more than 300 miles per jump to ensure fighter cover for bombers. It discusses the use of combined arms task forces to undertake this with naval support under the cover of an air umbrella. U.S. National Archives and Records Administration (hereinafter NARA), Record Group (hereinafter RG) 407 98-GHQ1-3.2, SWPA GHQ G-3 Journals and Files Box 566 (no. 1), April 1942–30 May 1943.

and an army that possessed a well-trained amphibious force. The bedrock of these operations during 1942–43 was the cooperation between Australian and United States militaries; a partnership that continues to be Australia's most important strategic relationship to this day.

Yet in 1942 Australia was neither well organized nor experienced in amphibious warfare. After Gallipoli, amphibious warfare did not form a part of Australia's operations in the rest of the First World War. In the inter-war period, little thought or attention was paid to this mode of warfare by the military. During the period 1939–41, the Australian Army had concentrated on the deployment of the Australian Imperial Force to the Middle East and mastering the art of desert warfare, while the Militia station in Australia had concentrated on continental defense. This meant that in 1942 the Australian Army had almost no institutional knowledge in amphibious operations and certainly no cultural tradition.[11]

Centers of Gravity in the SWPA and the Adoption of a Maritime Strategy

The victories at Coral Sea and Midway had provided Australia with security and allowed MacArthur to plan for offensive operations in the SWPA. These plans were greatly assisted by the U.S. Marine Corps (USMC) landing at Guadalcanal in August 1942 which forced the Japanese command at Rabaul, who controlled operations in both the Solomon's and New Guinea, to reduce both air and naval strength around Papua. Despite this diversion of Japanese assets from MacArthur's region, his relative lack of air and naval strength meant that he was unable to establish sea control or air superiority.

The Japanese had reacted to their defeat at the Coral Sea by landing Major General Tomitarō Horii's South Seas Detachment on the north coast of Papua on 21 July 1942 in order to attempt an overland assault on Port Moresby via the Kokoda Trail. The Japanese underestimated both the condition of this mountainous track as well as the logistics and force structure required for a successful assault, and under mounting Australian resistance and continued setbacks at Guadalcanal, Horii's forces were eventually ordered to withdraw to the north coast beachheads at Buna, Gona, and Sananada.

By the time these bastions had been wiped out in January 1943, the Japanese had lost more than 13,000 soldiers killed in Papua while the Australians had lost 2,000 killed and the Americans more than 600. A further 20,000 Australian and American troops had been wounded or succumbed to one of the many diseases

---

[11]Evans, *The Tyranny of Dissonance*, pp. ix, 35. Evans also argues that a "genuine maritime consciousness [has] [n]ever dominated the mainstream of Australian strategic thought," p. 34. For an overview of Australian amphibious operations, see Russell Parkin, *A Capability of First Resort: Amphibious Operations and Australian Defence Policy, 1901–2001*, Land Warfare Studies Centre, Working Paper No. 117 (Canberra: Land Warfare Studies Centre, 2002).

found in the harsh tropical climate.[12] By way of contrast, when the Japanese had finally abandoned Guadalcanal in February 1943 they had lost approximately 20,000 killed. U.S. losses had amounted to 1,769 killed on land, some 60% of the losses in the battle for Papua, however "for every United States soldier or marine ashore who died almost three sailors and marines in ship companies had perished."[13]

The relative balance between the opposing forces in the Kokoda and Beachhead operations had, for much of 1942 and early 1943, prescribed an attrition-style campaign. Furthermore, the absence of suitable numbers of landing craft and a lack of training in amphibious operations meant that MacArthur's forces had been unable to undertake any landings from the sea to outmaneuver the Japanese.[14] The SWPA's lack of naval ships and landing craft as well as the Navy's unwillingness to operate in the dangerous waters around the north coast of Papua at Buna, Gona, and Sananada had an almost crippling effect on Allied supply lines. One major consequence was the restrictions on the use of artillery and tanks in these battles.[15] It was critical, therefore, that the Allied forces in the SWPA develop adequate amphibious forces to take advantage of their increasing air and naval superiority in order that they not repeat the costly mistakes of the Papuan campaign.

Preparing the Force: Amphibious Warfare Training in the SWPA

The need for amphibious forces had been recognized early in 1942. In March, the Deputy Chief of the General Staff in Australia, Major General Sydney Rowell, recommended the acquisition of specialist equipment and the establishment of a school of combined operations.[16] Rowell had, however, recognized that the

---

[12]See Dudley McCarthy, *South-West Pacific Area—First Year: Kokoda to Wau,* Australia in the War of 1939–1945, Series I, Vol. V (Canberra: Australian War Memorial, 1959), pp. 527–28; Samuel Milner, *Victory in Papua,* U.S. Army in World War II, The War in the Pacific (Washington, DC: U.S. Army Center of Military History, 1989), pp. 370–72.

[13]This figure comes to 4,770 lost at sea. This figure does not include the Australian and New Zealand casualties nor the casualties from the U.S. Navy, Marine Corps, and Army. Richard B. Frank, *Guadalcanal: The Definitive Account of the Landmark Battle* (New York: Penguin, 1992), pp. 613–14.

[14]Australian War Memorial (hereinafter AWM), 52 1/5/51, War Diary, Adv NGF HQ, Appreciation for GOC NGF by DCGS at Adv NGF, 1 January 1943.

[15]This lack of artillery was seen as crippling to the campaign and one of the reasons for the high casualty rate. See NARA, RG 319, Records of the Army Staff, Center of Military History, Victory in Papua, 270/19/7/1, Box 1–3, Albert W. Waldron to Orlando Ward, 5 March 1951. Waldron was artillery commander of the U.S. 32nd Division at Buna.

[16]National Archives of Australia (hereinafter NAA), B6121 289, "Long Range Planning for offensive Action—Landing operations," 13 March 1942. See also Ross Mallett, "Together Again for the First Time: The Army, the RAN and Amphibious Warfare 1942–1945," in

raising of such a force would be a long-term endeavor. It was also realized that this would eventually be a combined operation between the U.S. and Australian forces in the area. The story of the development of Australia's advances in its amphibious warfare capability is really an Allied story, built around U.S.–Australian cooperation.

With the establishment of the SWPA on 30 March 1942, MacArthur's General Headquarters South West Pacific Area (GHQ), as well as the Australian Commander-in-Chief and Commander Allied Land Forces, General Sir Thomas Blamey's Headquarters (LHQ), set about the enormous task of planning for the directive set by the U.S. Joint Chiefs of Staff (JCS) to retake Rabaul.[17] After discussions between MacArthur and Blamey, the decision was made to establish a Joint Overseas Operational Training School (JOOTS). This school was designed as a joint U.S.–Australian, Army-Navy-Air Force, operation and MacArthur described its mission as "Training of Land Forces in overseas operations in conjunction and cooperation with Naval Forces and Air Forces, both land and carrier based. . . . The task will involve the combat loading of ships, the disembarkation of troops and supplies in small boats in the face of an enemy, the landing on hostile shores, a rapid and strong thrust inland, and the occupation of hostile coast lines with continued operations into the interior, all tasks in coordination with Naval and Air support."[18]

JOOTS was just one aspect of a multi-pronged plan to train the U.S. and Australian forces in the SWPA. Key to this was the acknowledgement that amphibious operations are not just about the transportation of troops, equipment, and supplies over water. Rather, an amphibious operation requires a well-defined doctrine, detailed operational planning, tactical acumen, complex ship-to-shore movement, and landing operations. As USMC General Holland Smith noted, it was "not a ferry boat ride, but [rather] a tactical movement, culminating in an assault."[19]

Senior commanders and planners in the SWPA realized that the early Japanese successes owed a great deal to their prowess in this form of warfare. It was therefore paramount that the Allies develop a high level of expertise for their troops through appropriate training. Allied commanders in the SWPA were exceptionally

---

David Stevens and John Reeve, eds., *Sea Power Ashore and in the Air* (Sydney: Halstead Press, 2007), pp. 118–32. In the British, and thus Australian military, amphibious landings were referred to as "combined" operations as opposed to the U.S. terminology of "joint" operations.

[17]Milner, *Victory in Papua*, p. 20.

[18]NARA, RG 496, Box 667, Memo "Combined Training for Offensive Operations," MacArthur to Blamey, 4 June 1942.

[19]Smith as quoted in M. H. H. Evans, *Amphibious Operations: The Projection of Sea Power Ashore* (London: Brassey's, 1990), p. 9.

lucky that they had ready-to-hand the inter-war developments in doctrine from both the U.S. and UK.[20]

Doctrine

Despite facing a number of critical infrastructure and equipment difficulties, the training schools set up in the SWPA were to benefit from established doctrine and training methods and experienced staff. Although neither the doctrine nor the training staffs were to prove themselves perfect in 1942–43, they did provide a critical foundation for success and innovation. Instructors were sourced from the U.S. and Australian armies and navy, the U.S. and Royal Marines (RM), and the Royal Australian Air Force. The Australians initially relied heavily on instructors who had attended the British Combined Training Centre (CTC) at Kabrit, Egypt.[21] Additional instructors were added to JOOTS in February 1943 when a number of 9th Australian Division officers, who had received instruction at Kabrit, returned from the Middle East.[22] The original draft in 1942 included one British Army officer and Lieutenant Colonel Thomas K. Walker of the RM. Walker was a particularly valuable addition to the team. A highly experienced officer, Walker had previously been the Deputy Assistant Adjutant General for the RM, and in 1940 had prepared a detailed paper on "Preparation[s] for Combined Operations on a Major Scale." This document outlined the requirements for successful amphibious operations, including the organization of a staff, formations, equipment, training establishments, and naval and air support. As Donald Bittner has noted, Walker's paper "proposed the correct solution, but their implementation would be over a long period of time and on a scale hardly envisioned in 1940."[23]

Doctrine on which these preparations were to be based was originally a mixed affair with the Australians relying on *British Combined Operations Doctrine* (1942) and *Combined Operations for Unit Commanders* (1941), and their own Australian Military Forces (AMF) *Combined Operations Pamphlet No. 1* (Provisional—1942). The U.S. military did, of course, prefer their own approach to joint operations.

---

[20]See Kenneth J. Clifford, *Amphibious Warfare Development in Britain and America from 1920–1940* (Laurens, NY: Edgewood, 1983) and William F. Atwater, "United States Army and Navy Development of Joint Landing Operations, 1898–1942," Ph.D. dissertation, Duke University, 1986.

[21]AWM, 54 943/16/1, p. 1, A Summary of Combined Operations Training in Australia (Amphibious) 1942–1945.

[22]U.S. Naval History and Heritage Command, Washington, DC (hereinafter NHHC), Barbey Papers, Correspondence, Box 2, Landenberger to Barbey, 26 April 1962.

[23]Donald F. Bittner, "Britannia's Sheathed Sword: The Royal Marines and Amphibious Warfare in the Interwar Years—A Passive Response," *The Journal of Military History* 55:3 (July 1991), p. 359.

Their doctrine was contained in the manuals: FM 31 *Landing Operations on a Hostile Shore* (1941); FM 31-5 *Landing Operations on a Hostile Shore—Air Operations* (1941); USN *Joint Action Doctrine* (1927); and USN *Landing Operations* (1938). This meant that, at least initially, as each unit rotated through the various training establishments set up in the SWPA, the doctrine and methods were adapted to the unit's nationality. This attempt to mesh together a combination of British and U.S. Navy, Army, and Air Corps doctrine did cause some tensions. JOOTS' original Commanding Officer, Colonel B. Q. Jones (U.S. Army), found these difficulties tiresome and complained about his problems with Australian/British methods to MacArthur's GHQ operations chief, Brigadier General Stephen Chamberlin. He noted: "We constantly have to deal with our Allies. Their system, their methods and their line of thought are different from ours. In many cases measured by our own standards, these methods appear most inefficient."[24]

Conversely, Lieutenant Colonel Walker (RM) believed that the Americans were too rigid theoretically, and their training methods were based on "rather out of date theory from American Army textbooks."[25] A lot of the initial tension at JOOTS seemed to be the result of the unsuitability of Jones to his command. The British Army liaison officer at JOOTS, Lieutenant Colonel Hope, who had been brought out from the UK to provide expert advice on amphibious operations, had been allocated to a junior role at the school and he found his position so untenable that he requested that he be sent back to the Middle East. The Australian Deputy Chief of the General Staff,[26] Major General Frank Berryman, noted that Jones "is not all together suitable . . . he has been trained as a staff officer, but appears to be academic and unable to handle his staff . . . it is inevitable that there should be difficulties when a school is first established, but making allowances for this, I feel that under his [Jones] direction the place will never be a success."[27]

Three days after Berryman made these observations to the C-in-C Allied Land Forces, Jones was replaced by Brigadier General Robert H. Van Volkenburgh (U.S. Army),[28] and Lieutenant Colonel Hope was transferred to the First Australian Army's amphibious training school at Toorbul Point, Queensland.[29] From this

---

[24]Douglas MacArthur Memorial Archives (hereinafter MMA), Norfolk, VA, RG36, Box 1, Chamberlin Papers, B. Q. Jones to S. Chamberlin, 17 September 1942.

[25]Walker as quoted in Parkin, *A Capability of First Resort,* p. 20.

[26]The DCGS position was effectively Blamey's chief of staff and head of operational planning.

[27]AWM, PR84/370, Papers of Lieutenant-General Sir Frank Horton Berryman, item 11, Berryman to Blamey, "Amphibian Training," 22 September 1942.

[28]NARA, RG 496, Records of GHQ SWPA, G-3 General Correspondence, 1942–1954, Box 665, Special Order No. 146, GHQ SWPA, 25 September 1942.

[29]NARA, RG 496, Records of GHQ SWPA, G-3 General Correspondence, 1942–1954, Box 665, Chamberlin to Sutherland, 1 March 1943.

point onwards, relations between the two armies at the Port Stephens School improved significantly.

Throughout 1942, the initial amphibious warfare training had come under the command of a senior U.S. Army officer appointed by GHQ,[30] but at the beginning of February 1943, this would change with the creation of the 7th Amphibious Force (7th AF). The 7th AF was formed in December of 1942 and on 10 January 1943 it was placed under the command of Rear Admiral Daniel E. Barbey. But when Barbey arrived in Australia, his force existed in name only. His immediate tasks were to create an amphibious force, train his amphibious ships, and train the U.S. and Australian Army units in amphibious operations.[31] As such, 7th AF was given control over all amphibious warfare training from 8 February 1943.[32]

Barbey was to become one of the most successful amphibious warfare commanders in history, and by the end of 1945 he had commanded fifty-six successful amphibious operations. One of his first moves was to appoint Commander "Red" Jamison (USN) as the commanding officer of JOOTS, which along with the First Australian Army Combined Operations Training School, was absorbed into 7th AF Amphibious Training Command. In addition, Barbey added more U.S. and Australian Army and Navy officers to the staff of these two training centers, mainly combat veterans with experience at Guadalcanal, North Africa, New Guinea, or Malaya.[33]

Barbey's desire was to standardize equipment and doctrine for all units, Australian and U.S., on U.S. Navy Fleet Training Publication FTP 167 *Landing Operations Doctrine.*[34] This was a relatively smooth transition, noting that there was "little difference in the two techniques [British and American] and they were readily integrated into the training program."[35] The only difficulties were "a few

---

[30]During 1942, relations between senior U.S. and Australian officers was at an all-time low point, often driven by cultural imperialism on both sides. This relationship was also greatly affected by the divisions and conflicts within MacArthur's headquarters. Relations improved significantly during 1943 and had proven to be close at the tactical and operational level from late 1942. For details, see Peter J. Dean, *The Architect of Victory: The Military Career of Lieutenant-General Sir Frank Horton Berryman* (Melbourne: Cambridge University Press, 2011), Chapter 9, pp. 209–18.

[31]United States Navy, *Command History: Seventh Amphibious Force, 10 January 1943–23 December 1945,* Section I, p. 4.

[32]NAA, B6121, 194B, p. 2, Combined Operations—RAN Beach Commandos.

[33]Ibid.

[34]NAA, MP742/1, 246/1/164, Annexure 2, Appointment of Commander Amphibious Force, SWPA. This order included the training of Australian troops in "Combined Operations Pamphlets" from the British War Office.

[35]Daniel E. Barbey, *MacArthur's Amphibious Navy: Seventh Amphibious Force Operations, 1943–1945* (Annapolis: United States Naval Institute, 1969), p. 36.

differences in communication procedures [that] cropped up and caused minor problems."[36] He was, however, insistent that if Australian troops were to be mounted in American vessels, "they must follow the entire American technique."[37]

Despite Barbey's insistence, the SWPA was so low in the allocation of amphibious (and all other) resources in 1943 that the 7th AF had to modify standard U.S. Navy doctrine to account for a force based mainly around smaller, often improvised landing ships and landing craft with little support from APDs (high-speed destroyer transports) and large landing ships. Subsequently, Commander Jamison and his Allied staff at the Amphibious Training Centre at Port Stephens were charged by Barbey with "developing this new and untried technique."[38] Thus, Port Stephens became not only the major school for the training of instructors in amphibious warfare, but also the specialists in the development of doctrine in the SWPA. These new approaches were formalized through the establishment of Standard Operating Procedures (SOP) for the 7th AF; for instance, the adoption on 31 August 1943 by GHQ of the SOP for *Boat Teams in Small Boats* (*LCVP and LCM3*)[39] that had been developed by Jamison and his team. Twenty-four other SOP documents were produced by the ATC team during the war. These include SOPs for air-amphibious communications, medical services, loading of Landing Craft Tanks (LCT), and various documents on operating with different classes of amphibious ships and craft. These documents were based on USN FPT 167 and 211 and U.S. Army FM 31-5, British doctrine, and the experience gained in the SWPA.[40]

Until HQ Amphibious Training Command at Port Stephens was up and running and in charge of all training and doctrine development, Barbey had to improvise training programs. One of the first was for the U.S. 1st Marine Division that was stationed in Melbourne, recovering from the Guadalcanal operations. To do so, he drafted the RAN ship HMAS *Manoora* (Landing Ship Infantry—LSI) and USS *Henry T. Allen* (Amphibious Attack Ship—APA), as well as a Free French and a U.S. destroyer to provide naval gunfire support. Barbey used a USMC regiment as a demonstration unit and brought a large group of Allied observers of all ranks and formations, including the staff at Amphibious Training Command at Port Stephens, down for a training demonstration and post-operation analysis.[41] Soon

---

[36]Ibid.

[37]NARA, RG 496, Box 667, "Amphibious Training Oct 1942–Oct 31 1943," Notes from Conference, GHQ, 18 January 1943.

[38]Barbey, *MacArthur's Amphibious Navy*, pp. 48–49.

[39]NARA, RG 496, Box 667, "Amphibious Training Oct 1942–Oct 31 1943," Memo Commander Seventh Amphibious Force to C-in-C SWPA, 31 August 1943.

[40]NAA, MP742/1, 246/1/164, Annexure 2, Training Carried Out by Seventh Amphibious Force.

[41]NAA, MP729, 50/401/430, Barbey to MacArthur, "Amphibious Training," 11 March 1943.

afterwards, he made available for training operations at Port Stephens the converted armed merchant cruisers HMAS *Manoora, Westralia,* and *Kanimbla,* which had been refitted as British-model LSIs.[42]

With a new doctrine to initiate and with new ships containing highly inexperienced crews, Commander Jamison and his Allied training staff at the ATCs in Port Stephens, Toorbul Point, and a new facility that was soon to be opened at Trinity Beach in Cairns on the far north coast, had a daunting task. However, within two months the staff had expanded from approximately 400 to 2,500 U.S. and Australian personnel and they had in training some 20,000 combat personnel made up initially of the 7th Australian and U.S. 32nd Infantry Divisions.

Joint Training

Barbey's initiatives seemed a long way off in the dark days of 1942. Before he took control of amphibious training, the first moves had been undertaken by First Australian Army in April 1942. In March, Lieutenant Colonel M. W. Hope (British Royal Artillery), who had previously been the Senior Army Instructor at the British Combined Training Centre in Kabrit, Egypt,[43] along with Commander F. N. Cook, RAN, who also had considerable experience in the UK and Middle East in amphibious operations and had been the commanding officer of HMS *Tormentor* and the senior naval officer in charge of the Bruneval commando raid on France in 1942, arrived in Australia.[44] These were the first officers to begin serious planning in combined (amphibious) operations in the SWPA.

Lieutenant Colonel Hope and Commander Cook, along with Major A. L. Rose, Australian Imperial Force (AIF),[45] undertook reconnaissance down the length of

[42]Barbey, *MacArthur's Amphibious Navy,* pp. 36–38.

[43]AWM, 54 943/16/1, p. 1, A Summary of Combined Operations Training in Australia (Amphibious) 1942–1945.

[44]AWM, PR00631, Papers of HMAS *Assault,* Record of Commander Fredrick Norton Cook. See also W. N. Swan, *Spearheads of Invasion: An Account of the Seven Major Invasions Carried Out by the Allies in the South-West Pacific Area during the Second World War, as Seen from a Royal Australian Naval Landing Ship Infantry* (Sydney: Angus and Robertson, 1953), p. 18.

[45]During World War II, the Australian ground forces were split between the all-volunteer Australian Imperial Force (AIF) that was raised for service in Australia and overseas and the Militia or conscripted forces that were raised for the defense of Australia and its mandated territories. AIF units were prefixed with a 2/ to denote the 2nd AIF (the 1st being raised for service in World War I). Only the AIF saw service in the Middle East, while both the AIF and the Militia served in the SWPA. However, the Militia were restricted to Australia's mandated territories (Papua, New Guinea, Solomon Islands) and thus did not serve in Borneo or the Philippines.

the eastern seaboard of Australia in search of appropriate sites for training facilities. The subsequent Hope-Cook report recommended a center for the training of staff at Port Stephens on the NSW north coast and a training school for units at Bribie Island (Toorbul Point) in Queensland. These later became the sites for the Joint Overseas Operational Training School, also known as the Combined Training Centre (CTC) and the First Army Combined Training School (CTS).[46]

LHQ gave approval for these two schools to be established on 28 June 1942, but the major difficulty at this time was the fact that any such planning could not be carried out in isolation, and as all Australian forces fell under GHQ, they would also have to be coordinated with, and include, the U.S. forces based in the SWPA. GHQ was only just becoming interested in the development of an amphibious force, and though GHQ held a number of conferences in June and July, final approval for JOOTS was not given until 9 August 1942.[47] By the time GHQ approval was forthcoming, the Australian initiatives meant that the CTS at Toorbul Point had already opened four days prior to GHQ's approval and the CTC at Port Stephens began operations six days later, on 15 August 1942.[48]

Combined Training School—Toorbul Point, QLD

CTS began its first course on 5 August 1942 under the direction of I Australian Corps HQ with the newly-promoted Lieutenant Colonel A. L. Rose (AIF) as its chief instructor. He had arrived in July and within three weeks roads, telephone lines, water supply, accommodation blocks, and a pier had been constructed. Rose was supported by Lieutenant J. M. Band, Royal Australian Naval Volunteer Reserve (RANVR), the senior naval officer, who would command the landing craft used at the school.

Initially, new units arrived at CTS every week to receive six days of introductory instruction. In order to ensure that these units got the best out of their short stay at the CTS, they all participated in preliminary training at their regular training camps. The object of this training was to give all ranks an introduction in amphibious operations through boat management and towing, company and battalion level exercises, and training with equipment. For the battalion exercise, the

---

[46] AWM, 54 422/7/8, Combined Training School, Toorbul Point: Origin of the School. A joint board made up of U.S. and Australian officers was appointed by GHQ in late 1942 and undertook a second reconnaissance of the eastern seaboard of Australia and concluded that as per the Hope-Cook report no other suitable site existed. See AWM, 54 943/16/1, p. 1, A Summary of Combined Operations Training in Australia (Amphibious) 1942–1945.

[47] NARA, RG 496, Records of GHQ SWPA, G-3, General Correspondence 1942–45, Box 667, Directive—Joint Overseas Operational Training School, 9 August 1942.

[48] AWM, 54 943/16/1, p. 2, A Summary of Combined Operations Training in Australia (Amphibious) 1942–1945.

unit was given a beachhead to secure and a more distant objective to reach. The major restriction at this time was the fact that training at CTS had to begin with folding boat equipment until the new Landing Craft Vehicle Personnel (LCVP), Landing Craft Mechanized (LCM), and Landing Ship Tank (LST) became available in November.[49]

Despite the lack of landing craft, the school did everything possible to ensure a realistic training environment. This was helped by the liberal use of explosives as simulations and cooperation with aircraft from the RAAF. 32nd Squadron RAAF, commanded by Squadron Leader Hampshire, provided "whole hearted cooperation" with CTS from the very beginning. He was provided with a senior Air Liaison Officer (ALO), Major Hopkins, and they coordinated dive-bombing and low-flying attacks for most of the daylight landing exercises. The RAAF was provided with targets near the landing troops and these were attacked with either 8lb smoke bombs or live 112lb or 250lb bombs. The ALO was at all pilot briefing and debriefing sessions and was often able to observe the exercises from one of 32nd Squadron's aircraft. Pilots and observers from the squadron also regularly visited CTS to participate in debriefings and undertake liaison work. It was reported that "a valuable feature of this work . . . was the mutual understanding that it fostered" between the RAAF, Army, and Navy, all of whom were working together at the CTS.[50]

In order to facilitate the training, a captain-level school instructor was allotted to each sub unit on the course. In September, officers from the U.S. 32nd Division arrived on temporary assignment as instructors for when their division's units arrived, and later, officers from the U.S. 41st Division were attached more formally to the staff of the school. Two Australian Independent (Commando) Companies also were trained at CTS with a slightly longer training program. They then stayed on at CTS to conduct their regular training while proving enemy parties for other units passing through as well as demonstration troops for the school. Their pres-

[49]AWM, 54 943/16/1, p. 3, A Summary of Combined Operations Training in Australia (Amphibious) 1942–1945. Originally, MacArthur had asked for specialist landing craft to be made available to the Australians via Lend-Lease, but he was notified that sufficient craft could not be produced to meet the needs of Australia in addition to the requirements in the USA and UK. However, with the cancellation of some operations in Europe in 1942, 284 Landing Craft Vehicle Personnel (LCVP) and 172 Landing Craft Mechanized (LCM) were allocated to the SWPA. The majority of these were shipped to Australia in parts and a construction plant was built in far north Queensland. Australian engineers did design and build a number of British Landing Craft Assault (LCA) in 1942, but this design was deemed inferior to the U.S. craft and they were mainly used in training. The cancellation of further operations in Europe in 1942 provided the SWPA "another windfall" with seventy-six Landing Ship Tanks, thirty-seven Landing Craft Infantry, and seventy-two Landing Craft Tanks being transferred to the SWPA in December 1942.

[50]AWM, 54 422/7/8, p. 7, Combined Training School, Toorbul Point: Origin of the School.

ence allowed for much more imaginative training for the units rotating through the battalion and company exercises. They also undertook their own specialized course which involved water and land movement by day and night, camouflage and concealment, and air resupply by parachute.[51]

Cooperation between the services, as well as the two national armies, was also a prerequisite to success. The quick establishment of the schools had only been possible through cooperation with the RAN. On 18 July, Lieutenant J. M. Band, who was on his way from HMAS *Hobart* to command of a new corvette, was seconded to command the naval elements of the CTS. He was instrumental in overseeing the development of the naval facilities and the establishment of crews and craft for training.[52]

Lieutenant Band labored under the problem of not only proving craft and crews for the Army, but also having to train his own men from scratch. Band struggled, however, to keep up with the demands of his command, especially as initially he had so few men and resources at hand. This changed, however, after the first visit of the naval officer in charge of the Brisbane area, Captain Thomas, RN. He visited CTS on 17 September and after his inspection remarkable progress was achieved. Thereafter, Band's naval personnel rose from forty to eighty and after the Chief of the RAN, Admiral Sir Guy Royle, visited on 30 November 1942, naval support became "almost prodigal." By December, the naval personnel at CTS peaked at 300, including eighteen officers. In the same month, two officers and forty men from the U.S. Navy arrived to help train the RAN members on the new landing craft. Thereafter, U.S. Navy numbers greatly increased when the U.S. 7th Amphibious Force took over the base in May 1943.[53]

From August 1942 through March 1943, CTS trained all three battalions of the 162nd, 163rd, and 186th U.S. Infantry Regiments, along with the infantry battalions of the 25th Brigade (7th Australian Division), 2/3rd and 2/8th Independent Companys, 2nd and 4th Field Regiments, 2/4th Armoured Regiment, and the 22nd and 57th/60th battalions. The majority of these troops being infantry until the arrival of the LCVP, LCM, and LST craft, which allowed armor and artillery units to be trained. This continued in the transition to 7th Amphibious Force control with the 113th Light AA Regiment, 6th, 9th, and 13th Armoured Regiments, and the 1st Tank battalion all receiving training during early 1943.[54]

---

[51] Ibid., p. 10.

[52] AWM, 54 943/16/1, p. 3, A Summary of Combined Operations Training in Australia (Amphibious) 1942–1945.

[53] Ibid, p. 5.

[54] Ibid, pp. 4–5.

The Joint Overseas Operational Training School

While the CTS at Toorbul Point in Queensland had been established by First Australian Army specifically for the training of units in amphibious operations, JOOTS at Port Stephens on the New South Wales mid-north coast had been established as an officer training school and was seen as the senior of the two organizations. JOOTS' locality was also to be the base area for the RAN's training school for boat handling in amphibious warfare, the aptly named HMAS *Assault,* under Commander F. N. Cook. The first students graduated from JOOTS on 25 September 1942.

Initially, JOOTS was heavily theoretical and designed to run courses to train commanders and their staffs rather than units. These officers at division, brigade, and battalion level would then return to their formations and units to pass on relevant information and establish amphibious warfare training programs; others would be posted to the CTS at Toorbul Point. The Deputy Chief of the General Staff, Major General Frank Berryman, noted that the course would "Standardise the training of the Allied forces in Combined Operations and train selected representatives from each of the services in order that they in turn may form a nucleus of trained instructors for further training of their formations."[55]

The staff at JOOTS consisted of: one U.S. Commander (initially Jones, then Van Volkenburgh); eight U.S. Army instructors; seven Australian Army instructors (including two British officers: Lieutenant Colonel Hope and Lieutenant Colonel Walker, RM); two U.S. Army Air Corps Instructors; two RAAF Instructors; and one RAN Instructor.[56]

In addition, a considerable number of support units were made available to the school. These included 3rd Battalion, 128th U.S. Infantry Regiment; 2nd Battalion, 129th U.S. infantry Regiment; 19th Battalion AMF; and the 2/11th Field Company. These units were used for demonstrations in the different courses. The Royal Australian Engineers also maintained a detachment under Major Gilmore to undertake experimental developments in landing operations.

The senior course at JOOTS was the Staff Officers Course run over twenty days for twelve senior officers. It used a combination of lectures, staff studies, demonstrations, and exercises on: functions of Navy, Army, and Air Force in combined operations; staff duties, including planning and preparation of tactical and administrative orders; signal and communications and procedures; beach organization; and supply and deployment.

The second course was a much larger unit officer's course normally attended by approximately fifty officers from the rank of captain to lieutenant colonel. This course covered: cooperation between the three services in combined operations; practical

---

[55]NAA, MP508/1, 323/701/804, DCGS to First Australian Army, "Training in Combined Ops," 26 September 1942.

[56]Ibid.

demonstrations of a combined operations landing by a battalion; theory of planning and preparation; fire support; and intercommunication and administration.[57]

In addition to these two major courses, JOOTS was quickly expanded to run a range of specialist programs for all ranks consisting of courses on communications, navigation, reconnaissance, logistics, and shore parties.

As with the CTS at Toorbul Point, the worldwide shortage of equipment meant that the early training at JOOTS was completed with the bare minimum of amphibious craft. This meant that troop training was largely restricted to small boat operations. Improvization was the major feature up until this time, with ship's cargo nets hung over cliffs and the construction of simulated ramps and ships in order to undertake realistic training.[58]

Despite these limitations, the operations of JOOTS were expanded in November 1942 when it was realized that Toorbul Point could not handle the volume of training required for the formations based in Australia. As such, JOOTS took on running an initial orientation course for combat units that was based on five key areas: 1) An outline of amphibious operations concepts and doctrine; 2) Organization; 3) Mission; 4) Planning; and 5) Training and Rehearsals.

This orientation course was capped off with a joint operations exercise in order to practice and assess the newly-taught skills. The major instructors of each of these courses and their various components were a combination of U.S. and Australian personnel. During the rehearsal and exercise phase, regimental officers from the combat units were paired with naval commanders down through the ranks. After the rehearsals and training exercises, a debriefing and critique was provided by both army and navy officers—a system which was deemed "very effective."[59]

The completion of this program was not the end of the unit's involvement with the school. In addition to developing their own training programs, mainly based around dry-shod (away from the sea) training, units were often required to attend one of the training establishments for advanced training, especially when they were preparing for or refitting after a major operation, returning from a tasking in a different operational setting, or after the unit had absorbed large numbers of replacements due to combat losses or the transfer of personnel. DCGS, Major General Frank Berryman, noted that by 1943 the Army's policy on combined (amphibious) operations laid down that in order to achieve a high degree of proficiency and readiness and to "meet and overcome difficulties in advance," units needed continuous and ongoing training.[60]

---

[57]NARA, RG 496, Box 667, Directive—Joint Overseas Operations Training School, GHQ SWPA, 9 August 1942.

[58]Barbey, *MacArthur's Amphibious Navy,* p. 35.

[59]NHHC, Barbey Papers, Correspondence Box 2, Jamison to Barbey, December 1960.

[60]AWM, Papers of Lieutenant-General Sir Frank Horton Berryman, PR84/370, Item 40, Combined Operations (General).

HMAS *Assault*

Commander Cook, coauthor of the Hope-Cook report and a highly experienced combined operations naval officer, was originally slated to be posted to JOOTS. However, the RAN decided that in order to meet its needs in providing crews for amphibious craft for combined operations, a naval training center had to be established and Cook, the RAN's most experienced officer in this field, was given command. In order to facilitate cooperation with the Army, HMAS *Assault*[61] was co-located with JOOTS.[62] Cook's biggest initial obstacle was the construction of the base which required piers, refueling points, accommodation, slipways, and maintenance facilities. Cook was supported in this endeavor by the allocation of one the RAN's three Landing Ships Infantry, HMAS *Westralia,* to provide accommodation while the base facilities were established, and to support training.[63] In addition, HMAS *Ping Woi* was provided as a tender ship. Once these facilities were sorted, the next biggest obstacle was the acquisition of suitable craft. Initially, like the CTS, HMAS *Assault* had to make do with a small number of landing craft, but it was eventually to build up its strength to be the base for the three Australian LSIs (*Manoora, Westralia,* and *Kanimbla*); one U.S. Amphibious Attack Transport (AKA), USS *Algorab;* one LST; twelve LCI; four LCT; seven LCM; two APC; sixty-seven LCVP; nine ALCA; three LCP(R); and seven LCS. These vessels were crewed by RAN personnel and the U.S. Navy's Landing Force Equipment Depot, which was also established on the site in March 1943.[64]

HMAS *Assault* was responsible for the training of the landing craft crews for the three Australian LSIs as well as the crews for the amphibious craft at CTS at Toorbul Point. The school was steadily increased in size and functions. After the initial amphibious operations in 1943, it was realized that beach assault groups needed to be formed and HMAS *Assault* became the training ground for the Navy Beach Commandos that formed a core component of the two Beach Groups that were raised. In the end, a total of 100 officers, 100 coxswains, 120 Navy Beach Commandos, forty signalmen, 453 boat crews, and 250 stokers were trained at the facility.[65]

---

[61] As in the British tradition, shore bases in the RAN are referred to as per a naval vessel. For example, HMAS *Assault* was a shore-based establishment.

[62] NHHC, Barbey Papers, Correspondence, Box 2, Landenberger to Barbey, 26 April 1962.

[63] This role meant that HMAS *Westralia* underwent modifications to accommodate the additional sailors and officers and her ability to embark troops for amphibious operations was reduced.

[64] AWM, 54 943/16/1, p. 8, A Summary of Combined Operations Training in Australia (Amphibious) 1942–1945.

[65] Ibid.

When 7th AF took over amphibious training in February 1943, control of the three Australian LSIs passed to Admiral Barbey, and the training activities of HMAS *Assault* were integrated with the new Amphibious Training Centre (ATC). However, as there was a requirement for ongoing training of RAN crews, HMAS *Assault* continued to function within the ATC command structure. Thus, when on 1 October 1943 Barbey ordered the facilities at Port Stephens shut down and moved north to Toorbul Point, HMAS *Assault* remained in place and took on the role of a reserve pool depot for trained personnel.[66] But by 1944, with the provision of landing craft crews complete, the reduced role of the Australian Military Forces in the SWPA, and the U.S. Navy's vastly increased role in amphibious warfare, it was decided that the Beach Commandos would be attached to the Australian Army training facility in Cairns and HMAS *Assault* wound up its activities in July 1944 and was transferred to the Royal Navy in 1945.[67] By the end of its life, HMAS *Assault* has assisted in the training of 2,000 Australian and 20,000 U.S. soldiers in addition to the officers, crew, and mechanics that it had trained for service with the Australian landing ships and amphibious transport squadrons.[68]

7th Amphibious Force and the U.S. Navy

This increased U.S. Navy involvement was predicated on the 7th Amphibious Force's control of amphibious warfare training, which occurred in February 1943 with the arrival of Admiral Barbey. With the raising of his command flag, it was decided by GHQ that all amphibious warfare training would come under his direction. Thus on 1 March 1943 JOOTS, HMAS *Assault*, and the U.S. Advanced Base Unit at Port Stephens were all combined to form the Amphibious Training Centre (ATC).[69] This transition process was relatively smooth as almost all of the officers posted to CTS and JOOTS were retained in the new ATC formation in 1943. In May the CTS at Toorbul Point also came under control of the ATC at Port Stephens. The ATC standardized training of units and formations. JOOTS, and later the ATC, ran a number of specialist officer training courses. These included courses designed to train division-, regimental-, and battalion-level staff officers. In addition, other more specialized courses on areas such as intelligence, operations, logistics, engineering, and medical problems were also provided. These officer training courses were scheduled so that at their completion the units from

[66]The RAN did not wish to forgo this base given its heavy investment in the facilities at Port Stephens. It also had in mind the post-war requirements of the RAN in amphibious warfare.

[67]Mallett, "Together Again for the First Time," p. 123.

[68]AWM, PR 00631, Papers of HMAS *Assault,* RAN's part in Combined Operations Training in Australia.

[69]*Seventh Amphibious Force: Command History, 10 January 1943–23 December 1945* (N.p.: United States Navy, 1945), p. II-11.

which the officers on course had been drawn arrived at the ATC to undertake group training. This enabled these units to leverage these newly-acquired skills in their unit-level amphibious training exercises.[70]

One of the most pressing problems in 1943–44 was the location of these different schools and centers. The original Hope-Cook report had stated that Port Stephens was an ideal location given its protected harbor, access to suitable beaches, as well as troop training grounds. Toorbul Point was not viewed in the same regard, but was seen as the next best option given the absence of suitable locations identified along Australia's eastern seaboard. However, this did not solve the problem that the vast majority of the Allied troops in Australia were based farther north and that considerable time and expense were being taken up moving between training grounds. This had prompted GHQ to instigate another review of suitable training grounds in late 1942.[71] The board reported in 4 November 1942, but it struggled to find any of the areas in northern Queensland suitable.[72] However, a decision was made to establish an ATC in Cairns and this center started operations on 25 June 1943 under Captain P. A. Stevens, USN. This was the location of the training ground for the U.S. Army's 2nd Engineering Special Brigade (2ESB), which was also incorporated into the 7th AF at this time.[73] The Cairns ATC was to become the major training ground to Australian formations from mid-1943 until the end of the war.

A few months later it was decided that the Port Stephens facility was to close in October 1943 and move to Toorbul Point, and from there it was eventually moved to Milne Bay in New Guinea. On 5 February 1944 the personnel, equipment, and records from Toorbul were moved north to establish the Milne Bay ATC in Papua, the remaining facilities being handed back to the Australian Army. In March 1945, GHQ directed that the ATC at Milne Bay be moved to Subic Bay in the Philippines.[74]

Mobile Training Units

The realization in late 1942 that the Port Stephens locality would have to be moved farther north was, however, never going to solve all of the amphibious training problems in the SWPA. As the front line moved, so did the troops requiring training, and by 1943 a considerable number of formations were concentrated in

[70]Ibid., p. II-13.

[71]NARA, RG 496, Box 667, Selection of Alternative Site for Joint Overseas Operations Training School, 24 October 1942.

[72]NARA, RG 496, Records of GHQ, SWPA, Box 667, Captain M. R. Kelly USN to MacArthur, RE: Selection of Alternative Site for Joint Overseas Operations Training School, 4 November 1942.

[73]*Seventh Amphibious Force*, pp. II-13–14.

[74]AWM, 54 943/16/1, A Summary of Combined Operations Training in Australia (Amphibious) 1942–1945.

the rear combat areas in New Guinea and the Solomon Islands. In order to alleviate this problem, in December 1942 GHQ instigated a review of training requirements and processes and raised the possibility of the development of mobile training teams.[75] Soon after, the JOOTS/ATC at Port Stephens was charged with the training and development of these teams. Based on the original one-third/two-thirds split of responsibility between the Australian and U.S. forces, three mobile training units, one under the command of an Australian Army officer and two under the command of U.S. Army officers, were established.

Initially, it proved very difficult to get suitably qualified and experienced officers to man these teams, with parent units and formations, including the ATCs at Port Stephens and Toorbul Point, reluctant to part with their best officers. However, pressure from GHQ's operations chief, Major General Chamberlin, and then the C-in-C, MacArthur, meant that highly qualified personnel were eventually forthcoming.[76] These units, a highly integrated mix of U.S. and Australian army and navy personnel, trained troops based on availability and schedule, not nationality. For instance, the 3rd Mobile Training Unit under Major C. L. Woodcliff (U.S. Army) trained the 6th, 7th, and 9th Australian Divisions in Cairns during 1944, while the 1st Mobile Training Unit under Lieutenant Colonel C. T. Barton (AIF), trained the 33rd, 43rd, and 40th U.S. Infantry Divisions, as well as the 112th U.S. Cavalry Regiment and 6th U.S. Ranger Battalion in New Guinea and New Britain in 1943–1944.[77]

U.S.-Australian Amphibious Training Cooperation

One of the major keys to the effectiveness of JOOTS, the CTS, and then the ATCs was that they had developed from a largely army-controlled operation to integrated joint services training schools. They included RAN and USN ships and amphibious crews and craft and the demonstration troops. The RAAF contributed by providing meteorologists as well as a RAAF photography team, which took pictures of the training program, demonstrations, and exercises to produce training slides. In addition, RAAF air liaison officers, including personnel and aircraft from the 32nd Squadron and 5th Army Cooperation Squadron, were brought in to provide theoretical and practical training on the use of air power in amphibious operations. The only service from which cooperation was not secured was from the U.S. Army Air Corps.[78] Although they provided two officers to the JOOTS/ATC staff, they argued that scarce resources and other priorities precluded their involvement

[75]NARA, RG 496, Box 667, Memo RE Amphibian Training, GHQ 4 December 1942.

[76]See NARA, RG 496, Box 667, correspondence on the formation of the Mobile Training Teams, 25 December 1942–15 January 1943, GHQ SWPA.

[77]NHHC, Barbey Papers, Correspondence Box 2, Jamison to Barbey, 14 November 1961.

[78]See NARA, RG 496, Box 667, Wellington to Chamberlin, RE: Joint Overseas Operational Training School, 19 August 1942.

beyond this level. This situation was to lead to difficulties in cooperation between this service and the U.S. and Australian Army and Navy planners and formations in operations throughout 1943–44.

One of the most significant factors was the high level of cooperation not just between the three services, but between the United States and Australia. While initial relations had been strained, culminating in the relief of Colonel Jones and Lieutenant Colonel Hope, from this period on the two nationalities had been especially cooperative. Commander of Allied Naval Forces, Admiral Carpenter, noted that "the Australians delivered all that they could," and while there were natural differences of opinion on matters of policy which were sometimes heated, Australian cooperation was always wholehearted. Carpenter noted his particularly good relations with the Australian Army, especially the 9th Australian Division which undertook the amphibious assaults in New Guinea in 1943.[79]

Admiral Barbey and the CO of JOOTS/ATC, Commander Jamison, were particularly complimentary about working with their Australian counterparts, as were the Australian personnel. Captain J. Louis Landenberger, USN, one of Barbey's key staff officers who was also "very closely associated with the training operations at Port Stephens, Toorbul Bay and other places," noted that the "comradeship between the Australians and the 7th Amphibious Force was most cordial at all times," adding that "I am convinced that this period of training [JOOTS/ATC] was well worth while."[80]

Australians in Action: The Amphibious Assault on Lae

By early 1943, MacArthur and his subordinate headquarters had begun planning for offensive operations in the SWPA. The bloody battles in Papua had meant that the key Australian and U.S. strike formations, the 6th and 7th Australian Divisions and the U.S. 32nd Division, had to be withdrawn from Papua and allowed to absorb replacements, and be provided with time to retrain. For the Australians, this included the 9th Division that had recently returned from the Middle East, while significant reinforcements in the form of the U.S. 1st Calvary Division and 24th Infantry Division were just arriving in theater and also needed time to acclimatize and complete their training.

For the Australians, this period was used to reorganize the force structure of the Army's key divisions to meet the challenges of jungle and amphibious warfare.[81] This included the adoption of the "Jungle Division" structure. The AIF

---

[79]NAA, Naval Intelligence Division, B6121 3R, p. 4, Interview with Admiral Carpenter, July 1945.

[80]NHHC, Barbey Papers, Correspondence Box 3, Landenberger to Barbey, 24 January 1961.

[81]John Moremon, "No 'black magic,'" in Peter Dennis and Jeffrey Grey, eds., *The Foundations of Victory: The Pacific War 1943–1944* (Canberra: Australian Army History Unit, 2003), pp. 70–80.

infantry divisions, along with the 3rd, 5th, and 11th Militia Divisions, were modified to this new structure, which removed the division's cavalry regiment along with HQ Royal Artillery, two field regiments, the anti-tank regiment, and the survey battery. Supply, transport, and medical services were also reduced. The brigade's battalions were similarly affected, and overall, a jungle division had approximately 4,000 fewer soldiers than a standard British infantry division.[82]

In addition to these reforms, on 15 March 1943 Lieutenant General Leslie Morshead, General-Officer-Commanding (GOC) I Australian Corps, which included all the AIF divisions, oversaw the transition to the jungle warfare establishment and instigated a corresponding training program. This program was augmented on 6 May by the U.S. Army's 2nd Engineering Special Brigade (ESB).[83] After initial discussions between GHQ, Australian land HQ, Admiral Barbey, and the ESB's commander,[84] Morshead and his troops worked out the details of a two-week training program for each brigade in the 9th Division starting 16 June.[85]

The 2ESB had been formed in Massachusetts from a few regular Army engineering officers and with men made up of a motley collection of yachtsman, small boat engineers, garage mechanics, and volunteers. They had lots of "enthusiasm and little military know-how." The brigade consisted of three Boat and Shore Regiments made up of one boat and one shore battalion, which operated the small Landing Craft Personnel Ramp (LCP-(R)); Landing Craft Vehicle Personnel (LCVP); Landing Craft Mechanized (LCM); and the 2½-ton amphibious truck known as the DUKW. In addition, the brigade included a maintenance battalion; a medical battalion; and signals, maintenance, headquarters, and transport companies. These brigades were originally designed for ship-to-shore movement after the Navy had transported Army units into an operational area, and also for shore-to-shore operations for short-range maneuver.[86]

The 2ESB started to move to Australia in February 1943 and formed up in Cairns and Rockhampton. The major problem was that there were no landing craft awaiting their arrival. An assembly plant in Cairns had to be built and the material for the landing craft, as well as the appropriate tools, had to be found amongst the

---

[82] Albert Palazzo, "Organising for Jungle Warfare," in Dennis and Grey, eds., *The Foundations of Victory*, pp. 183–86.

[83] This was confirmed in an order from GHQ to Commander Allied Land Forces, 15 May 1943. AWM, 54 389/3/10, Operation Cartwheel—Correspondence—Messages and Plans.

[84] AWM, 54 213/3/20, Conference Notes—Cartwheel Postern, Planning, May–July 1943, Berryman, Points for discussion with G3 GHQ—Commander Engineer Amphibian Brigade—Commander Amphibious Force 19 May 1943.

[85] AWM, 54 589/3/11, Part 1, Correspondence, Memos and Operation Instructions for Operation "Cartwheel," April–December 1943, Berryman to Kruger, 6 May 1943.

[86] William F. Heavey, *Down Ramp! The Story of the Army Amphibian Engineers* (Landisville, PA: Coachwhip, 2010), p. 68.

supply ships that had split up and docked all over Australia. It was not until 9 April 1943 that the first LCVP was launched; however, by June 1943 more than 500 landing craft per month were being produced. Soon after its establishment, the plant in Cairns was able to deliver seven armor-plated landing craft per day: enough for training the Australian troops on the Atherton Tableland; to send a number of LCMs to New Guinea to operate supply runs; and to prepare for preliminary landings in support of the Australian diversionary operation before the capture of Lae.[87]

Blamey's concept of operations was to use the town of Salamaua, to the south of Lae, as a magnet to draw the Japanese forces away from the town before he launched a two-division assault. The 9th Division would launch an amphibious landing; and the 7th Division, supported by a U.S. parachute regiment, would capture an abandoned Japanese airfield; with the Australians to be air-landed. These two assaults would then converge by opposite axes on Lae. The amphibious assault would drive on the Japanese base in Lae while the 7th Division's assault would put a large Australian force across the Japanese lines of communication to Madang. Once Lae was secure, these divisions could then advance up the Markham Valley and along the coast to allow 5th AF to establish airfields.[88]

With limited fire support available from the Navy, and no carrier-based aircraft to support these operations, the planning directed that "9th Division with the U.S. 2nd Engineer Amphibious Brigade [ESB] will conduct a shore to shore operation" in an area that will provide for an "unopposed landing . . . outside the range of field artillery in [Lae]." 9th Division was to concentrate at Milne Bay with the final assault being undertaken "as close to the landing as possible, but NOT more than 60 miles" from the objective—the range of the ESB's landing craft.[89]

One of the major limiting factors of this plan was that the 9th Division had only just returned from the Middle East and had not done any training at either Port Stephens or Toorbul Point. Priority, therefore, was to get this division together with the 2ESB as soon as possible. In order to coordinate these efforts, a training team from the ESB began working with Lieutenant Colonel T. K. Walker, RM, who had recently been transferred from JOOTS to be GS01, Combined Operations at I Australian Corps HQ. As soon as the details were ironed out, the 9th Division units undertook dry-shod training, and this was followed by an interchange of personnel from the 2ESB with the 9th Division down to brigade level. Training began with the 24th Brigade AIF under Brigadier Evans. His battalions did a day and night landing followed by three brigade-level exercises, each after a twelve mile close to shore ocean movement. After nineteen days of training, the troops proved efficient at using the craft;

---

[87] Ibid., pp. 67–72.

[88] Dean, *Architect of Victory,* p. 170.

[89] AWM, 54 589/3/7, Planning "Postern" Reference Advanced LHQ, letter to GOC New Guinea Force, Initial inquiries "G" factors influencing replies and sequence planning, Berryman to Herring, Future Operations—New Guinea, 17 May 1943.

however, nothing had been learned in the areas of supply and maintenance issues, some of the most important and complicated areas in amphibious operations.

Brigadier Windeyer's 20th Brigade joined the 2ESB next, but its training was greatly curtailed as the 2ESB's commander noted that to continue the training regime of the 24th Brigade with two more brigades would mean that his craft and crews would not be serviceable for operations. This meant that 20th Brigade received only ten days training, while the last of the brigades, the 26th under Brigadier Whitehead, only received seven days training in Cairns with no time for a brigade-level exercise. In August a single LCI arrived in the area and all troops from the 24th and 26th brigades received training in loading and landing practices from this vessel.

While this training was underway, the plan for the 9th Division's landing had been the subject of a staff exercise in Brisbane on 16–17 June by I Corps HQ with the GOCs of 6th, 7th, and 9th Divisions, Brigadier W. F. Heavy (CO 2ESB); Captain Roy C. Hudson (U.S. Navy liaison officer at GHQ); and Lieutenant Colonel A. L. Rose, AIF, also formerly of JOOTS and now the Australian Army liaison officer to the ESB. At this conference it was realized that the nature of the operations was beyond the capabilities of the 2ESB. As such, GHQ directed that the 2ESB come under the command of the 7th Amphibious Force and that Admiral Barbey and his staff assume responsibility for both training and the operation. Barbey's HQ analyzed the plan, and agreeing with the staff study, proposed that the 9th Division be moved by the 7th AF's ships and craft and that the 2ESB be used for logistical maintenance of the division once ashore. The plan called for two of the 9th Division's brigades to be landed on D-Day and the third on D+1.[90]

This meant that besides the limited training undertaken at Cairns, the most obvious problem was the fact that a significant portion of the troops in the 9th Division would not be arriving at the Lae beachhead from the LVCPs of the 2ESB, but rather from APDs, LCP(R)s, and LCIs from the 7th AF. Training with this unit could only be undertaken after the troops had been moved from Cairns to their staging point at Milne Bay in Papua, just prior to the operation commencing.

The planning for the operations was undertaken at the 9th Division's HQ in Milne Bay with the staff of the 7th AF, representatives from the 7th Fleet, and the Allied Air Forces. Barbey had twelve new officers, recently arrived from the U.S., to form the core of his planning staff, and this was supplemented by a number of Australians under Brigadier Ronald Hopkins. While planning was underway, the 9th Division familiarized themselves with the ships and crew of the 7th AF, but training was limited with "only two weeks allocated to assemble the ships, distribute plans, hold rehearsals and load troops and supplies for the operations."[91]

---

[90]AWM, 589/3/11, Amphibious Operation Postern: Outline Plan, 14 July 1943, and Amphibious Force Seventh Fleet; Amphibious Operations for Operation II, Memos and Orders—Cartwheel—April–September 1943.

[91]Barbey, *MacArthur's Amphibious Navy*, p. 74.

For the assault, the Australians were to be landed in waves with the first echelon of 560 troops landed from APDs; the second from eighteen LCIs, carrying some 3,780 troops; and the third from the 2ESB to organize the beachhead. Given the pauperdom that the SWPA found itself in during 1943, the assault was to be carried out without the benefit of air cover, no prolonged naval bombardment, and from a "hodgepodge of ships" that comprised the 7th AF.[92] The lack of air cover meant that Barbey was unwilling to risk his three large Australian LSIs in the operation while his only U.S. AKA was not in a fit condition to take part. Thus the troops were to be transported in the APDs, LST, and LCTs. Barbey considered the plan more like a "commando raid, except that it would be daylight and the troops that would go ashore would stay there."[93] When the operation was complete, some 17,000 troops and 12,000 tons of supplies, carried in some 156 ships, had been landed.

The landing on 4 September was "unopposed" and Australian Deputy Chief of the General Staff, Major General F. H. Berryman, recorded in his diary that it "all went according to plan . . . a vindication of [the Australian] C-in-C [Blamey] and my judgement in adhering to [the] plan to bypass Salamaua." The following day the paratroopers of the 503rd PIR and the soldiers of the 7th Australian Division opened their assault on Nadzab against "practically no opposition."[94] At the landing site for the 9th Division, five destroyers from the 7th Fleet laid down a short bombardment while the APDs lowered their boats to land the first wave. The arrival of the second wave in eighteen LCIs coincided with the appearance of three Japanese bombers escorted by six fighters. The attack rendered two LCIs inoperable, killing seven Australians and leaving another twenty-eight wounded.[95] This attack, however, did not halt the landing or its schedule. After four hours, more than 8,000 men and 1,500 tons of supplies were landed. A second air raid would cripple two LSTs later in the day, but again this did not disrupt the landing timetable and over the next few days the 9th Division undertook the sixteen-mile advance on Lae against steadily increasing Japanese opposition.

Yet within a week of the landing, the Australian high command had concluded that the Japanese opposition was "less than we formerly expected." With the bulk of their troops committed to the defense of Salamaua farther south, and with Lae under pressure from two directions, coupled with Allied control of the sea and air superiority, the commander of the Japanese defenses, Lieutenant General Hidemitsu Nakano, decided to abandon Lae. This was confirmed on 14 September when the Australians captured a copy of the Japanese order to withdraw and by 16 September the town had been occupied by elements of the 9th and 7th Divisions.[96]

---

[92] Ibid.

[93] Ibid., pp. 74–77.

[94] AWM, PR89/370, Berryman, diary, 4–5 September 1943.

[95] AWM, 54 589/7/27, Report on the Naval Aspects of the Lae Operation.

[96] AWM, PR89/370, Berryman, diary, 6–16 September 1943.

Lessons from Lae

The landing of the 9th Division outside Lae in September 1943 had been the first time that the Australians had undertaken a major amphibious operation since Gallipoli in 1915. This divisional-size landing may, in many respects, pale in comparison to many other amphibious operations carried out during World War II, but as Barbey noted, "the landing at Lae was not likely to make much of a dent on history. But in the Southwest Pacific it was a matter of great importance. . . . Failure or success would determine the pattern of MacArthur's forces."[97]

The landing had gone well and the assault troops had performed admirably, but it was undertaken against no opposition save for a small number of Japanese air raids. It was only after the first waves had gotten ashore that the limitations of the Australian's preparations and training became evident. One of the greatest difficulties in amphibious operations is logistics and the organization of the beach area. Soon after the landing got underway it became apparent that the one shore battalion from the 2ESB was not sufficient for the operation. This unit was designed to support one brigade, not a whole division, and the 9th Division had completely miscalculated their logistical requirements in the beachhead. The 9th Division's lack of experience in amphibious operations had only been made worse by their limited preparations and training in Cairns and Milne Bay, which had concentrated on the landing of troops as opposed to supply and logistics arrangements.

The 9th Division's miscalculation had been so great that they had even rejected an offer for the 2/1st Ordnance Beach Detachment to be allocated to the division.[98] The lack of troops for unloading the landing craft and organizing supply dumps meant that the beaches became confused.[99] Troops from a pioneer battalion and two infantry battalions had to be diverted to help with the unloading of ships, but this meant that stores and equipment simply piled up on the beaches and blocked exits. Supplies remained unsegregated, with fuel and ammunition dumps located in extremely close proximity, causing a major concern in relation to the sporadic, but at times effective, Japanese air raids on the beachhead.[100] The problem with unloading ships and the Japanese air raids had a cascading effect on Army-Navy relations, with the Army demanding that ships and craft remain on the beach until unloaded, but the Navy demanding that they leave in order to run the gauntlet of Japanese air raids in greater safety. This dispute simply led the Navy to load fewer supplies in each ship.

---

[97] Barbey, *MacArthur's Amphibious Navy*, p. 74.

[98] Mallett, "Together Again for the First Time," p. 77.

[99] Ross Mallett, "Logistics in the South-West Pacific 1943–1944," in Stevens and Reeve, *Sea Power Ashore and in the Air*, p. 113.

[100] Ibid.

The operation at Lae revealed a number of critical shortfalls: principally, the need for a beach master with overall authority in the landing zone; the requirements for improvements in training, especially for non-combat forces in amphibious landings; modifications to standing doctrine; improved command and control arrangements between the navy and army; and better cooperation with Allied air forces. Significant improvements in these areas would be critical, especially if future landings were to be opposed by the Japanese.

Balikpapan

While Lae in 1943 was the first large-scale Australian amphibious operation, Balikpapan in 1945[101] was to be the last Australian amphibious landing of the war and the last major action of the Allies in World War II. Like Lae, this landing was to be a divisional-size operation undertaken this time by the 7th Australian Division, but the difference in scale, preparation, and execution was to demonstrate just how far the Australians had come in terms of mastering the amphibious craft.

By 1945, the Allied forces in the region around Borneo had acquired air and sea supremacy. These factors, along with the combat and logistics capacities of the U.S. and Australian forces, meant that unlike the landing at Lae, Balikpapan would be provided with overwhelming support. This meant that the planning staff at I Australian Corps and 7th Australian Division headquarters was able to consider a wider range of options for the invasion. Given the overwhelming fire support, the availability of armored landing vehicles tracked (LVT), and the requirement from GHQ that the "advance inland be made without delay" and that prompt steps be taken to "secure the airfield . . . and seize the port area," the planners decided that a full-scale assault into the Japanese defenses was more viable than landing in a more remote spot and fighting through the jungle to the objective.[102]

The confidence to conduct a direct assault into the Japanese defenses was also born of the fact that the 7th Division had twelve months to prepare for amphibious operations, was combat proven, and was exceptionally well trained and led. It was also provided with a significant number of support troops, which meant that the division would go into action with some 33,446 troops under command, including

---

[101]The AIF conducted no major operations between April 1944 and the Oboe landings in Borneo in 1945. Originally, the AIF had been slated to participate in the operations to liberate the Philippines, but for a variety of political and strategic reasons they remained as MacArthur's only major reserve for the landing on Luzon. With the rapid progress of U.S. forces on Luzon, the I Australian Corps was not required to undertake operations and was allocated to the Oboe landings, which had been originally slated for U.S. forces. See David M. Horner, *High Command: Australia and Allied Strategy, 1939–1945* (Sydney: Allen & Unwin, 1982), pp. 382–95.

[102]MMA, RG-3 Records of Headquarters, Southwest Pacific Area (SWPA), 1942–1945, Box 158, G-3 operations, Oboe II, Staff Study, Operation Oboe-Two, GHQ, 21 March 1945.

more than 2,000 U.S. and NEI troops. This included a full complement of artillery support, a full nine infantry battalions plus a commando regiment, an armored regiment, a pioneer battalion, an amphibious tractor battalion, and a machine gun battalion. Of most significance was the allocation of the two Australian Beach Groups, which consisted of two Royal Australian Navy Beach Commandos, a pioneer battalion, a transport company, an armored salvage unit, a workshop, and an engineering field company. Additional support included no less than forty-eight corps units or detachments that were temporarily allocated to the division for the assault.[103]

Fire support came in the form of the entire 1st Australian Tactical Air Force, which included four B-24 heavy bomber squadrons, as well as support from all the heavy bombers in the 13th AF, and a wing from the 5th AF. Additional fighter cover for the landing was to be provided by a division of U.S. Navy escort carriers. The 6,000 Japanese defenders, with their eighteen coastal defense guns and 104 heavy and medium anti-aircraft guns, were pounded for twenty days prior to the landing. The naval bombardment started sixteen days before the landing during the minesweeping operations. During this period, 3,000 tons of bombs, 7,361 rockets, 38,052 shells from 3–8 inch, and 114,000 rounds of automatic fire from five cruisers and eleven destroyers hit the Japanese defenses. This support was to continue throughout the operations with 25-pounder artillery fire providing 41,800 rounds of support in the first ten days. The one brigade commander noted that the operation was a "lesson on the use of firepower," reflecting on the fact that on the "Kokoda Trail the division had no artillery . . . [at] Gona . . . we received only 250 rounds . . . [of] 25-pounder [artillery] . . . for the final assault," and in the Markham valley, "one tank would have saved 200 lives if only we had it."[104]

The first two waves landed on 1 July 1945 in LVTs at 08.55. The first battalion met little opposition killing thirty-one Japanese for the loss of only one killed in action (KIA) and eight wounded in action (WIA). By the end of the day, the Australians had advanced 2,000 yards for twenty-two KIA and seventy-four WIA. Later that night, the Japanese counterattack was smashed with more than 300 Japanese dead being counted the next morning. Despite the success of this initial action, Japanese resistance was to continue until the end of the war.[105] The tragedy for the Australian and Allied servicemen killed and wounded during this operation was the fact that it had all been in vain.

MacArthur hoped to have crushed the Japanese in Java by 1 August 1945 so that he could announce to the U.S. Joints Chiefs that his task in the SWPA, set

[103]AWM, 52 1/5/14/74, War Diary 7th Australian Division, June 1945, Operation Orders, Oboe Two.

[104]Gavin Long, *The Final Campaigns: Australia in the War of 1939–1945*, Series 1—Army, Vol. VII (Canberra: Australian War Memorial, 1963), p. 531.

[105]Ibid., pp. 517–20.

down in 1942, was completed and that the command could be dissolved.[106] While these priorities have some legitimacy, MacArthur's other arguments for the Oboe landings, and especially Balikpapan, are more tenuous. With an effective naval blockade off Japan in place, and with U.S. forces having landed at Okinawa and Iwo Jima, Borneo was effectively by-passed. Arguments that the Borneo oil supplies could be secured for the Allied war effort were also disingenuous. Allied supply lines were secure; the war in Europe was winding down and it would take months if not years to make the Borneo oilfields operational. The claim that Brunei (Oboe Six) was needed as a base for the British Pacific Fleet was rejected by the British Admiralty. The rationale for Balikpapan was the most dubious of all the Oboe operations. The landing was only ever designed as a staging post for further operations against Java and, after the U.S. Joint Chiefs released their directive on 3 April 1945, MacArthur knew that operations against Java would not occur as the Allied war effort in the Pacific was now to concentrate on preparing for an invasion of Japan. The Australians were well aware of these limitations, and Blamey had argued against the landing to MacArthur and the Australian government, but in the end, MacArthur's will power triumphed over the opposition from Blamey, the U.S. Joint Chiefs, and the reservations of the Australian government.[107]

At the purely tactical level, however, the operation had been an outstanding success and had vindicated all of the lessons learned by the Australian and U.S. forces from the operations in 1943. The planning timetable used for the operation by the 7th Division was very similar to the one used in training. The naval and air bombardment went as planned, tanks and artillery were landed in the initial waves of the assault as planned, and the first day's objectives were all secured, as was the high ground over the beach area. The area of critical failure at Lae, logistics, had also been overcome. Five hundred and fifty-two vehicles and more than 900 tons of stores were landed on the first day, and it was noted that the bulk loading of specialist equipment, such as engineering stores and bridging equipment, worked exceptionally well. The only major criticisms of the amphibious elements of the operation by 7th Division HQ was that many of the attached units to the division had not received adequate training in amphibious operations; of particular note was the lack of loading tables for these units. The only other point of note from Corps and Division HQs was that the ever-present "friction" of war had caused a number of unforseen operational issues that had restricted the planning process for the operation.[108] In the end, while Lae had been a success based on limited, but

[106] AWM, PR89/370, Berryman, diary, 13 March 1945.

[107] Dean, *Architect of Victory,* pp. 219–20.

[108] See AWM, 52 1/5/14/87—October 1945, War Diary 7th Australian Infantry Division, October 1945, Report on operations, Oboe Two (original report completed September 1945); AWM, 52 1/4/1/74—1945, part 1, appendices, Report on Operations, Oboe Two and 1st Australian Corps Report on Operations Borneo Campaign, 15 September 1945.

adequate training, doctrine, and preparations utilizing experimentation and maneuver, Balikpapan had left nothing to chance with training, doctrine, support measures, and operational planning reaching the pinnacle of excellence for the U.S. and Australian forces involved.

Conclusion

Difficulties in Australia in 1942 with respect to amphibious warfare were hard to fathom. The Australian Army had no tradition or experience in this type of warfare, and preparation and organization had to be undertaken with coalition partners, which meant adapting to different doctrine, equipment, and techniques in a region during World War II that topped the Allies "list of 'have nots' and 'won't gets.'"[109] What eventuated in 1943 was a hybrid doctrine that mixed British and U.S. methods, and adapted itself to new equipment under a rather unique set of circumstances.

Problems in procuring equipment had severely hampered both training and operations in 1942–43. Almost inevitably, the first units and formations that went through the training schools set up in Australia were sent to the frontline where their amphibious training held no relevance and distracted from much-needed training in conventional operations and jungle warfare. Meanwhile, new units with little training and experience were thrust into the forefront of landings on hostile shores. The 9th Division's training in Cairns with the 2ESB did not prepare them well for Lae, especially in the area of logistics, and this lack of familiarity and experience caused major problems for the Australian landings in and around the Huon Peninsula in 1943.

The operations in Borneo in the last year of the war were to prove a very different experience for the Australians. The Allied superiority in logistics meant that they were able to provide the forces in the SWPA with more than adequate supplies, and while shipping and landing craft were always in short supply, their availability dwarfed that which had been used two years earlier. The improvements in training and organization, in particular the advent of the mobile training teams in the SWPA, proved highly effective, and, combined with the greatly reduced operational tempo for the Australians in 1944, this meant that there was an abundance of time available for training and preparations. Furthermore, the experiences and lessons of 1943 had led to the establishment of specialist units for the support of Australian amphibious landings, the most critical being the raising of two Beach Groups allocated to I Australian Corps HQ, which greatly improved and enhanced logistics arrangements.

---

[109]Samuel Eliot Morison, *Breaking the Bismarcks Barrier, 22 July 1942–1 May 1944: History of United States Naval Operations in World War II*, Vol. VI (Boston: Little, Brown, 1950), p. 32.

In 1945, the overwhelming Allied superiority on the battlefield allowed I Australian Corps to plan the assault on Borneo utilizing two amphibious forces, backed up by the resources of the First Australian Tactical Air Force, 5th and 15th U.S. Air Forces, the 7th Fleet, and carrier aviation. On board the amphibious ships were Australia's most experienced infantry divisions that had first seen action in the Middle East in 1941, and had been in action at Tobruk, El Alamein, the Kokoda Trail, the beachhead battles, Lae, the Markham Valley, and the Huon Peninsula.

By 1945, the Australians in the SWPA had been the beneficiaries of a long learning curve and had thus developed into highly proficient practitioners, if not masters, of the amphibious craft. The critical enabler to this success, however, was their cooperation and combined operations with U.S forces, without which the Australians would never have been able to undertake such assaults. However, all of this experience and knowledge was soon to be forgotten and by 1947, the last of the major amphibious forces in the Australian military had been dismantled; all that remained were a few obsolete landing craft and a single naval vessel able to take part in amphibious operations. Now, more than seventy years on from those fateful first steps in 1942, the modern Australian Defence Force is trying to relearn these valuable lessons from the past.

8

Naval Gunfire Support at Iwo Jima

The Perils of Doctrinal Myopia

Donald K. Mitchener

During the 1930s, the United States Navy and Marine Corps developed a doctrine for training amphibious troops and for planning an attack from the sea against heavily defended objectives. A key component of that doctrine consisted of guidelines for the use of naval gunfire in support of troops going ashore with a special emphasis on the softening up of enemy beach defenses over a period of time preceding a landing. Naval and Marine commanders applied this doctrine in each of the island invasions of the Central Pacific Campaign leading up to Operation Detachment, the invasion of Iwo Jima, on 19 February 1945.[1] Both the official and scholarly consensuses on the Pacific war agree that, while the Navy made mistakes, it learned lessons and applied those lessons along the way. It is generally agreed that the most important lessons regarding naval gunfire were those learned at Tarawa in November of 1943 (see Appendix 8.1). Adjustments were made to the doctrine and in the techniques for its application in each subsequent Central Pacific operation of 1944, the Marshall Islands in January–March and the Mariana Islands in June–August, with the "Lessons of Tarawa" in mind. These adjustments resulted,

[1]The Central Pacific Drive consisted of the following major operations: Galvanic (Gilbert Islands, 13 November–8 December 1943); Flintlock (Marshall Islands, 19 January–2 March 1944); Forager (Mariana Islands, 11 June–10 August 1944). Additional operations that grew out of and were related to this Central Pacific Drive were: Stalemate II (Peleliu, 6 September–27 November 1944); Detachment (Iwo Jima, 19 February–16 March 1945). Gordon L. Rottman, *U.S. Marine Corps World War II Order of Battle: Ground and Air Units in the Pacific War, 1939–1945* (Westport, CT: Greenwood Press, 2002), pp. 300–58. The six components of the amphibious warfare doctrine that was developed during the 1930s were: command relationships, naval gunfire support, air support, ship-to-shore movement of troops, the securing of a beachhead, and logistics. Department of the Navy, *Tentative Landing Operations Manual, 1935* (Washington, DC: Headquarters U.S. Marine Corps, 1935). Descriptions of each of these components are found, in the order of the elements given above, on the following pages of the *Tentative Manual:* pp. 42–51, pp. 145–85, pp. 186–96, pp. 81–144, pp. 51–55, and pp. 250–300.

historians argue, in the development by February 1945 of a method of beach preparation that the Japanese were unable to counter successfully.[2]

But is this consensus accurate? In the case of Iwo Jima, it would appear that some modification is in order. Disagreements between the Navy and Marine Corps regarding the length of time to be allotted for the preliminary bombardment produced a naval gunfire plan for the initial assault that was inadequate to the task. The traditional history takes this internecine conflict and its results into account, but generally does not carry the analysis to its logical conclusion. Just what is that conclusion? It would be this: that the defensive preparations made by the Japanese not only dissipated much of the destructive capacity of the American advantage in firepower, but that they led the Americans to apply one tenet of their own doctrine in a manner that, while doctrinally correct, was ultimately ineffective and counterproductive.

The Failure of Air Bombardment: Foundation for the Great Debate

Preparations for the assault of Iwo Jima began in June 1944 when the U.S. Navy carried out the first large-scale carrier raid against the island. Regularly scheduled air strikes began in August and the pre-invasion bombing campaign against Iwo got underway on 8 December. For seventy-four consecutive days, American bombers based in the Marianas dropped payload after payload on islands in the Bonin and Volcano groups. On 31 January 1945, air operations against Iwo entered a new phase, the number of sorties per day increasing in preparation for the start of the preliminary naval gunfire bombardment on 16 February (see Appendix 8.2). The stated purpose for the increased bombing was the neutralization of the airfields and other installations on the island, the destruction of gun positions and various fixed defenses, and the unmasking of additional targets for the bombarding ships. On 16 February, the same day on which the preparatory fire began, the daylight air attacks increased in intensity, and were augmented by night harassing missions for good measure. To top off the air campaign, fighter sweeps shot up every airplane and living thing the pilots could find on the island.[3]

[2]There are three histories that are most important in this respect. They are as follows: Various authors, *History of U.S. Marine Corps Operations in World War II,* 5 vols. (Washington, DC: GPO, 1958–68); Samuel Eliot Morison, *History of United States Naval Operations in World War II,* 15 vols. (Boston: Little, Brown, 1947–62); Jeter A. Isely and Philip A. Crowl, *The U.S. Marines and Amphibious Warfare: Its Theory and Its Practice in the Pacific* (Princeton: Princeton University Press, 1951).

[3]George W. Garand and Truman R. Strobridge, *History of U.S. Marine Corps Operations in World War II,* Vol. IV, *Western Pacific Operations* (Washington, DC: Historical Branch, HQMC, 1971; reprint, Nashville: The Battery Press, 1994), pp. 485–86; United States Navy, *Amphibious Operations: Capture of Iwo Jima, 16 February–16 March 1945* (COMINCH P-0012), p. 3–1, Marine Corps University Archive, Historical Amphibious Files, Box 11B, Folders 255.

Photographic reconnaissance monitored the damage done by these air strikes. A comparison of photos taken between 3 December 1944 and 10 February 1945 showed that, while they had done appreciable damage to aircraft on the ground and had indeed neutralized the airfields, they had not thwarted the Japanese attempts to strengthen Iwo's defenses. A study submitted to the Chief of Staff of Fifth Amphibious Corps on 9 February 1945 demonstrates the high level of concern on the part of Marine Corps planners for this lack of results: "Photographic coverage of IWO JIMA to 24 January, 1945, indicates that damage to installations resulting from bombing strikes between 3 December, 1944, and 24 January, 1945, was, on the whole, negligible. These strikes have apparently not prevented the enemy from improving his defensive position and, as of 24 January, 1945, his installations of all categories had notably increased in number. The island is now far more heavily defended by gun positions and field fortifications than it was on 15 October, 1944, when initial heavy bombing strikes were initiated."[4] With the air campaign falling short of accomplishing its goals, the preliminary naval gunfire bombardment grew in importance proportionately.[5]

Location and Physical Description of the Objective

American photo analysts and planners were witnessing the indirect evidence of a massive undertaking on the part of the Japanese to transform a small island into as close to an impregnable fortress as they could construct in the short time the fortunes of war had allotted to them. But just what was this place called "Iwo Jima?" The name translates as "Sulfur Island" from the Japanese. It was, and is once again, a Japanese possession. It is located in a chain of islands, the Nanpo Shoto, stretching from just beyond the entrance of Tokyo Bay to within 300 miles of the Mariana Islands. There are three island groups making up the chain. The northernmost group, known as Izut Shichito, is composed of six main islands and a number of smaller ones. The center group, named the Ogasawara Gunto but more commonly referred to by Americans during World War II as the Bonins, consists of four islands: Muko Jima, Nishino Shima, Chichi Jima, and Haha Jima. Iwo Jima is found in the center of the third group, the Kazan Retto ("Volcano Islands"). American planners tended to lump this third group together with the Bonins and use the latter name for all of them, but, technically, the Volcanoes and the Bonins are not the

---

[4]Combined Arms Research Library Digital Library, World War II Operational Documents, Headquarters, Expeditionary Troops, Task Force 56, Enclosure C, G-2 Report of Intelligence, Iwo Jima Operation, dated 1 April 1945, p. 4. The referenced quote is included in this post-operational report. The authors of the G-2 Report on Intelligence state that it came from the "Bomb Damage Assessment, IWO JIMA, Report No. 2" that was submitted to the Chief of Staff and G-3 of Task Force 56 on 9 February 1945.

[5]Garand and Strobridge, *Western Pacific Operations,* pp. 486–87.

same.[6] As far as Pacific Ocean distances are concerned, its location (24°44' N 141°22' E) places Iwo in close proximity to a number of objectives important in the American war against Japan: 670 miles south of Tokyo, 625 miles north of Saipan, and 700 miles north of Guam.[7]

Iwo Jima's appearance from the air has been described variously over the years. To this writer's eyes the most accurate analogy is to a pork chop. Covering a total of seven and one-half square miles, the island stretches four and two-thirds miles to the northeast from Tobiishi Point (just southwest of Mount Suribachi) to Kitano Point, and it expands from a width of 700 yards at the landward base of Suribachi to two and one-half miles across the widest part of the northeastern portion. The profile of the island from seaward has been likened to a partially submerged whale with Suribachi, an inactive volcano rising 546 feet above sea level, providing a zoologically inaccurate vertical tail and the northeast plateau gently rising above the surface giving the appearance of a whale's head and back.[8]

The presence of Suribachi is a clue as to the origin of the island: Iwo Jima is the result of relatively recent volcanic activity. There is very little vegetation, so the landscape is rather moon-like in appearance. The soft volcanic rock allows for the rapid construction of tunnels as well as the effective dissipation of the explosive power of artillery rounds. There is no fringing reef like American forces had encountered so often before in the Pacific and the surrounding waters are mostly free of rock outcroppings. The island affords no natural harbors or protected anchorages of any kind. This is the main reason that Iwo was rarely ever mentioned in the prewar Orange plans, the lack of harbor facilities making the island useless as a base in a pre-aviation-centric era naval campaign. The seabed rises to the beaches very steeply so that within just a few feet of the shoreline the water is ten to twelve feet deep. Without the protection of a reef or harbor, ships attempting to offload their cargo are pounded by unrelenting open ocean surf that makes a mockery of the efforts of their crews. American forces found just how difficult these conditions were to overcome on 19 February 1945 when it was too late to do anything but stay the course and hope to God that they could persevere.[9]

---

[6]It can be assumed that when the documents mention the "Bonins," the planners were usually referring to the Bonin and Volcano Island groups together. References to the "Bonins" in this chapter reflect the verbiage of the primary documents.

[7]Gordon L. Rottman, *World War II Pacific Island Guide: A Geo-Military Study* (Westport, CT: Greenwood Press, 2002), pp. 417–21; Whitman S. Bartley, *Iwo Jima: Amphibious Epic* (Washington, DC: Historical Division, HQMC, 1954), pp. 4–5; Garand and Strobridge, *Western Pacific Operations,* p. 447.

[8]Rottman, *World War II Pacific Island Guide,* pp. 421–22.

[9]Ibid., pp. 422; Edward S. Miller, *War Plan Orange: The U.S. Strategy to Defeat Japan, 1897–1945* (Annapolis: Naval Institute Press, 1991), p. 101.

The Planning Process

If Iwo Jima had never been considered an important objective prior to World War II, what happened to change the minds of U.S. planners? To understand the evolution of American strategic thought in this regard one must look into the inner workings of the Joint Chiefs of Staff (JCS) during the war. Robert S. Burrell argues in *The Ghosts of Iwo Jima* that their decision to take this obscure piece of volcanic flotsam was influenced more by interservice rivalry than by realistic strategic considerations.[10] An argument can be made for such a case, and Burrell does a workmanlike job of putting that case forward, but the process was more intricate and open to various and extraneous influences than his explanation implies.[11] Stripped of all but the essentials, the process was as follows. Islands in the Nanpo Shoto were first mentioned during a debate amongst the Joint Chiefs that took place in August of 1943, and Admiral Ernest J. King, Commander-in-Chief of the U.S. Fleet and Chief of Naval Operations (CominCh and CNO), was ultimately responsible. On 6 August 1943, in preparation for the upcoming Quadrant Conference, the Joint Chiefs began discussing a report submitted to them by the Rainbow Team of the Joint War Plans Committee (JWPC). Designated JCS 446, the report outlined 1943–44 operations in the Pacific. The five projected for the Central Pacific

---

[10] Robert S. Burrell, *The Ghosts of Iwo Jima* (College Station: Texas A&M University Press, 2006), pp. 123–26.

[11] Burrell originally presented his argument in an article published in the October 2004 issue of *The Journal of Military History.* The article elicited a rebuttal on the part of Brian Hanley, published along with Burrell's reply in the July 2005 issue of the journal. Hanley criticizes Burrell for what he sees as a most selective use of primary documents in making his case and for exhibiting the same parochial attitude in reverse that Burrell attributes to World War II-era Navy and Army Air Force officers. Hanley is on target in many of his criticisms, but there are others that need to be mentioned as well. First, Burrell never cites, nor does he refer to, Hayes' *History of the Joint Chiefs of Staff in World War II: The War Against Japan.* Second, Burrell never cites JCS 446. Not referencing Hayes's work is understandable. She presents a well-documented argument concerning JCS planning for the war in the Pacific that is at odds with Burrell's own. Not mentioning JCS 446 is also understandable because its contents would make it harder for him to sustain his own argument that interservice rivalry, and only interservice rivalry, ultimately explains why the Joint Chiefs decided to invade Iwo Jima. Interservice rivalry did play a part in the decision, but to believe Burrell's argument one also must believe that men such as Ernest King and Henry Arnold were no more capable of thinking about highly consequential issues like the possible slaughter of thousands of men than if they were a couple of twenty-year-olds whose rival fraternities were trying to best one another in some insignificant intramural competition. As Hanley points out, evidence does not support such a conclusion. Robert S. Burrell, "Breaking the Cycle of Iwo Jima Mythology: A Strategic Study of Operation Detachment," *The Journal of Military History* 68 (October 2004), pp. 1143–86; Brian Hanley, "The Myth of Iwo Jima: A Rebuttal," *The Journal of Military History* 69 (July 2005), pp. 801–09.

Campaign were: 15 November 1943—Gilbert Islands (Nauru, Tarawa, Makin); 1 January 1944—Marshall Islands (Kwajalein, Wake, Kusaie); 1 June 1944—Ponape (in the Carolines); 1 September 1944—Truk (in the Carolines); 31 December 1944—Palau Islands.[12]

This schedule was tight, the joint planners admitted, but it had two important advantages. First, it allowed strategic momentum against the Japanese to be maintained. Second, it never allowed any Allied assets to remain idle for long since forces and equipment in theater would be transferred where and when needed for prosecution of the necessary operations. If conflicts over men and materiel arose, JCS 446 stated that "due weight should be given to the fact that operations in the Central PACIFIC promise more rapid advance."[13] Admiral King was the only member of the Joint Chiefs to offer any objections to the contents of the report. He was not happy that mention of the Marianas had been left out. He reminded his fellow Chiefs that occupation of the Marianas would be considered a serious threat by the Japanese while the Palaus were nothing more than a staging area on the way to the Philippines. At his recommendation, then, the following was added to the report: "It may be found desirable or necessary to seize Guam and the Japanese MARIANAS, possibly the BONINS, in conjunction with the seizure of the western CAROLINES, and in particular with the attack on the PALAUS. The MARIANA-BONIN attack would have profound effects on the Japanese because of its serious threat to the homeland." This statement left open the possibility of heading in the direction of the Marianas once the conquest of the Carolines was completed.[14] King insisted on including the Bonins in these operational calculations because their inclusion strengthened his argument for what he saw as the more important and necessary invasion of the Marianas. Shortly after the CNO won his argument for including this broadening statement in JCS 446, Army Air Forces planners in Washington recommended to the JWPC that the timetable for seizing the Marianas be accelerated. They also argued that it would be a good idea to establish air bases in the Bonins. Fighters operating from these bases could fly escort missions for the bombers that would be operating from the Marianas. The air planners thereby supported and strengthened King's position—after the fact.[15]

The Joint Chiefs presented the revised JCS 446 at the Quadrant Conference as the combined paper CCS 301. The British Chiefs asked their American counterpart only one question concerning the Central Pacific during discussions of the paper: Would it not be a better idea to limit operations in the Japanese Mandates in order to release resources for use in the upcoming European Theater Operation

---

[12]Grace Person Hayes, *The History of the Joint Chiefs of Staff in World War II: The War Against Japan* (Annapolis: Naval Institute Press, 1982), pp. 427–30; list is from p. 430.

[13]Ibid., p. 429.

[14]Ibid., pp. 430–32; quote is from p. 431.

[15]Ibid., pp. 431 and 496.

Overlord? The American Chiefs argued that the requisite forces for operations in both the Central and Southwest Pacific were already in theater and could not be used effectively in Europe. Therefore, it made better sense to use the men and materiel where they were and treat Overlord as something entirely separate. Admiral King defended the American strategy of attacking along two paths in the Pacific by arguing that each route was complementary to the other and that both were essential for ultimate victory. The existence of the two paths provided Allied forces with options: either the two routes could converge on the Philippines or the Central Pacific forces could turn north from Truk to attack the Marianas. The Americans further assured the British that the planned operations for the Pacific would not interfere with the invasion of Nazi-held Europe. The British appeared to be satisfied with the American responses to their questions and gave CCS 301, with a paragraph concerning China added, their tacit approval.[16] The Quadrant Conference, upon its closing on 24 August 1943, signaled the demise of the idea that the Allies should hold the line against the Japanese until such time as Germany was defeated.

Another issue that influenced American interests in the Bonins was the need to push their British allies to support an invasion of Japan. The American joint planners by mid-1944 believed that the preponderance of air, sea, and land power built up by the U.S. in the Pacific argued for the mounting of an invasion as early as was practicable. They presented this argument to the Joint Chiefs on 30 June 1944 in a Joint Staff Planners' report entitled "Operations against Japan Subsequent to Formosa" (JCS 924). This paper defined the objective of the war with Japan as "invasion of the industrial heart of Japan and the seizure of objectives therein, in order to force the unconditional surrender of Japan." The intermediate objectives on the way to the Home Islands should be:

> (a) Concurrent advances through the Ryukyus, Bonins, and Southeast China coast for the purpose of intensifying the blockade and air bombardment of Japan and creating a situation favorable for:
>
> (b) An amphibious assault on Kyushu for the purpose of further reducing Japanese capabilities by engaging and fixing major enemy forces and establishing a tactical condition favorable to:
>
> (c) A decisive stroke against the industrial heart of Japan by means of an amphibious attack through the Tokyo plain assisted by continued pressure from Kyushu.[17]

---

[16]Ibid., pp. 432–33. The British appeared to be suggesting that the majority of the "savings" in the Pacific be at MacArthur's expense. King defended American actions in the Southwest Pacific because of his belief in unremitting pressure on the Japanese and because he did not want to give back any gains in the European Theater vs. Pacific Theater logistical contest.

[17]Quote found in Hayes, *History of the Joint Chiefs*, p. 627.

In addition, the JWPC recommended dates from about 1 April to 30 June 1945 for air attacks on the Home Islands and the invasion of the Bonins and the Ryukyus.[18]

The Committee went even further: the members believed that invasion and unconditional surrender needed to be made the officially sanctioned basis for all future Pacific war planning. Such a fundamental change in military strategy, however, required British agreement. With persuasion of the British Chiefs to their way of thinking in mind, then, the American joint planners suggested that the Combined Chiefs undertake at the Octagon Conference (12–16 September 1944) a restatement of the overall objective for the Pacific. They recommended the following wording: "[The overall objective of the war in the Pacific will be t]o force the unconditional surrender of Japan by: (1) Lowering Japanese ability and will to resist by establishing sea and air blockades, conducting intensive air bombardments, and destroying Japanese air and naval strength. (2) Invading and seizing objectives in the industrial heart of Japan."[19]

The British Chiefs, upon receiving a copy of the recommended change, voiced concerns that such a modification to the stated overall objective in the Pacific might jeopardize the already agreed-to priority of "Germany first." The JWPC suggested that the JCS send a reply to the British assuring them that it was not the intent of the United States to jeopardize the war in Europe by undertaking an expansion of the war effort in the Pacific. The JCS sent this reply to the British Chiefs on 4 August, but the British did not answer. Finally, after the Octagon Conference had convened and discussions of overall strategy had begun, the British registered their official approval of the restated objective by offering no arguments against the Combined Chiefs signing off on the change.[20] The American Joint Staff Planners included the restated objective in a brief paper summarizing U.S. intentions regarding Japan which the U.S. Chiefs presented at the conference. Included in this paper was a tentative schedule of major operations: (1) 15 October 1944—Talaud; (2) 15 November 1944—Sarangani Bay; (3) 20 December 1944—Leyte-Surigao Area; (4) 1 March 1945—Formosa-Amoy Area, or 10 February 1945—Luzon. The fourth operation could not be specified at the time because American planners had made no firm decision as to whether Formosa or Luzon should be captured. The former had become important in Admiral King's strategic vision while the latter was integral to that of General Douglas MacArthur, Supreme Allied Commander in the Southwest Pacific (ComSWPac).[21] The tentative nature of the schedule did not distress Admiral King, however. Regardless of the problems that still required solution, the Marianas were already in American hands and the invasion of Japan was now the stated ultimate goal of the Pacific war.

---

[18] Ibid., pp. 627–28.

[19] Quote found in ibid., p. 628.

[20] Ibid., pp. 628–29.

[21] Ibid., p. 629.

A year after the debate over JCS 446, the idea of capturing an island in the Bonins caught the fancy of more than just the Army Air Forces' planners, and for different reasons. Admiral Raymond A. Spruance, U.S. Fifth Fleet commander, upon returning to Pearl Harbor from the Marianas in early September of 1944, met with Admiral Chester W. Nimitz, Commander-in-Chief of the Pacific Fleet and Commander-in-Chief of the Pacific Ocean Areas (CinCPac/CinCPOA), to discuss future operations. Nimitz told Spruance that the next operation designated by the Joint Chiefs would be Formosa and Amoy, the latter a port on the Chinese coast. Spruance, who already had been studying the question of where U.S. forces should go next, immediately told his boss that he did not like Formosa as an objective. Nimitz asked him his preference and Spruance said that he would prefer seizing Iwo Jima and Okinawa. Both men knew that Admiral King had decided that Formosa needed to be invaded in preparation for the imposition of a naval blockade of the Japanese Home Islands. Such a blockade, King believed, would make what promised to be a most difficult and bloody invasion of Japan unnecessary. Spruance and Nimitz were well aware that persuading the CNO to change his mind would be difficult. This challenge did not deter Spruance, however. He knew that even though King had convinced the other members of the JCS to give his Formosa strategy consideration, the tentative schedule of operations listed above being proof, there were others beside himself who had doubts about that strategy. One of those people was Chester Nimitz.[22] When Admirals King and Nimitz met in San Francisco from 29 September through 1 October 1944 for one of their periodic conferences, they discussed, as usual, the results of recently completed operations as well as other issues relating to the continuation of the war against Japan. The most important of this latter group was the issue of future operations, and it was in regard to them that Nimitz and his staff put before King their arguments against invading Formosa. Faced with determined and logical opposition to his treasured Formosa strategy from the men he had most trusted to carry the fight to the Japanese, King finally relented and agreed to recommend the invasions of Iwo Jima and Okinawa to the Joint Chiefs.[23]

---

[22]Thomas B. Buell, *The Quiet Warrior: A Biography of Admiral Raymond A. Spruance* (Boston: Little, Brown, 1974; reprint, Annapolis: Naval Institute Press, 1987), pp. 331–32; Garand and Strobridge, *Western Pacific Operations,* p. 465. Edward S. Miller, on pp. 362 and 363 of *War Plan Orange,* explains how the invasion of Formosa fit into King's strategy.

[23]Burrell, *The Ghosts of Iwo Jima,* p. 35.; E. B. Potter, *Nimitz* (Annapolis: Naval Institute Press, 1976), pp. 326–27; Buell, *Quiet Warrior,* pp. 330–35; Hayes, *History of the Joint Chiefs,* p. 623. Burrell describes Spruance at this meeting as simply agreeing, after prodding from Nimitz, with the latter's plan to forego Formosa. Burrell does grant that it was originally Spruance's idea to go into Iwo Jima and Okinawa and not Nimitz's, but he seems to want to shield Spruance from any real responsibility for the decision. It is hard to see how an argument can be made that Spruance only reluctantly recommended the invasion of Iwo. According to his biographer, the Fifth Fleet commander's early reasoning was based upon the classic strategic concept of

Upon his return to Washington on 2 October, King circulated amongst the Chiefs a draft of a directive that was to be sent to Nimitz and to General MacArthur. This directive embodied all the ideas that he had discussed with Nimitz in San Francisco. On 3 October, the JCS sent a directive to MacArthur telling him to make preparations to take Luzon on 20 December and they sent another to Nimitz telling him to provide cover and support for MacArthur's operation. Additionally, Nimitz was to make preparations to take Iwo Jima on 20 January 1945 and Okinawa on 1 March 1945. With the successful completion of all of these operations, a lodgment on Formosa or the China coast would be unnecessary for the prosecution of a naval and air blockade of Japan. King, Nimitz, and Spruance hoped that it also would mean that there would be no reason for an invasion of the Home Islands. Admiral William F. Halsey, as Commander of the U.S. Third Fleet, began planning for the operation that would provide cover for MacArthur. Meanwhile, Spruance and his Fifth Fleet staff began planning for the move into the Bonins.[24]

Tying Operation Detachment so closely to the invasion of Luzon affected CinCPOA scheduling almost immediately. Pacific Ocean Areas planners recognized by mid-November that the specified one-month interval between the Luzon and Iwo assaults was too short a timeframe to allow for the transference of shared amphibious and support shipping between commands. Nimitz, accordingly, recommended that the invasion of Iwo be put off until 3 February and that of Okinawa until 15 March in order to compensate. As American forces became bogged down on Leyte, the Japanese fighting tenaciously to hold on to every square inch of the island as long as possible and nature conspiring with the enemy to slow U.S. troop progress and airfield construction to a standstill, MacArthur decided that the Luzon operation would have to be postponed to 9 January. Once again, Nimitz was forced to change the dates of his operations to coordinate with ComSWPac. In early December Nimitz sent a message to the JCS requesting that the Iwo operation be postponed until 19 February and Okinawa until 1 April. The Joint Chiefs agreed and the dates were reset a second time, this time for good.[25]

---

interior versus exterior lines of communication. Since interior lines can be more easily protected than exterior lines, it behooves the strategist to strive to establish the former. According to this logic, Spruance saw Iwo Jima as the focus of an arc running through Tokyo, Kyushu, and the Ryukyus, its capture allowing the U.S. to establish an airbase that would augment American carrier-based air in protecting the fleet from the attacks of thousands of aircraft the Japanese were believed to have hoarded in the Home Islands. In other words, the capture of Iwo Jima would allow the U.S. to operate along interior lines of communication as it tightened its grip on Japan. As such, a lodgment in the Bonin or Volcano Islands was a military necessity.

[24]Hayes, *History of the Joint Chiefs*, pp. 623–24; Maurice Matloff, *United States Army in World War II: The War Department, Strategic Planning for Coalition Warfare: 1943–1944* (Washington: Office of the Chief of Military History, Department of the Army, 1959), pp. 530–31. Halsey's cover operation would lead to the seizure of Peleliu in September of 1944.

[25]Bartley, *Amphibious Epic*, p. 23; Garand and Strobridge, *Western Pacific Operations*, p. 465.

Nimitz's staff at Pearl Harbor could not wait for every JCS planning possibility to work itself out before beginning to make their own plans for Detachment. Accordingly, they went right to work upon receipt of the 3 October JCS directive, aiming for the original target date of 20 January. The central planning agency of CinCPOA issued a study on 7 October 1944 that was meant to serve as the preliminary planning foundation for all subordinate commands. It was not issued as a directive, so Nimitz was not restricted by it to a particular course of action. Tentative as it was, however, the description of the purposes of Detachment contained therein was rooted in the ideas of America's top-level planners and as such is instructive. These ideas were:

> To maintain unremitting military pressure against JAPAN.
>
> To extend our control over the Western Pacific.
>
> To establish a base from which we may: (1) Attack the Japanese Empire. (2) Protect our bases in the MARIANAS. (3) Cover our naval forces and conduct search operations in the approaches to the Japanese Empire. (4) Provide fighter escort for VLR [Very Long Range] operations. (5) Initially stage heavy bombers from the MARIANAS for direct assaults against the EMPIRE. (6) Eventually base heavy bombers for direct assaults against the EMPIRE.[26]

As far as specific tasks were concerned, the study envisioned the following:

> (a) Reduce Japanese naval and air strength and production facilities in the EMPIRE.
>
> (b) Destroy enemy naval and air strength in the BONINS.
>
> (c) Capture, occupy, and defend IWO JIMA and develop an air base on that island.[27]

It would be the first of these specific tasks that Spruance and Vice Admiral Richmond Kelly Turner would use as their ultimate trump card in their argument with the Marine Corps over the amount of preliminary naval gunfire to provide at Iwo.[28]

Nimitz issued the directive for Detachment on 9 October 1944. He designated in that document the following men to be his top-level commanders at Iwo: Operational Commander, Admiral Raymond A. Spruance; Joint Expeditionary Force

---

[26]Steven T. Ross, ed., *U.S. War Plans, 1938–1945* (Boulder: Lynne Rienner Publishers, 2002), p. 336.

[27]Ibid., p. 337.

[28]Bartley, *Amphibious Epic,* pp. 23–24; Garand and Strobridge, *Western Pacific Operations,* pp. 465–66.

Commander, Vice Admiral Richmond Kelly Turner; Commanding General Expeditionary Troops, Lieutenant General Holland M. Smith, USMC; and Second in Command of the Joint Expeditionary Force, Rear Admiral Harry W. Hill. All levels began operational planning at once. Spruance issued his Operation Plan No. 13–44 on 31 December 1944. Given the importance of the question of preliminary naval gunfire support as the planning process unfolded, knowing what Spruance understood—at the time—to be his ultimate objectives is enlightening:

> (a) The occupation of SAIPAN, TINIAN, and GUAM has served to establish our shore based air forces in locations from which continuing air attacks are being made against enemy positions in the NANPO SHOTO. Continual reconnaissance from the MARIANAS of the sea areas in the southern approaches to the Japanese Empire is being conducted. The MARIANAS also provide bases from which long range bombers can operate against enemy positions in JAPAN. In order to operate with the greatest effectiveness and with a minimum of attrition, long range bombers should be provided with fighter support at the earliest practicable time. IWO JIMA provides three sites for airfields and is admirably situated as a fighter base for supporting long range bombers between the MARIANAS and the Empire.
>
> (b) The results of carrier attacks on PALAU, MINDANAO, the VISAYAS and LUZON have revealed a weakness of Japanese aviation in their forward areas which will permit more rapid advances. To exploit this situation it is necessary to establish ourselves in positions from which we can engage the enemy in the Empire itself.
>
> (c) Several divisions are now deployed in the South PACIFIC and the Central PACIFIC, from which positions they can be readily mounted for an assault on the NANPO SHOTO. The interval between the LUZON operation and the earliest subsequent operation of comparable size offers the possibility of sufficient time to complete an assault on the NANPO SHOTO without delaying or interfering with other scheduled operations.
>
> (d) This assault will be executed in a single phase, with two divisions employed simultaneously in such a manner that the complete subjugation and occupation of the island will be quickly effected, after which all major assault troop units will be evacuated, leaving the development and defense of the island in the hands of garrison forces.[29]

---

[29]Marine Corps University Archive, Iwo Jima Project, Fifth Fleet Operations Plan, 13–44, 31 December 1944, p. B-1.

Two observations can be made here. First, Spruance concentrated his attention on overall strategic questions rather than on the operational problems associated with taking the objective. One must not read too much into his behavior in this instance, however, because Spruance himself admitted that he was a better strategist than a tactician and, also, he had operated successfully in this way since the very beginning of the Central Pacific campaign in late 1943.[30] Second, Spruance initially believed that the "complete subjugation and occupation" of Iwo would not be too difficult a proposition. Consequently, he believed that Detachment would fit in neatly between the two larger operations in the Philippines and the Ryukyus. Not realizing the difficulty of taking Iwo did not necessarily have to lead to the expenditure of large numbers of American lives. The combination of the two in Spruance's mind, however, would prove deadly to the Marines tasked with assaulting the island because this combination heavily influenced Spruance when it came time to decide how many days of preliminary naval gunfire bombardment to provide the assault troops.

There are other questions that must be asked pertaining to the naval gunfire planning process beyond those dealing with Spruance's view of the overall strategic picture for the Iwo operation. One of the most important of these questions concerns what the Americans knew about the Japanese defenses on the island prior to 16 February 1945, the date on which the naval gunfire bombardment began. A persistent story associated with the history of the assault on Iwo Jima is that American planners and commanders were not aware of the nature of those defenses until it was too late to do anything about them. This was not true, at least not in certain particulars. A study submitted on 9 February 1945 shows that at least some of the operational planners were very well aware of the fact that Iwo was going to be a tough fight. The question for them was what to do about it. There is other evidence as well. Consider the following description found in the final version of the naval gunfire annex of Admiral Turner's Operation Plan: "IWO JIMA is a relatively small but heavily defended island presenting a series of integrated defense installations which will require a carefully coordinated scheme of fire from all attacking agencies. It must be emphasized that advanced intelligence cannot be expected to disclose more than a fraction of existing defenses. Close observation, both visual and photographic, plus imagination stimulated by a close study of the terrain, are of the utmost importance in locating others."[31]

In his book *Storm Landings,* Joseph Alexander states that Nimitz's intelligence section was aware that Japanese defensive tactics were changing but that a bulletin from his command forecasting Iwo Jima's defenses to be more along the lines of Peleliu's than Roi-Namur's or Saipan's did not receive much attention from the plan-

[30] Buell, *Quiet Warrior,* p. 332.

[31] Marine Corps University Archive, Iwo Jima Project, Commander Amphibious Forces (CTF 51) Operation Plan No. A25–44, Iwo Jima Operation, Ships' Gunfire Support Plan, pp. 4–5.

ners of the Iwo assault.[32] Turner's description quoted above is dated 27 December 1944, fourteen days prior to the CinCPac/CinCPOA bulletin of 10 January 1945 alluded to by Alexander and over a month prior to the Bomb Damage Assessment of 9 February 1945 quoted earlier. This description of the conditions Turner expected to find on the island, a description that predates even the CinCPac/CinCPOA bulletin, indicates that someone on his staff had given the photographic evidence more than just a cursory glance and had come to conclusions on his own that paralleled the conclusions being reached at the same time by photo analysts in other commands. All this evidence shows that planners appreciated the dangers, at least to an extent, in time to do something. The real question, once again, was just what could be done.

The Great Gunfire Support Debate: Navy vs. Marine Corps

It was out of appreciation of these dangers that the great debate over how much preliminary naval gunfire bombardment to provide first arose. Central Pacific experience taught that several days of slow, thorough preparation would be necessary, and Admiral Nimitz's headquarters included eight days of preliminary gunfire support in their original plan with this idea in mind.[33] Spruance, strategic concerns uppermost in his mind as explained above, decided that eight days were too many given the need to coordinate Detachment with simultaneous carrier air raids on the Japanese Home Islands and with the subsequent invasion of Okinawa. In addition, he determined that he would need more gunfire support ships for an eight-day bombardment than he could obtain at the time because of General MacArthur's continuing need for bombardment vessels to support his operations in the Philippines. Spruance explained his solution to the gunfire dilemma in his After Action Report: two battleships and two cruisers from the Fast Carrier Force were sent to participate in covering the landings at Iwo on 19 February after they had finished providing protection for the carrier raids on the Home Islands (see Appendix 8.3). He determined that the best that could be done, given all these

[32]Joseph H. Alexander, *Storm Landings: Epic Amphibious Battles in the Central Pacific* (Annapolis: Naval Institute Press, 1997), p. 129. The report was CINCPAC-CINCPOA Bulletin No. 9–45, "Iwo Jima: First Supplement to Nanpo Shoto Information Bulletin No. 122–44," dated 10 January 1945. Isely and Crowl seem to allude to the same document on p. 436, but they do not name the source. They also report on the same page that records captured in the Marianas outlined the basic defensive plan for Iwo and that aerial reconnaissance photographs showed work continuing on island defenses despite almost daily American air and periodic sea bombardment.

[33]Ross, ed., *U.S. War Plans, 1938–1945*, p. 338. The referenced quote is as follows: "Surface ship bombardment by cruisers will be commenced on about D-8 and will be continued by the full fire support group from D-3 until no longer required by the assault force."

relevant considerations, was to limit the period of preliminary bombardment to three days.

Lieutenant General Holland M. Smith and Landing Force Commander Major General Harry Schmidt, USMC, knew from their own bitter experience that a target as tough as Iwo Jima required massive amounts of naval gunfire preparation over several days in order to soften it up sufficiently for assault. They also knew that Spruance and Turner were set on three days of naval gunfire, so they wrote numerous letters to the two admirals in the desperate hope that they could get them to authorize more. In one of the first of these, dated 24 October 1944, Schmidt listed the basic assumptions from which he and his planners were working in making their recommendations:

> (a) That all known defensive installations must be destroyed.
>
> (b) That a minimum of 50% more targets will be uncovered during preliminary bombardment and that to discover these targets it will be necessary to fire a considerable amount of ammunition for the purpose of blasting away camouflage and concealment.
>
> (c) That a methodical "working over" with major caliber ammunition of the slopes of the crater on the south flank and the terrain formation on the north flank will be required to uncover and destroy concealed emplacements and casemates.
>
> (d) That from one to four major caliber direct hits will be required to destroy known defensive installations and that 10% hits is the average expected occurring in direct fire.[34]

With these and other considerations taken into account, Schmidt recommended:

> (a) That a minimum period of ten days be allotted for naval bombardment of (IWO Jima).
>
> (b) In addition to the Cruiser Division now allotted, that a minimum of one division of BB's [*sic*] be added for preliminary bombardment.
>
> (c) That a staff trained and equipped to evaluate results be present to assess the results of preliminary bombardment and to make target assignments; that this staff not only include gunfire personnel but also support air personnel and that this staff be present on SAIPAN by D-20 to study results of 7th AAF bombing.
>
> (d) That the date of D-Day be dependent on the successful prosecution of the destruction of enemy defensive installations. . . .[35]

---

[34]Marine Corps University Archive, Box 35, Folder 649, Collection of communications between parties concerning naval gunfire requirements for the invasion of Iwo Jima, p. 1.

[35]Ibid., p. 6.

Smith forwarded Schmidt's request to Turner on 26 October with his recommendation and the comment that "[i]n order to insure the quick seizure of the objective with a minimum of casualties, the increase in the duration of intensity of naval preliminary bombardment over that now contemplated is considered a necessity."[36]

Turner sent his reply on 15 November. He turned Schmidt down, explaining himself in similar terms as Spruance would have used:

> Limitations on the availability of ships, difficulties of replenishment of ammunition and lose [*sic*] of surprise interpose serious obstacles to meeting your recommendations. However, it is felt that the following plan will accomplish all the objectives desired:
>
> (a) It is understood that (IWO Jima) will be bombarded by one Cruiser Division on or about 15 December and at irregular intervals thereafter. Bombardment will be for the purpose of accomplishing such destruction as is possible, and for establishing a flow of intelligence of target destruction supplementing that expected from air strikes during the same period.
>
> (b) On Dog minus three, a methodical and thorough bombardment will be instituted by the Amphibious Support Force. Competent evaluation of results and corresponding adjustments in the bombardment plan will be assured through control and coordination by an Amphibious Group Commander embarked in an AGC (with representative of your command aboard if desired). The increased number of bombardment ships and local control and evaluation of results will enable more ammunition to be delivered at the target in three days than the estimated requirements listed [in your request] for ten days. . . . It is believed that the excess of 5," 14," and 16" capabilities [in this plan] over your estimate of requirements will more than offset the apparent shortage of 6" and 8" capabilities.
>
> (c) Emphasis will be placed on the securing and utilization of all information of target destruction.[37]

Turner's statement that "[t]he increased number of bombardment ships and local control and evaluation of results will enable more ammunition to be delivered at the target in three days than the estimated requirements listed [in your request] for ten days. . . ." struck the Marines the hardest blow of all. They believed that the Navy had learned the "Lessons of Tarawa." One of the most important of these, as far as the Marine Corps was concerned, was that quantity of ammunition

---

[36]Ibid., p. 8.

[37]Ibid., pp. 9–10.

delivered did not make up for accuracy of placement on target. Turner, like so many Navy men before him, continued to place faith in the sheer quantity of ordnance fired at an objective. In addition, he speaks of "destruction," but he does not appear to understand, at least as far as the wording of his letter is concerned, that destruction can only occur when that massive quantity of ammunition is delivered slowly and methodically, a point the Marines would continue to make.

In a letter to Admiral Turner dated 24 November, General Schmidt conceded that he would not receive the ten days of preliminary naval gunfire support that he had requested. Believing that more support than Turner allotted to be absolutely necessary, however, he provided evidence to support a request for four days instead of three:

> 2. As has been stated in previous correspondence, the preliminary bombardment should provide for the deliberate and methodical destruction of all known or subsequently located enemy gun positions and other installations that can interfere with the successful execution of the landing. The question is whether the time now allotted is adequate for deliberate destruction. The amount of ammunition allotted appears to be adequate.
> 3. Since this question may be determined accurately in advance, two alternatives suggest themselves:
>   (a) To provide for an additional day of bombardment prior to the landing, if it is found during the preliminary bombardment that all known targets can not be destroyed prior to D-Day. This involves obtaining early information by the Expeditionary Force Commander in order that orders could be issued in time to task forces at sea as to the postponement. Under conditions of radio silence, and the dispersion of forces involved, this would appear difficult. The target date in this case would be postponed one day.
>   (b) Allot four days for the preliminary bombardment. This would allow for more deliberate firing, particularly after necessary minesweeping has been effected. It theoretically would involve no increase in the expenditure of ammunition, but rather a slower rate of fire. Any increase in ammunition requirements would have to come from that allocated to call fire.
>   (c) It is recommended that, if consistent with other considerations, the plan outlined in paragraph 3(b) above be adopted and that the preliminary bombardment begin on D-4.[38]

General Smith sent Schmidt's latest request on to Turner along with his own recommendations:

---

[38]Ibid., p. 25.

1. From lessons learned in previous operations and continued study and analysis of the proposed Operation (against IWO Jima), it is the considered opinion of this Headquarters that four (4) full days for preliminary bombardment is the absolute minimum necessary for success.
2. This objective lends itself to a strong defense with elaborate installations, and unless these defenses are destroyed or at least neutralized we may expect casualties far beyond any heretofore suffered in the Central Pacific, or possibly jeopardize the success of the entire operation. It is considered that three (3) days for preliminary bombardment are insufficient.
3. A comparison of casualties suffered on SAIPAN after an approximate three (3) days of preliminary bombardment with the casualties received on TINIAN and GUAM after many days of preliminary bombardment further convinces this Headquarters of the necessity for a minimum of at least four (4) days of preliminary bombardment. . . .[39]

Both Smith and Schmidt referred to current doctrine in explaining their request for more naval gunfire and Smith pointed out that the lessons learned thus far in the Pacific supported their opinions. The Marines believed that their argument to their Navy superiors was based upon irrefutable grounds.

It was at this point that Spruance himself weighed in on the debate. On 2 December 1944, he sent a letter to Turner refusing to allow the additional day of bombardment. He drew the greater strategic picture for Turner so that the latter could leave Schmidt and Smith with no illusions as to what the overall operational commander thought:

1. The recommendation of Commanding General, V Amphibious Corps, that surface ship bombardment of (IWO Jima) be commenced on D minus 4 Day is not approved for the following reasons:
   (a) The initial surface bombardment must be simultaneous with the initial carrier attack upon the TOKYO area. The carrier attacks will be continued there for three days if practicable, but it may prove desirable or necessary to discontinue attacks after two days or less. If the attacks were made on D minus 4 and D minus 3 days, time would then be available to the enemy for recovery and reinforcement early enough to initiate threatening air operations at the objective by D day.

[39]Ibid., p. 26.

(b) The situation at (IWO Jima) will be different from that encountered at SAIPAN for reason that by D day the enemy personnel and fixed defenses at (IWO Jima) will have been under heavy shore based air attack for a considerable period of time. This prolonged air bombardment, which was not undertaken at SAIPAN, may be considered at least as effective as the recommended additional day of surface ship bombardment.

(c) The trench and pill box defense system at IWO Jima is strong and must be taken by assault, supported, until shore artillery can be emplaced, by intense naval gunfire on D day. There will be no early opportunity for replacement of naval ammunition, a large proportion of which must be saved for support of D day. There is a limit to the quantity of ammunition which can be made available for pre D day bombardment and no advantage is seen in delivering that quantity in four days rather than three.[40]

His comments show that Spruance placed great faith in the effectiveness of the air bombardment. Also, just like Turner, Spruance emphasized quantity of ammunition delivered rather than accuracy in placement of rounds.

The two Marine generals tried again on 6 January 1945, summarizing some of the deficiencies of the bombardment plan and pointing out that photographic reconnaissance was not bearing out Spruance's predictions concerning air bombardment:

2. The preliminary bombardment plan for (Operations against IWO Jima), indicates the following deficiencies:
   a. A lack of sufficient time for destruction of known targets. . . . Total bombardment allotted does not exceed an average of 18 hours per ship of which not more than six hours is at ranges sufficiently short to insure destruction.
   b. Within the time now allotted, the lack of sufficient ships on the targets located on and adjacent to the preferred landing beaches.
3. Since the overall time element is an important factor in the capture of (IWO Jima) it is believed that if sufficient time is allotted for preliminary bombardment, the overall time will be reduced, with a consequent reduction in the time that fire support ships and other vessels will be required to remain in the area. An adequate, methodical preliminary bombardment can reduce the island defenses to a point where a quick capture is assured. The preliminary bombard-

[40] Ibid., p. 28.

ment now planned will not only increase the overall time necessary to complete the operation, but will require an unnecessary expenditure of lives during the initial assault phase.

4. As to the effect of heavy horizontal bombing attacks on the objective, photographic interpretation to date indicates a negligible destruction of the critical pin-point targets which are a particular menace to assault troops. Although a stepped up schedule of bombings will occur prior to D-Day, it is not anticipated that the final result will equal that which can be expected from an additional day of preliminary naval bombardment. Attention is invited to the fact that studies of photographs flown in mid-December, when compared with sorties of October, November and early December show that considerable new construction of block houses and caves has been completed and that additional new construction has begun.[41]

Wanting to bring the debate to an end, Turner sent one last letter to Smith and Schmidt on 9 January 1945 in which he took them to task for sticking so closely to the "lessons" the Marines thought everyone had learned thus far in the Pacific war. First, he pointed out that, contrary to Marine contentions, short-range fire was not always most effective for destruction. He emphasized the high number of ricochets that occurred when firing from short ranges at flat targets to support his conclusion.[42] Second, he criticized the Marines for constantly trying to change the gunfire support schedule, pointing out that the schedule had already been approved by both of their commands.[43] Third, he took exception to the Marine emphasis on

[41]Ibid., pp. 32–33. These comments are further proof that some American commanders were aware of the strengthening of Japanese defenses prior to 19 February 1945.

[42]Ibid., pp. 34–35. The referenced quote is as follows: "It is not agreed that short ranges are always the most effective for destructive fire. The best experience, as clearly set forth in the PhibsPac Circular letter AL15-Rev. 1, indicates that short range pointed fire is best for targets that present an appreciable vertical surface. On the other hand, short range firing at flat targets wastes ammunition because so many ricochets and duds occur. Therefore, a considerable amount of long and medium range firing must be done, to take care of the many flat targets on (IWO Jima). For example, it would be quite impossible to hit any targets on the tops of the hills by means of short-range fire."

[43]Ibid., p. 35. The referenced quote is as follows: "The firing schedules were originally prepared in collaboration with representatives of ComGenVPhibCorps [Schmidt] and ComGenFMFPac [Smith]. Then after a final check, before issue, they were completely revised at the request of representatives of ComGenVPhibCorps in order to allow every ship the maximum possible firing time. The allotment of ammunition to targets was also changed radically. More time for general targets could be allotted than as directed in the present plan if it were not considered necessary to perform UDT [Underwater Demolition Team] operations on the Alternate Beaches. If UDT reconnaissance of the Alternate Beaches shows that

gunfire coverage of the landing beaches, pointing out that there were numerous gun and artillery positions inland from the beaches that were just as deadly to the assaulting troops as those on the beaches themselves. These positions needed to be fired on as well. Emphasis on the landing beaches, given the shortage of time and ammunition, would therefore unwisely dilute the overall gunfire support effort.[44] Finally, Turner allowed the Marines some flexibility by granting the naval commander of the support forces during the assault, Rear Admiral William H. P. Blandy, the discretion to modify the gunfire and air support schedules if he found it necessary to do so.[45]

Smith and Schmidt could argue effectively against the first two of Turner's points. While it was true that the flat trajectories of most naval guns made it exceedingly difficult for them to hit non-vertical targets, it was also true that hitting targets from longer ranges required more firings and, therefore, more time. This would mean that Turner would have to allow more than three days for the massive quantity of ordnance he stressed so heavily in his 15 November letter to be used effectively. He did not seem to appreciate this apparent contradiction at the time. As to Turner's second point, the two Marines could argue that their original estimates for gunfire were based on incomplete photographic reconnaissance and that they were modifying their request based upon the latest evidence at their disposal, rather than any wish to be obstinate. Turner was aware of the new evidence as well, but his knowledge of it did not seem to affect his decisions.

---

no beach demolitions are necessary, then there will be more time and ammunition available for bombardment of the Preferred Beaches. However, it should be recognized that fire for covering UDT operations is *all* [underline in original] directed at targets that will particularly threaten troops."

[44]Ibid. The referenced quote is as follows: "The concentration of more firing vessels off the Preferred Beaches as desired by ComGenVPhibCorps would result in very thin coverage of the other portions of the objective. Guns in such locations can most probably be used to fire on the Preferred Beach area after How Hour. Therefore they too should be destroyed. If the recommendations in the basic letter are approved, the coverage of the Alternate Beaches would be only about 25% of that on the Preferred Beaches, and this is believed to be too small. Most assuredly, it is not proposed to fire on coast defense batteries only on Dog minus THREE, and thereafter concentrate only on what are spoken of in [your] basic letter as 'troop targets.' Any such procedure would be unsound to an extreme degree. It is essential that coast defense and anti-aircraft batteries receive a great deal of attention."

[45]Ibid. The referenced quote is as follows: "ComPhibGrp ONE [Rear Admiral William H. P. Blandy] as, Commander Support Force, will be in charge of the preliminary bombardment. He has been given full authority to change schedules of fire to meet eventualities as they develop. He is fully cognizant of the features brought out in the basic letter and endorsements, and has been instructed to exercise discretion in modifying schedules to get the best results as a whole from the air and ship bombardments."

Turner's third and fourth points were not open to direct Marine refutation, however. As to his third point, his concern with the Marine Corps insistence on gunfire coverage of the landing beaches proved to be prescient. As to the fourth, it would be a Marine, ironically enough, who would offer words of advice that, when acted upon, would raise Turner's concerns to the level of prophecy. Turner's grant of change authority to Admiral Blandy gave the Marines an opportunity to influence the modification of the firing schedules based upon their accumulated experience and their developed doctrine. They would, in the end take advantage of the opportunity.

The Japanese Defensive Scheme

As it turned out, the Marines were going to need all of the naval gunfire support allowed by Turner and more. Lieutenant General Tadamichi Kuribayashi, commanding general of the 109th Division of the Imperial Japanese Army in the Volcano Islands, had been given the responsibility of preparing Iwo Jima for an anticipated American invasion. He knew that it would be a difficult assignment. Studying the tactics Japanese forces had used in earlier Pacific island battles and paying special attention to the defense of Biak, Peleliu, Angaur, and Luzon, Kuribayashi concluded that standard Japanese doctrine would not suffice. He would have to approach the situation from a purely defensive perspective if he expected his forces to withstand the onslaught of the American juggernaut. The defensive side of warfare, however, was generally loathed by the Japanese.[46] His decision to throw away convention, therefore, stirred controversy among his more militarily conventional subordinates.

Disregarding their objections at first, he issued orders that put the Iwo garrison on a thoroughly defensive footing. First, he chose to establish his headquarters on Iwo rather than on the larger island of Chichi Jima. Second, all civilians on Iwo were evacuated, including "comfort girls," and all alcohol consumption was banned. Third, all facilities were moved underground and mining engineers were brought from Japan to design the necessarily extensive tunnel system needed for the implementation of his defensive scheme. The engineers produced a plan that integrated existing caves with constructed tunnels, providing the Iwo defenders with adequate ventilation and protection from naval gunfire. Fourth, conceding that the Americans would take the landing beaches, he concentrated his defenses on the higher ground of the central and northern sectors of the island. This decision proved to be his most controversial because it was seen to negate the all-important warrior spirit that was believed to set the Japanese soldier apart from his enemies. Rear Admiral Toshinosuke Ichimaru, in particular, pressured Kuribayashi to

[46]United States War Department, *Handbook on Japanese Military Forces* (London: Greenhill Books, 1991; reprint, Baton Rouge: Louisiana State University Press, 1995), p. 99.

modify his plan in this regard. The general eventually bowed to the pressure and, in a compromise, ordered that 135 pillboxes be constructed along the most obvious landing beaches in the southern part of the island. The Americans, as it turned out, overran these pillboxes within three hours. Finally, large-scale *banzai* attacks were strictly forbidden. This was another point of contention with his subordinates. Dying a glorious death in battle when the odds were totally against success was a Japanese tradition. Contrary to this tradition, all counterattacks were to be carefully considered and efficiently carried out. With this concept of efficient counterattack in mind, small groups of troops were to be sent out at night into the American lines to gather intelligence, damage equipment, and kill Marines.[47]

Kuribayashi issued his final defensive plan just before the battle began. His emphasis was on maintaining a flexible defense that would wear down the enemy. To that end, he ordered his artillery to remain silent during the preliminary bombardment of the island. He did not want their positions given away too early in the battle. He stressed that he did not want the landings opposed by any of his men. Let the Americans land! Once the invading forces had come ashore and advanced up the beach approximately 500 yards, automatic weapons in the vicinity of Motoyama Airfield #2, along with automatic weapons and artillery on the high ground north of the landing beaches and around Mount Suribachi to the south, would rain concentrated fire down on them. The general's purpose was not necessarily to defeat the Americans. He knew that U.S. command of the sea surrounding Iwo Jima, principally as a result of the effectiveness of their submarine force, insured that he most likely could not win.[48]

If victory was to be denied him, then he instead would inflict the maximum number of casualties on the American invasion force as was humanly possible. Once the artillery had done all the damage that it could do from its original positions on the high ground above the beaches, it was then to move northward to continue its mission. Infliction of casualties was the rule. Perhaps he could cause the Americans to lose heart and give up, or, barring that outcome, perhaps massive casualties on Iwo Jima would cause the American public to think twice about carrying the fight all the way to the Home Islands.[49]

---

[47]Bartley, *Amphibious Epic,* p. 11; Garand and Strobridge, *Western Pacific Operations,* pp. 453–59; Alexander, *Storm Landings,* p. 132.

[48]Joseph Alexander points out on p. 61 of *Storm Landings* that the U.S. submarine force was placing increasing pressure on the Japanese ability to send supplies out to its garrison forces in the Pacific. The ease with which they had been able to supply Betio would not be the norm for 1944 and 1945.

[49]Garand and Strobridge, *Western Pacific Operations,* pp. 456–57; Alexander, *Storm Landings,* pp. 131–32.

The American Response

This was Kuribayashi's plan. It was Kelly Turner's job to produce a plan for an offensive from the sea that would overcome whatever defenses the Japanese put in place. Turner was known to be a workaholic, but even workaholics need help and he was no different. The nuts and bolts of the various sections of an invasion plan were developed and then torqued together by a multitude of officers from a multiplicity of backgrounds and specialties. One such officer involved in the process that produced the Detachment plan was Lieutenant Colonel Donald M. Weller, USMC. As Fifth Amphibious Corps naval gunfire liaison officer at Iwo Jima, it was his job to work with the Navy to make sure that Marine Corps gunfire support concerns were addressed as the operation unfolded. Weller was present during the preliminary bombardment from D-3 to D-1 and he was instrumental in the making of decisions concerning the application and the altering of the original gunfire plan. His after action comments are instructive because of his intimate knowledge of what alterations were made and the reasons for making them. Weller reports that ships were to engage targets during the preliminary bombardment according to an established priority system. The assignment of targets in this way was in line with established doctrine. Priority A targets consisted of installations that could threaten ships and aircraft of the invasion force and underwater demolition team operations. Examples included: any battery that opened fire on ships or aircraft of the invasion force; all coast defense guns; all antiaircraft guns; all covered or open artillery emplacements where the presence of weapons was confirmed; and all antitank guns. Priority B targets consisted of installations that threatened the Landing Force in the ship-to-shore movement. Examples of these targets included: blockhouses; covered artillery; pillboxes; and various targets such as command posts, unspecified unidentified installations, earth-covered structures, machine guns, and areas of heavy growth. Priority C targets consisted of installations that could oppose troops after the landing had taken place. This was a broad category that included such targets as caves near the beach, radio and radar stations, ammunition storage areas, fuel dumps, and bivouac areas.[50]

Targets were addressed according to this priority system on D-3 and D-2, but photographic assessments of damage caused by the bombardment on those two days indicated that the level of destruction achieved was unsatisfactory. The weather on D-3 had limited the effectiveness of that day's bombardment and, while the weather had been good on D-2, the skillful use of camouflage by the Japanese increased the number of rounds necessary to take out island installations. An increase in the number of rounds per target meant an increase in the amount of time needed to work through all the targets that presented themselves, time the preliminary bombardment team did not have. These concerns caused Weller to

[50]Marine Corps University Archive, Iwo Jima Project, CG VAC, VAC Landing Force SAR, Annex CHARLIE, Appendix 2, 30 May 1945, Comments of Naval Gunfire Officer, pp. 7–8.

begin thinking of what needed to be done to insure that as effective a bombardment as possible could be delivered in the short time left. The Japanese reaction to UDT operations on D-2 offered proof that gunfire coverage, while knocking out the obvious Priority A targets, had been woefully inadequate in other respects (see Appendix 8.4). Additional action needed to be taken, but what? The solution to the problem was dictated ultimately by yet another event on D-2. Weller states:

> A hitherto undiscovered four-gun battery situated in *Idaho's* area was disclosed in TA 166 VW on top of the bluff in a portion which enfiladed the entire landing beach. [This position was located to the right of the 4th Marine Division landing beaches.] This discovery of additional enemy defensive positions as well as this [*sic*] full realization of the heavy firepower the enemy was able to bear, was of paramount importance. Officers conducting the preliminary bombardment were impressed far more by this development than by negative results of photographic assessment of the day's firing; it brought home the large amount of damage yet to be inflicted on the enemy installations, and impressed all concerned with the absolute necessity of placing the bulk of the remaining fire power in the vicinity of the preferred landing beaches.[51]

To address the lack of destruction in the vicinity of the landing beaches, Weller

> recommended that a maximum concentration of bombardment be placed on and near the preferred landing beaches; the recommendation was approved by the Commander, Amphibious Group 1 [Admiral Blandy]. The schedule for Dog Minus One Day was accordingly abandoned and the following plan substituted. . . . This arrangement placed the fire of four battleships and one heavy cruiser on, and in close vicinity to, the main landing beach. . . . Ships delivering this fire were directed to close the range as early as practicable and to execute maximum destruction. All unexpended ammunition allotted for the preliminary bombardment was to be expended if practicable.[52]

In other words, he advised Blandy to do exactly what doctrine taught should be done: expend ammunition on the landing beaches and adjacent areas because that is where the weapons that are most deadly to the landing force are located. This was a rule of thumb that had served the Marine Corps well for most of the war, but

[51] Ibid., p. 12.

[52] Ibid., p. 15.

by following it in this situation Donald M. Weller fell into Kuribayashi's unintended trap.[53]

The Marine Perspective

Admirals Spruance and Turner had argued that strategic considerations had to take precedence over what Generals Smith and Schmidt saw as overwhelming tactical realities when determining how much preliminary naval gunfire support should be allocated for Operation Detachment. In the end, the Navy won the argument. The Marine commanders finally realized that further remonstrance was futile and that three days was all they were going to get. Upon inspecting Japanese fortifications after the fall of the island, Spruance believed himself justified in limiting the bombardment to three days. Japanese defenses were so thoroughly hardened, in his estimation, that only Marines with rifles and flamethrowers could have gotten them out, regardless of the number of aerial bombs dropped on and naval shells fired at them.[54] For Spruance, naval gunfire doctrine was essentially irrelevant in the face of such strength and organization.

Did the Marines who were given the task of using those rifles and flamethrowers agree with Spruance? Are we able to ascertain their views? Some have left us a record of their personal reminiscences. Charles Lindberg, one of the men involved in the first flag-raising on Suribachi, recalled years later: "We were sitting there [aboard ship] watching [the H-Hour bombardment]. Big battlewagons were sitting back there [firing], and aircraft were coming in. The whole island looked like a big dust bowl. They told us that the thing was bombed for seventy-two days, around the clock. I thought to myself, 'Oh, a couple of days, and we'll be out of here. Nothing to it!'"[55]

Another Marine's memory tended more toward the somber and morbid when he thought back to that day:

> We got out of our bunks that morning and had breakfast. We were anchored off of Iwo Jima. I remember coming out of the ship and walking along the rail. I looked out, and we weren't too far off of the island. It was my first view of it, and it was everything that the map said—but not quite. I thought maybe it was a place with palm trees, something like Saipan, Tinian, and the Marshalls. When I first saw it, the only thing I could see was death. . . . It was dead. I've often described it as something that nobody

[53]Isely and Crowl, on p. 470, praise Admiral Blandy's decision to go with Weller's recommendation.

[54]Buell, *Quiet Warrior*, p. 366.

[55]University of North Texas Oral History Program, Charles W. Lindberg, interview by Ronald E. Marcello and John Chalkley, 19 February 1998, interview OH 1224, transcript, p. 27.

> wanted, and they burned it up. All that was left was ashes. It was gray, and there was smoke coming up from different places where they had bombed. There couldn't be any life on it. It was a picture that one would call Hell.[56]

But what about Marines whose job it was to analyze the results of the operation? The Action Reports of the Fourth and Fifth Marine Divisions provide the official conclusions of various officers attached to the regiments used in the initial assault. Their assessments of the preliminary bombardment are not especially complimentary. According to the Fifth Division report:

> The Japs did not elect to defend the beach. First troops ashore reported occasional enemy small arms fire from the base of Mount Suribachi, mortar fire from a defiladed area in that same vicinity and from the north; they also found coast defense guns on Suribachi out of action and nearly every pillbox on the beach and ground in rear of beach destroyed. The troops crossed the island in about [3½] hours. . . . [The preliminary NGF bombardment was intense and well executed, but it did not]: 1. succeed in taking out a large percentage of the enemy fortifications; 2. prevent the enemy from being able to place accurate artillery and mortar fire from prepared positions on any part of the island.[57]

The Fourth Division report was not so generous:

> It is estimated that approximately [10%] of the beach defenses were knocked out and all enemy troops neutralized by scheduled bombardment in RCT 23 zone of action. Destruction was principally accomplished by heavy caliber shells. High velocity guns were not knocked out on the flanks and automatic weapons from emplacements along abutments of the airfield to the front were active. Mortars from the flanks and reverse slopes exacted heavy casualties on the beaches. Construction of emplacements, in conjunction with terrain, rendered area fires largely ineffective, although observation revealed it possible to close small caves with 5" projectiles. . . . In RCT 25 zone of action large pillboxes on beach Blue 1 were knocked out by heavy caliber fire prior to landings. Effect of 5," rocket, and mortar fires on beaches was not visible due to the sandy nature of the beach. The rolling barrage screened initial waves sufficiently to enable them to organize on the beach but enemy riflemen had held

[56]University of North Texas Oral History Program, Donald Boots, interview by Ronald E. Marcello, 17 May 2001, interview OH 1405, transcript, p. 156.

[57]Marine Corps University Archive, Iwo Jima Project, CG 5th Marine Division, 5th Marine Division Action Report, Annex HOW, Naval Gunfire Support, pp. 4, 12, and 13.

> their positions and were able to fire through the barrage from the right flank. Most emplacement observable from seaward were destroyed and the fire received came from reverse slopes and emplacements further inland.[58]

Donald Weller provides us with a more sanguine analysis. He states that "the Dog-Day bombardment, in conjunction with the destruction accomplished in the preliminary firing, enabled a successful landing to be made against one of the enemy's most heavily and skillfully defended possessions, with acceptable casualties."[59] There are a number of problems with this conclusion. First, the person expressing it was integrally involved in the naval gunfire planning process and it can be argued that he was hardly unbiased in his assessment of its results. Second, the number of American casualties was not considered acceptable by everyone. The extent to which the casualty figures affected later American decisions will be dealt with presently. Third, and most importantly, Weller's conclusion does not take into account Kuribayashi's decision to grant the landing to the invasion force—a decision made long before the first gunfire support shell struck Iwo on D-3. Their emphasis on the "Lessons of Tarawa" and their study of the application of those lessons in operations over the course of 1944 insured that the Navy and Marine Corps planned to fight a different battle at Iwo Jima than the one they were forced to fight. Even with photographic proof of the depth and complexity of Iwo's defenses before them, the naval gunfire emphasis on D-1 and on D-day was still on the landing beaches. Marines over the intervening years have argued that a longer pre-assault bombardment would have cut down the number of U.S. casualties. They are conceivably correct, but just how much more damage to the enemy could have been done given that the Japanese were buried underground and their emplacements well concealed? American Central Pacific amphibious warfare planners failed to understand fully the changes that occurred in Japanese defensive strategy and, consequently, U.S. naval gunfire doctrine failed to adapt sufficiently to address this changed strategy. In failing to adapt, it also failed to meet one of its most important objectives: to minimize American casualties.

Weller has spent his life since the invasion of Iwo Jima analyzing the use of naval gunfire in support of amphibious operations and explaining away the planning and execution anomalies that plagued that operation. He was not the only Marine artillery officer present during that hellish month on Sulfur Island, however, and at least one of them has left us a record of his own considered opinions as to the efficacy of the gunfire plan. Lieutenant Colonel Kenyth A. Damke served as the Executive Officer of the 13th Marines, the artillery regiment of the Fifth Marine Division, at Iwo.

---

[58]Marine Corps University Archive, Iwo Jima Project, CG 4th Marine Division, 4th Marine Division Operation Report, Annex CHARLIE, Operations, pp. 18–19.

[59]VAC Landing Force SAR, Comments of Naval Gunfire Officer, p. 34.

He took the Senior Course in the Amphibious Warfare School after the war and wrote a paper in which he analyzed the pre-D-day naval gunfire bombardment at Iwo Jima. The question he attempted to answer: did pre-D-day naval gunfire accomplish its mission in Operation Detachment? He reviewed Action Reports from several levels of command and considered everything from the stated mission of naval gunfire and the priority of targets set forth in the plans to the hydrographic conditions in the waters surrounding the island. His conclusions:

> Adopting a very strict interpertation [*sic*] of the mission of the pre-D-day bombardment of an objective by naval gunfire, as given in U.S. F. 63, the over all [*sic*] mission was not accomplished at IWO JIMA.
>
> Hundreds of targets remained; naval gunfire was capable of destroying them. At the rate of destruction achieved on D-1, another five (5) days of bombardment, under conditions similar to those existing for that day, would have been required to have reduced the fortifications.
>
> The destruction of defensive installations in the prefered [*sic*] beach areas was generally very good. A landing force had been placed ashore which had captured the island; in doing so, it had suffered a very high percentage of casualties. Who can say what the casualties might have been in this force and in other forces with in [*sic*] the theater, had the bombardment been extended until more of the enemy installations opposing the landing force had been destroyed? As in other major decisions in warfare which involves large forces in large theaters, the answer to this question will never be known. Suffice to say that insufficient time and ships were available with in [*sic*] the theater to allow additional bombardment of IWO JIMA before the landing force went ashore.[60]

[60]Marine Corps University Archives, Papers and Theses Collection, Kenyth A. Damke, "Was the Mission of the Pre-D-Day Bombardment by Naval Gunfire Accomplished at Iwo Jima?," paper presented in fulfillment of 1947–1948 Senior Course requirements at the Amphibious Warfare School, Marine Corps Schools, Quantico, VA, pp. 12–13. Isely and Crowl mention Damke's paper. They state on page 441: "An excellent post-operational study of the preliminary fires delivered during the three-day period at Iwo reveals that whereas the purely naval missions—protecting shipping from coastal defense guns, curtailing antiaircraft fire, and allowing minesweepers, underwater demolition teams, and hydrographic experts to do their work—were accomplished in the time allowed, the period required to prepare the island as thoroughly as was possible for occupation by assault troops would have been at least eight days. This study is premised upon the continuous employment of six old battleships and four heavy and one light cruisers, which, along with destroyers, were the vessels ultimately sent to fire on Iwo for three days before the assault date." Their summary is accurate, but they do not address Damke's concerns directly in their analysis. U.S. F. 63, Amphibious Instructions Landing Forces, 1947, was a distillation of amphibious lessons learned during World War II. It represented the latest official Navy statement of amphibious doctrine at the time Damke was writing.

He takes into account the strategic arguments made by Spruance and Turner that resulted in the limitation of gunfire support to only three days, but he still concludes that the amount and quality of the support were inadequate. Damke, a man who was there, a man who witnessed much of the carnage wreaked upon his fellow Marines, and a man who was trained in the proper use of artillery in support of land forces, had serious questions about the planning for and the effectiveness of pre-D-day naval gunfire at Iwo Jima.

Conclusion

Despite the inadequacies of the American naval gunfire plan, Operation Detachment was successful in the stated goal of taking Iwo Jima from the Japanese. Can the same be said of Kuribayashi and his goals? On the whole, he too can be said to have been successful. First, he made the Marines pay an extremely high price for the taking of Iwo Jima. Second, the sacrifice of himself and his command did affect U.S. thinking concerning the invasion of Japan. As mentioned earlier, Admiral King already had doubts about invading the Home Islands before the Iwo Jima operation commenced. He had agreed to preparations for an invasion as a contingency plan, believing that it was only prudent to make such plans because an invasion indeed might prove to be necessary. It could be called off easily, however, in the event that it was not. Chester Nimitz expressed support for Operation Olympic, the codename for the invasion of Kyushu, in a message to King on 28 April, but he withdrew his support in an "eyes only" message to King on 25 May. The heavy casualties suffered by U.S. forces at Iwo Jima, the slow progress of fighting on Okinawa, and intelligence reports of a buildup of Japanese defenses on Kyushu caused him to reconsider. King was about to make Nimitz's change of heart known to the Joint Chiefs of Staff when the dropping of the atomic bombs on Hiroshima and Nagasaki made the arguments concerning invasion moot.[61] Having accomplished his two major objectives, therefore, it can be argued that Kuribayashi's essentially hopeless defense of Iwo Jima was not totally in vain.

In conclusion and summary, reference must be made once again to Turner's third point in his 9 January 1945 communication to Generals Smith and Schmidt. American commanders at the time did not appreciate the accuracy of Turner's assessment of what the Marines would find upon going ashore at Iwo. Neither has it been understood fully in the years since, judging from the post-war writings of

[61] Richard B. Frank, *Downfall: The End of the Imperial Japanese Empire* (New York: Random House, 1999), pp. 37, 147, 212, and 276; John Ray Skates, *The Invasion of Japan: Alternative to the Bomb* (Columbia: University of South Carolina Press, 1994), pp. 21, 25–26, 29, 44–45, and 54. Skates does not mention Nimitz's change of heart indicated in the Admiral's 25 May "eyes only" communication to King.

men such as Donald M. Weller.[62] The Japanese, for their part, were not in a position to know of Turner's assessment. General Kuribayashi knew that American firepower would destroy anything he put on the beach. This realization led him to insist on a defense in depth. Weapons near the beach were not an integral part of his plan. Their presence represented a compromise that was forced upon him by his more conservative subordinates.[63]

Placing weapons near the beach, however, raised a most formidable defensive scheme to the level of sheer brilliance. Turner worried that concentrating naval gunfire on the invasion beaches would be counterproductive because to do so would draw fire away from other targets that needed attention as well. There was merit to his argument, but the pertinent reasons for that merit were apparent neither to him nor to any other American commanders at the time. The presence of the near-beach installations, afterthoughts though they were to Kuribayashi, insured that the Americans would spend their last precious day of preliminary bombardment blasting away at emplaced weapons that their doctrine told them were most dangerous to the assaulting troops and therefore should be given maximum priority in the firing schedule. The area where these weapons were located, regardless of what doctrine taught, was the one place on Iwo Jima where a massive expenditure of ammunition such as the Americans poured forth on D-1 was guaranteed to do the least amount of good for the invaders coming ashore. Though he did not mean to do so, Kuribayashi was able to use the Americans' own doctrine against them.

Appendix 8.1[64]

One of the best lists of these "lessons" is found in Rear Admiral Harry W. Hill's "Summary of Recommendations for Naval Gunfire Support," which he included as part of his Operation Galvanic After Action report. Hill recommended:

1. That immediate steps be taken to construct, at home base areas, dugouts, pillboxes and bomb proof shelters duplicating those found on [Betio] Island, and that aerial and naval gunfire bombardment be conducted on these for training and determination of effectiveness of bombardment methods and ammunition. . . .

---

[62] See, for example, Donald M. Weller, "Salvo-Splash!—The Development of NGF Support in World War II," *Proceedings* 80:8 (August 1954); Donald M. Weller, *Gunfire Support of Amphibious Operations: Past, Present, and Future* (Dahlgren, VA: Naval Surface Weapons Center, 1977).

[63] *Handbook on Japanese Military Forces*, p. 99; Bartley, *Amphibious Epic*, p. 11; Garand and Strobridge, *Western Pacific Operations*, pp. 453–59; Alexander, *Storm Landings*, pp. 131–32.

[64] NARA, RG 127, USMC Geographic Files, Box 26, File A3-1, Commander Task Force 53, Report of Tarawa Operations, dated 13 December 1943, Enclosure A, NGF Supplement n.d., pp. 41–42.

3. That fire support ships be strongly impressed with the necessity for delivering fire at close ranges for effective bombardment.

4. That landing plans definitely provide for destroying automatic weapons and small caliber batteries remaining in action after assault waves have landed.

5. That, in event of a pre-dawn approach, fire should commence as soon as earliest light permits and before the transport area can be taken under fire by shore batteries.

6. That naval gunfire must not be discontinued during delivery of aerial bombing strikes. . . .

8. That the bombardment allowance of ammunition be increased and include AP projectiles; that a bombardment allowance of common projectiles be established and made available for 5"/38 and 5"/25 batteries. Percentages of ammunition types must depend upon a study of the problem, but in general the delay action fuze will be required during all bombardment phases except Phase TWO. . . .

9. That, if fire support ships can be assigned early, the initial fire support plans outline only in broad form the overall bombardment effect desired, the general areas to be covered, and the type and amounts of ammunition to be expended for its accomplishment. That the details of fire schedules and their coordination with maneuvers required, track charts, and other details of section and individual ships problems, be delegated to Fire Support Section Commanders for planning and solution.

10. That gunfire support plans for future bombardment of enemy held islands similar to [Betio] embody the following:

- (a) Five general phases of bombardment as follows:
  - (1) Initial counter-battery fire against known strong points using HC and AP projectiles, and commencing at ranges producing an angle of fall of 15 degrees.
  - (2) General area bombardment using HC and AA common projectiles delivered at ranges between 10,000 and 5,000 yards.
  - (3) Destruction of heavy defenses, pillboxes, and dugouts along the landing beaches by slow, accurate, and deliberate fire, using AP and common projectiles. This fire is to be delivered at close ranges using pointer fire if practicable.
  - (4) Heavy concentration fire at close range on landing beach areas during the boat approach.
  - (5) Delivery of call fire by ships assigned after H hour [*sic*].
- (b) Use of a rolling barrage system of fire rather than radical shifting of target areas during area bombardment.
- (c) Restriction of air burst firing for use only as a weapon of opportunity against exposed personnel.

(d) Maximum use of 40[mm] batteries whenever range and other conditions permit and control of those batteries by 5-inch gun directors if fire control installation permits.

(e) Stationing of close fire support ships in a favorable position relative to the landing beaches, and at closest safe navigation range, so as to provide their continuous observation of assault boat waves, permitting these ships to decide at what time their fire must be ceased for safety of landing personnel.

(f) Destruction of hulks, latrines, and similar potential machine gun nests located in favorable beach enfilading positions.

(g) Plans for local supply of destroyer and light cruiser replacement ammunition, and depth charges.

(h) Readiness for immediate shift to radar controlled gunfire during all phases of bombardment except [numbers] 3 and 5 [of paragraph 10 (a) above].

Appendix 8.2[65]

Task Organization for Preliminary Naval Gunfire Support Group (Primary and Secondary armament in parentheses. Primary armament only for destroyers.)

54.1.1 Fire Support Unit ONE
BB 36 USS *Nevada* (10 x 14″/45; 16 x 5″/38)
BB 42 USS *Idaho* (12 x 14″/50; 10 x 5″/38)
@CA 35 USS *Indianapolis* (9 x 8″/55; 8 x 5″/25)
CL 86 USS *Vicksburg* (12 x 6″/47; 12 x 5″/38)
DD 586 USS *Newcomb* (5 x 5″/38)
DD 663 USS *H. L. Edwards* (5 x 5″/38)
DD 664 USS *R. P. Leary* (5 x 5″/38)
DD 662 USS *Bennion* (5 x 5″/38)

54.1.2 Fire Support Unit TWO
BB 43 USS *Tennessee* (12 x 14″/50; 16 x 5″/38)
CA 25 USS *Salt Lake City* (10 x 8″/55; 4 x 5″/25)

[65]Marine Corps University Archive, Iwo Jima Project, Information on Fire Support Unit assignments from Commander Amphibious Forces (CTF 51) Operation Plan No. A25–44, Iwo Jima Operation, Annex H, Change 2, p. (H)—1. Information on armament from Paul H. Silverstone, *U. S. Warships of World War II* (Shepperton: Ian Allan, 1965), and Stefan Terzibaschitsch, *Battleships of the U. S. Navy in World War II,* translated and adapted by Heinz O. Vetters and Richard Cox (New York: Bonanza Books, 1977).

*CL 60 USS *Santa Fe* (12 x 6″/47; 12 x 5″/38)
#CL 65 USS *Pasadena* (12 x 6″/47; 12 x 5″/38)
DD 597 USS *Wiley* (5 x 5″/38)
DD 513 USS *Terry* (5 x 5″/38)
DD 584 USS *Halligan* (5 x 5″/38)
DD 583 USS *Hall* (5 x 5″/38)

54.1.3 Fire Support Unit THREE
@BB 56 USS *Washington* (9 x 16″/45; 20 x 5″/38)
BB 35 USS *Texas* (10 x 14″/45; 6 x 5″/51; 10 x 3″/50)
BB 33 USS *Arkansas* (12 x 12″/50; 6 x 5″/51; 10 x 3″/50)
CA 37 USS *Tuscaloosa* (9 x 8″/55; 8 x 5″/25)
#CA 69 USS *Boston* (9 x 8″/55; 12 x 5″/38)
#CA 72 USS *Pittsburgh* (9 x 8″/55; 12 x 5″/38)
*CL 80 USS *Biloxi* (12 x 6″/47; 12 x 5″/38)
#CL 90 USS *Astoria* (12 x 6″/47; 12 x 5″/38)
DD 644 USS *Stembel* (5 x 5″/38)
DD 553 USS *J. D. Henley* (5 x 5″/38)
DD 590 USS *Hamilton* (5 x 5″/38)

54.1.4 Fire Support Unit FOUR
@BB 55 USS *North Carolina* (9 x 16″/45; 20 x 5″/38)
BB 34 USS *New York* (10 x 14″/45; 6 x 5″/51; 10 x 3″/50)
CA 27 USS *Chester* (9 x 8″/55; 8 x 5″/25)
CA 24 USS *Pensacola* (10 x 8″/55; 4 x 5″/25)
#CA 38 USS *San Francisco* (9 x 8″/55; 8 x 5″/25)
DD 665 USS *Bryant* (5 x 5″/38)
DD 551 USS *D. W. Taylor* (5 x 5″/38)
DD 591 USS *Twiggs* (5 x 5″/38)
DD 550 USS *Capps* (5 x 5″/38)
DD 481 USS *Leutze* (5 x 5″/38)
#CL 66 USS *Springfield* (12 x 6″/47; 12 x 5″/38)
CL 103 USS *Wilkes-Barre* (12 x 6″/47; 12 x 5″/38)

Ships also mounted 40mm and 20mm guns that could be used for gunfire support. The 40mm mounts were especially effective against soft material targets and exposed enemy personnel when support ships could approach close to shore.

@ = Attached from TF 58 DOG and DOG plus ONE
* = Attached from TF 58 DOG or DOG plus ONE
# = Attached from TF 58 DOG Day possibly

Appendix 8.3[66]

The referenced quote is as follows:

> The planning for and the actual execution of the Iwo Jima operation were affected to a considerable extent by the operations in the Philippines which immediately preceded it, and by the necessity of preparing for the Okinawa operation which was to follow it.
>
> The Philippine operations necessitated last minute changes and reduced the total number of ships which had been previously allocated to the Iwo Jima operation. This applied primarily to battleships, cruisers, and destroyers for the Joint Expeditionary Force, although other forces were also affected to a lesser extent. The only source from which additional gunfire support ships could be obtained was from Task Force 58 (Fast Carrier Force). In order to provide the necessary additional battleships for gunfire support, the Commander in Chief, United States Pacific Fleet, authorized the use of the *North Carolina* and *Washington* for this purpose. These two ships were, accordingly, loaded principally with bombardment ammunition, their service allowance of armor piercing ammunition being correspondingly reduced. Certain cruisers of Task Force 58 were also loaded with some additional bombardment ammunition.
>
> The Okinawa operation affected the Iwo Jima operation in two ways. First, it was felt that the threat of Japanese land based aircraft, while taking Okinawa, would be very great, both because of the great value of that island to us when we took it and because of its closeness to the Empire (325 miles). Therefore, anything which we could do during the Iwo Jima operation to reduce Japanese air strength, either in aircraft and, even more, in aircraft production facilities, would help the Okinawa operation as well as support the Iwo Jima operation. The second way in which the Okinawa operation and the Iwo Jima operation affected each other was in the close timing of the two operations. . . .
>
> After withdrawing from the Tokyo area to the westward of the Nanpo Shoto, Task Force 58 refueled to the westward of Iwo Jima, sending the *North Carolina, Washington, Birmingham,* and *Biloxi* to take part in the bombardment in support of the landings on 19 February (D-day), and furnishing air support.

The wording of Spruance's summary points to another of his concerns: how to keep the ships he could obtain supplied with large-caliber ammunition over the course of a long bombardment. Both Spruance and Turner are careful throughout

---

[66]COMINCH P-0012, p. 1–1.

the planning process to warn their ship commanders about the logistical problems associated with ammunition procurement. Isely and Crowl discuss on pages 444–45 how this problem could have been avoided.

A further point can be made in reference to this quote concerning Burrell's interservice rivalry thesis. While it is true that this is a summary written after Operation Detachment had been completed and, therefore, could have been an attempt by Spruance at justification after the fact, a comparison of this statement with the objectives specified in Spruance's 31 December 1944 operational plan shows that these ideas were uppermost in his mind from the very beginning.

Appendix 8.4[67]

Underwater Demolition Teams took samples of beach sand on D-2. According to the Operation Plan for the UDTs, among the types of information that they were to obtain was: "[the w]idth, gradient, and material of beach proper, with special attention to the following: (a) gradient and condition of beach at high water mark (b) contour of beach as to feasibility for attack forces to exit quickly with vehicles from high water mark (c) material of beach and an estimate as to what type of vehicle it will support."

Isely and Crowl report that, with a few exceptions, the UDTs "optimistically" deemed the beaches suitable for landings. One of those exceptions was Sergeant Frank E. Schnell. Sergeant Schnell was an observer with UDT #14. He filed a report of his observations after the operation. Included in this report was the following:

> [After UDT operations on 17 February were finished,] [w]e then headed back toward SAIPAN to meet the landing force. On the morning of 18 February 1945, I left the APD (USS WATERS), and went aboard the AGC #10 (USS AUBURN) to report to Admiral Hill. The following questions were ask [*sic*] by him: If the condition of the beaches would permit an LVT to climb over the terraces and I answered, "in most places they would." If rubber tired vehicles could climb them, and I said, "that I didn't think they could make it." Whether or not the beach was hard surfaced? I said no, they were soft sandy beaches.

---

[67]Marine Corps University Archives, Iwo Jima Project, Commander UDTs Amphibious Forces Pacific Operation Plan No. A1–45, 23 January 1944, Annex (A)—Daylight Reconnaissance Procedure, Appendix (I), p. (A)(I)-1; Isely and Crowl, *U.S. Marines and Amphibious Warfare*, p. 469; Marine Corps University Archive, Iwo Jima Project, CG V Amphibious Corps, VAC Report, Iwo Jima Campaign, 13 May 1945, Annex CHARLIE, Appendix 13 (Headquarters Commandant Report), Enclosure B, p. 1; VAC, Report, Annex CHARLIE, Appendix 10 (Shore Party Report), pp. 14–15, contains a description of the beach conditions; quote concerning the Japanese reaction to UDT operations found in COMINCH P-0012, p. 1–13.

Hill decided to go with the majority opinion. At the point when he received this information from Sergeant Schnell, there was not much else that he could do. It could be argued that an assessment of beach conditions should have been made earlier than two days before the landings were to occur, but putting men ashore on Iwo before the U. S. was prepared to launch an assault could have compromised the security of the operation. An accurate understanding of what the assault forces would be facing might have made a difference on D-day, but two days is precious little time to make any significant changes in the landing plan. Concerning the Japanese reaction to UDT operations on DOG minus 2 (17 February):

> By 1048 (K) all units were in position to commence the UDT reconnaissance set for 1100 (K). The last of the minesweepers was completing the sweep off the eastern beaches, these small ships having gallantly passed close along the eastern shore in precise formation, firing as they went, without deviation from their prescribed tracks although under occasional enemy fire. The UDT reconnaissance commenced exactly on schedule. As the LCI(G)'s [*sic*] moved in toward the beach, enemy fire began to concentrate on them. By 1105 (K), when they reached a point 1,000 yards off shore, enemy fire was intense from both medium and minor caliber weapons on the flanks and minor caliber along the beaches. The personnel of these little gunboats displayed magnificent courage as they returned fire with everything they had and refused to move out until they were forced to do so by materiel and personnel casualties. Even then, after undergoing terrific punishment, some returned to their stations amid a hail of fire, until again heavily hit. Relief LCI(G)'s replaced damaged ships without hesitation. Between 1100 (K) and 1145 (K) all 12 of the LCI(G)'s were hit. LCI 474 ultimately capsized after the crew had been removed, and was ordered sunk. Intensive fire from destroyers and fire support ships, and a smoke screen laid by white phosphorous projectiles, were used to cover this operation. Fire support ships took on board casualties from the LCI(G)'s as they withdrew, and CTG 52.3 in *Terror* most promptly and efficiently initiated emergency repairs for serious hull damage, as well as assisting in care of the wounded. At 1121 (K) *Leutze* was hit, the commanding officer receiving serious injuries, requiring his later transfer to *Estes,* but no extensive damage was sustained by the ship. By 1220 (K) all swimmers of the UDT's but one had been recovered, and the APD's and supporting destroyers moved out of the area. The reconnaissance had been accomplished.

9

Naval Power, Mao Zedong, and the War in China, 1926–1949

Francis Grice

The War in China began as a local struggle between the nationalist Kuomintang party and the Communist Party of China, became part of a regional conflict after the Japanese invasion of China in 1937, grew into an important part of the global Second World War following the outbreak of hostilities between Japan and America (and to a lesser extent Britain and Holland) in 1941, and then collapsed back into a primarily local civil conflict after the surrender of Japan in 1945.[1] The numerous interconnections that existed between events at a local, regional, and global level make the War in China an excellent example of a hub within a global-regional nexus. The links between these local, regional, and global tiers were facilitated to a large extent by the sea, and indeed naval power had a profound and lasting impact on the character and outcome of the War in China, from its beginning in the mid-1920s all the way through until its conclusion in 1949.

Despite popular conceptions to the contrary, the rise of Mao Zedong and the Communist Party of China was heavily influenced by naval-oriented events. As this chapter will demonstrate, naval power had a real and distinct impact on the efforts of the Communists to gain power throughout the period.[2] However, despite the frequency of their encounters with naval affairs, Mao made scarcely any mention of the topic within his myriad writings. This chapter is comprised of three main sections. The first provides an account of the impact of naval power on China during the first half of the twentieth century, as well as the historical context that preceded it, with particular emphasis on the interactions between naval power and the nascent Communist Party of China. This account is significant because it

[1]S. C. M. Paine, *The Wars for Asia, 1911–1949* (Cambridge: Cambridge University Press, 2012), pp. 3–12. This chapter was originally published in *Global War Studies* 11:1 (2014).

[2]This chapter uses the pinyin translations for the majority of names of Chinese people, ships, and locations. However, in some cases where names are commonly known within the older Wade-Giles system, these have been used instead. The Wade-Giles translations of Chiang Kai-Shek, Kuomintang, and Nanking are, for example, both widely known and used, whereas their pinyin translations of Jiang Jieshi, Guomindang, and Nanjing are far less commonly recognized.

represents the first real effort to appraise the interactions between naval power and the Communist Party throughout all of the struggles it was involved in during this period, rather than covering limited portions of the subject as part of a larger narrative. The second section exposes the dearth of references to naval power within Mao's works. It does this both by highlighting how rarely the topic was mentioned and by demonstrating how even the scarce references that are made portray little real interest in discussing naval power in a focused or serious manner. The chapter then argues that Mao neglected this area of warfare because he lacked the necessary strategic vision to recognize its significance and because his focus on providing a "victor" narrative repelled him from dwelling on naval power, an area which had primarily been filled with humiliating Communist failures. The third section highlights how Mao's omission of naval power from his writings helped to create an era of counterinsurgent thought that focuses almost exclusively on land power (and to some extent air power) and neglects the role of naval power, despite its demonstrated relevance for both insurgent and counterinsurgent forces in many modern conflicts.

A number of authors have discussed the conflicts which wracked China during the first half of the twentieth century. Some of these have focused on the Pacific War between Japan and America (and to a lesser extent Britain and Holland), while portraying the fighting between Japan and China after 1937 as a precursor to this conflict.[3] Others have concentrated on the Sino-Japanese War of 1937–1945, with the Pacific War presented as an important factor which influenced the outcome of this struggle.[4] A third group has honed in on the Chinese Civil War between 1945 and 1949, with earlier conflicts acknowledged briefly as the foundation for the final titanic clash between the Communists and the Kuomintang.[5] Recently, one scholar has endeavored to bring these three strands of research together by examining the overlapping conflicts holistically as an interconnected system rather than isolated struggles.[6] Other authors have considered the country during this period without concentrating on the wars being fought. There exist, for example, a variety of accounts that chart the rise of the Communist Party of China

[3]Andrew Wiest and Gregory L. Mattson, *The Pacific War: From Pearl Harbor to Hiroshima* (London: Spellmount, 2001); Dan van der Vat, *The Pacific Campaign* (London: Hodder and Stoughton, 1992).

[4]Frank Dorn, *The Sino-Japanese War, 1937–41: From Marco Polo Bridge to Pearl Harbor* (New York: Macmillan, 1974); James C. Hsiung and Steven L. Levine, eds., *China's Bitter Victory: The War with Japan 1937–1945* (London: M. E. Sharp, 1992); Mark Peattie, Edward Drea, and Hans van de Ven, eds., *The Battle for China: Essays on the Military History of the Sino-Japanese War of 1937–1945* (Stanford: Stanford University Press, 2011).

[5]Odd Arne Westad, *Decisive Encounters: The Chinese Civil War, 1946–1950* (Stanford: Stanford University Press, 2003); Lionel M. Chassin, *The Communist Conquest of China: A History of the Civil War, 1945–49* (London: Weidenfeld and Nicolson, 1966).

[6]Paine, *The Wars for Asia.*

from its creation in 1921 until its triumph in 1949.[7] There are also a number of works that explore political, social, and military themes as part of wider histories of China.[8] Biographies on specific historical figures such as the child emperor Pu Yi, Chiang Kai-shek, and Mao Zedong provide yet further information, as do writings which scrutinize specific events or specialist subjects such as the escape of HMS *Amethyst* after its impoundment in Nanking in 1949 or the development of citizen armies in history.[9] Finally, a number of scholars have examined the rise of Mao and Communism in China as part of the history of insurgencies and guerrilla warfare.[10]

In many of these works, the role of naval power in China is discussed to some extent, but always in a fragmentary fashion within a larger and less pertinent narrative. The result is that while many of the naval events which affected the Communist Party of China during its rise to power have been previously described, there has never before existed a comprehensive picture of the presence and impact of naval power on the rise of Chinese Communism. This has led to the relevance of naval power in this area being obscured and its significance being marginalized. Consequently, the first section of this chapter draws on a rich diversity of government reports, letters, newspaper reports, accounts written by witnesses and journalists shortly after the war, and pertinent secondary sources to provide, for the first time, a dedicated appraisal of the impact of naval power on the rise of the Communist Party of China. This examination demonstrates that, while naval power was by no means the most important influence on the conflicts in which the Communist Party was involved during this period, it nonetheless played a significant role.

---

[7]James P. Harrison, *The Long March: A History of the Communist Party of China, 1921–72* (New York: Praeger, 1974); Andres D. Onate, *Chairman Mao and the Communist Party of China* (Chicago: Nielson Hall, 1979).

[8]Hans J. van de Ven, *War and Nationalism in China, 1925–1945* (London; New York: RoutledgeCurzon, 2003); James E. Sheridan, *China in Disintegration: The Republican Era in Chinese History, 1912–1949* (New York: The Free Press, 1975); Jonathan Fenby, *The Penguin History of Modern China: The Fall and Rise of a Great Power, 1850–2008* (London: Penguin, 2008); Erik Durschmied, *Beware the Dragon: China—1,000 Years of Bloodshed* (London: André Deutsch, 2008).

[9]Brian Power, *The Puppet Emperor: The Life of Pu Yi, Last Emperor of China* (London: Corgi, 1986); Peter Carter, *Mao* (London: Oxford University Press, 1976); Jay Taylor, *The Generalissimo: Chiang Kai-shek and the Struggle for Modern China* (London: Harvard University Press, 2011); C. E. L. Phillips, *Escape of the Amethyst* (London: William Heinemann, 1957); Jock Haswell, *Citizen Armies* (London: Peter Davies, 1973).

[10]John Pimlott, ed., *Guerrilla Warfare* (London: Bison Books, 1985); David Rooney, *Guerrilla: Insurgents, Patriots and Terrorists from Sun Tzu to Bin Laden* (London: Brassey's, 2004).

I. Naval Power and the Rise of the Communist Party of China

During their rise to power, Mao and the Communist Party of China became enmeshed within a series of overlapping and interlinked struggles that pitted the Communists against asymmetrically stronger foes. These included a series of clashes with the Kuomintang prior to the Japanese invasion in 1937; a fluctuating matrix of warfare between the Communists, the Kuomintang, and the Japanese during the War with Japan; and the Civil War between the Kuomintang and the Communists that followed. These conflicts, along with the history that preceded them, provided the context in which Mao and the Communist Party developed. One common characteristic throughout all of these periods was the presence of naval power, either as an important factor in determining the nature and outcome of the conflicts, or at least as a feature with a clearly identifiable role and impact. This first section provides a detailed appraisal about the multiple ways that naval power impacted on the Communists, including its role in the general history that preceded their founding and its influence on the three broad periods of conflict that the Communists subsequently battled through.

Gunboat Humiliations: 1839 to the early 1920s

The background to the rise of both the Kuomintang and the Communist Party of China was rooted firmly in the context of centuries of encroachment by foreign powers within Chinese affairs. While the Manchu Empire had reached the zenith of its wealth and prestige towards the end of the eighteenth century, it was brought to its knees during the nineteenth century through a series of wars with the rising powers in the West, each of which was dominated by Western naval power.[11] Lin Tse-Hsu, who was the Imperial Commissioner at Canton at the start of the First Opium War of 1839–42, recorded his perspective about the impact of overwhelming British naval superiority: "The rebels' ships on the open sea came and went as they pleased, now in the south and now suddenly in the north, changing successively between morning and evening. If we tried to put up a defense everywhere, not only would we toil and expend ourselves without limit, but also how could we recruit and transport so many troops, militia, artillery, and ammunition, and come to their support quickly?"[12] He continued by outlining how the possession of naval power would have made a huge impact for helping the Chinese fight, and should remain an urgent priority for the Chinese empire:

---

[11]Robert Smith Thompson, *Empires on the Pacific: World War II and the Struggle for the Mastery of Asia* (Oxford: The Perseus Press, 2001), p. 6.

[12]Lin Tse-Hsu, "The Need for Western Guns and Ships, 1844," in William T. De Bary, ed., *Introduction to Oriental Civilizations: Sources of Chinese Tradition* (London: Columbia University Press, 1963), p. 671.

> I recall that after I had been punished two years ago, I still took the risk of recalling the Emperor's attention to two things: ships and guns. At that time, if these things could have been made and prepared, they still could have been used with great effect to fight against the enemy in Chekiang last fall. Now it is even more difficult to check the wildfire. After all, ships, guns and a water force are absolutely indispensable. Even if the rebellious barbarians had fled and returned beyond the seas, these things would still have to be urgently planned for, in order to work out the permanent defense of our sea frontiers.[13]

Whether because of a lack of political will or economic means, this advice was never followed, with tragic results. Writing in China nearly a century later, Chiang Kai-shek highlighted how defeat in this war had represented the first of many national humiliations.[14] For the remainder of the century, foreign powers from across the world swept in repeatedly to impose their will on China, often relying heavily on superior naval power to overwhelm their opponents and force them to provide ever more territorial, economic, and political concessions. By the end of the nineteenth century, China had ceded huge swathes of its territory to the Americans, British, French, Germans, Japanese, and Russians.[15]

The extent of China's impotence in the face of foreign maritime power was most acutely exposed at the turn of the century when the Movement of Harmonious and Righteous Fists—also known as the Boxers—rose up in revolt against the incumbent Manchu Dynasty. Whether seeking to use the uprising to solidify her authority, or impressed by its potential to rid the country of foreign influence, the Empress Dowager gave her support to the Boxer movement and approved their actions against the foreigners occupying China.[16] Led by the mysterious Yellow Lotus, the Boxers marched on first Tianjin and then Peking, laying siege to the former for a month and the latter for seven weeks. In response, the foreign powers put together a coalition force that drew from six European countries, the United States, and Japan.[17] Composed of fifty-one ships and 54,000 soldiers and marines, this combined force crushed the revolt swiftly and brutally, forcing the Empress

---

[13]Ibid., p. 672.

[14]Chiang Kai-shek, *China's Destiny* (New York: Da Capo, 1976), p. 24.

[15]Ibid., pp. 8–15.

[16]James H. Wilson, *Travels and Investigations in the "Middle Kingdom"—A Study of its Civilization and Possibilities. Together with an Account of the Boxer War: The Relief of the Legations and the Re-Establishment of Peace* (New York: D. Appleton, 1901), pp. 356–57; Sarah P. Conger, *Letters from China: With Particular Reference to the Empress Dowager and the Women of China* (Chicago: A. C. McClurg, 1910), pp. 139–42.

[17]Wilson, *Travels and Investigations*, pp. 359–60.

Dowager to flee the Forbidden City and enter into hiding.[18] A huge indemnity was imposed and the end of the last Chinese Empire was ushered into sight: three-year-old Pu Yi was removed from the throne in 1912 in the midst of a series of violent peasant uprisings and replaced by a Republican government.[19]

These events then presented the context in which the Communists in China began their rise to power. Foreign naval power had brought the country to its knees and the presence of special foreign territories ("the treaty ports"), protected when needed by state-of-the-art warships, acted as a constant reminder of this stark reality throughout much of the first half of the twentieth century, at least until the Japanese invasion in 1937.

Early Clashes: The late 1920s and 1930s

Cadets at the Kuomintang's Whampoa Military Academy enjoying the evening air at Changzhou Island on 18 March 1926 would have found their ordinarily peaceful surroundings disturbed by the unexpected arrival of *Zhongshan.* Sporting a formidable-looking turret, this prestigious vessel was not only the strongest ship in the recently-formed National Revolutionary Army, but she also drew on an esteemed history that included a leading role in Sun Yat-Sen's campaigns against the warlords Yuan Shikai in 1911 and Chenjiong Ming in 1925.[20] Chiang Kai-shek himself recorded how, under instruction from Li Chih-lung, the vessel had sailed to Whampoa from the southern sea port of Canton, only to turn around immediately upon arrival and return promptly to Canton.[21] The exact motivations that drove the gunship's movements remain a mystery—Chiang claimed that it represented part of a Communist plot to have him kidnapped and sent to Russia, while the Communists asserted that Chiang had orchestrated the incident in order to falsely accuse them of malicious activities—but the outcome was unambiguous.[22] Two days later, Chiang declared martial law and initiated the first substantial crackdown on Communist activities within the ranks of the United Front between

[18]Ibid., pp. 359–60; Conger, *Letters from China:* pp. 169–72; Fenby, *Penguin History of Modern China,* p. 90; Power, *Puppet Emperor,* p. 36; Thompson, *Empires on the Pacific,* pp. 16–17.

[19]Power, *Puppet Emperor,* pp. 46–47; Thompson, *Empires on the Pacific,* p. 17.

[20]van de Ven, *War and Nationalism in China,* p. 101.

[21]Chiang Kai-shek, *Soviet Russia in China* (New York: Farrar, Strauss and Cudahy, 1957), p. 39; Chiang Kai-shek, "Chiang Kai-shek," in The Committee for the Compilation of Historical Materials of the Chinese Kuomintang, eds., *Ke Min Wen Hsien* [Documents on Revolution], Vol. 9, pp. 1298–99, quoted in Chung-Gi Kwei, *The Kuomintang-Communist Struggle in China, 1922–1949* (The Hague: Martinus Nijhoff, 1970), p. 37–38.

[22]Chiang, *Soviet Russia in China,* p. 39; Zhou Enlai, *Selected Works of Zhou Enlai* (Beijing: Foreign Languages Press, 1989), p. 310.

the Kuomintang and Communists.[23] This "Zhongshan Warship Incident" generated seeds of distrust and hatred between the Communists and Kuomintang, which blossomed into open conflict and suppression during the workers strikes and White Terror that occurred in late 1926 and 1927. Its relevance here, however, is more that the incident demonstrated to the Communists the bitter truth about their military impotence. Regardless of who was behind the incident, the fact remained that for the Communists to be involved they would have had to use the Kuomintang's own military resources because they themselves possessed almost none. The Kuomintang had a strong military with powerful naval assets, the Communists did not. The difference was there for all to see.[24]

The next few years saw a number of confrontations between the Communists and foreign naval power. During the Communist-assisted strike in Shanghai in February 1927, for example, foreigners in the settlements called in reinforcements to guard against any trouble. Eventually, twenty-three Western warships fortified the river, a move which made any aggression towards the settlements impossible.[25] A further altercation occurred when a mixed Kuomintang and Communist force captured the city of Nanking from the warlord Zhang Zongchang the following month. The incoming army attacked a number of buildings owned by foreigners and killed several foreign officials, causing considerable alarm. British and American ships shelled the city in response and the attacks stopped.[26] This type of confrontation also occurred during the abortive Communist urban uprisings against the Kuomintang. The Communist takeover of Canton in December 1927, for example, was crushed after the Nationalists called upon the aid of foreign gunboats.[27] When the Communists captured Changsha in 1930, one of the Communist sub-commanders at the scene, Lo Ping-Hui, estimated that ten British and American gunboats remained within the city and that five of these opened fire on his troops.[28] Soon afterwards, a larger fleet of American, British, Japanese, and Italian gunboats steamed up the river and bombarded the city until the Red Army withdrew.[29] Similar

---

[23] Ah Xiang, "Zhongshan Warship Incident," http://www.republicanchina.org; van de Ven, *War and Nationalism in China*, pp. 94–104.

[24] Ah Xiang, "Zhongshan Warship Incident"; van de Ven, *War and Nationalism in China*, pp. 94–104.

[25] *New China Times*, 26 February 1927, quoted in Fenby, *Penguin History of Modern China*, p. 176.

[26] Frank Oliver, *Special Undeclared War* (London: Jonathan Cape, 1939), pp. 58–59.

[27] Harrison, *Long March*, p. 139; Tom Ryan, *China Rising: A Study in Revolution* (Melbourne: Historical Teachers' Association of Victoria, 2009), pp. 107–08.

[28] Nym Wales, *Red Dust: Autobiographies of Chinese Communists as told to Nym Wales* (Stanford: Stanford University Press, 1952), p. 124.

[29] *China Weekly Review*, September 6, 1930, quoted in Harold Isaacs, *The Tragedy of the Chinese Revolution* (Chicago: Haymarket Books, 2009), p. 300.

sequences of events occurred in several other urban hubs, including Swatow and Wuhan.[30]

The importance of naval power was again thrust into the path of the Communists during the Long March of 1934–35. Edgar Snow and Robert Payne relate several different accounts from survivors of the March that describe how, after the Communists had broken free from his Fifth Encirclement Campaign, Chiang Kai-shek used his navy to rapidly shift vast quantities of troops to fortify the Yangtse river against the Communists and draw all of the ferries onto the opposite bank of the river. Unable to cross at any of the nearest crossing points as a result, the Communists were compelled to undertake a grueling forced march along the river, until they could find a less guarded crossing point and hijack the necessary boats to cross.[31] Upon arriving at the river Tatu, they were unable to commandeer a sufficient number of boats and so were forced to traverse a partially destroyed bridge that was defended by a regiment of Kuomintang forces, armed with sniper rifles and machine guns.[32] The difficulty which both crossings posed was caused in no small part by the imbalance of naval power. The Kuomintang were able to use their naval assets to quickly fortify the two rivers whereas the Communists lacked the necessary naval strength to cross the rivers or clear the opposing river banks of enemy forces without a monumental struggle.

The Kuomintang were, however, distracted in their fight against the Communists by the application of Japanese naval power against their forces during an attack on Shanghai in 1932. During their assault, the Japanese deployed a squadron of thirteen warships—including two aircraft carriers—and 500 marines.[33] The impact was devastating, as an account of the Japanese bombardment of the Woosung Forts by an English-language newspaper described: "At 10:30 o'clock twelve Japanese warships commenced a bombardment of the Woosung forts. After ten minutes' firing, in which forty shells were fired by the Japanese warships, a lull in the fighting seemed to indicate that the forts had been silenced . . . No definite news of the result of the engagement could be got through to Shanghai last night, communications being in the utmost confusion. It was understood, however, that the superior Japanese armaments had taken their toll, and that the capture of the forts was simply a matter of time."[34]

---

[30] Carter, *Mao,* pp. 63–64; Harrison, *Long March,* p. 179.

[31] Edgar Snow, *Red Star Over China* (London: Middlesex, 1972), pp. 219–23; Robert Payne, *Journey to Red China* (London: William Heinemann, 1947), pp. 170–80.

[32] Snow, *Red Star Over China,* pp. 224–30; Payne, *Journey to Red China,* pp. 170–80.

[33] Fenby, *Penguin History of Modern China,* pp. 222 and 236–37; "The Sino-Japanese Clash at Shanghai 1932," *The Naval Review* 21:1 (1933), pp. 1–30.

[34] K. N. Lei, *Information and Opinion Concerning the Japanese Invasion of Manchuria and Shanghai from Sources Other than Chinese* (Shanghai: Shanghai Bar Association, 1932), pp. 210–13, quoted in Franz Schurmann and Orville Schell, eds., *The China Reader: Republican China: Nationalism, War, and the Rise of Communism 1911–1949* (New York: Vintage Books, 1963), pp. 187–89.

The fighting, which lasted roughly six weeks, was concluded by an international armistice that turned Shanghai into a demilitarized zone. In addition to losing control of a major urban center, the Kuomintang suffered many casualties and lost considerable prestige. This weakened them against the Communists and encouraged the likelihood of future aggression by the Japanese. The battle entered the national consciousness as a major defeat and news about it resonated across the country.

This overview of events has shown that naval power did have both a direct and indirect impact on the rise of the Communists during this early period. It helped to demonstrate their military impotence and inspire a corresponding build-up of arms. It contributed to their defeats during their urban uprisings and made their escape during the Long March considerably more challenging. Finally, it weakened their Kuomintang opponents at the 1932 Battle of Shanghai. However, it was not until the escalation of the conflict with Japan in 1937 that naval power really manifested itself as important in the struggle for China.

War with Japan: 1937–1945

During the War with Japan, naval power made a series of significant imprints on the nature and the outcome of the conflict. The first of these was that naval supremacy enabled the Japanese to physically engage in combat operations against both the Kuomintang and the Communists within China. The Japanese Navy included innumerable transports with the capacity to carry the country's soldiers into previously acquired portions of Southeast Asia (themselves secured through naval-supported actions), from which they were able to invade China on foot, as well as directly onto the Chinese coast itself. By the end of the war, roughly one and a half million Japanese ground troops had been ferried by boat into mainland China.[35] Transportation of troops into China itself was achieved both through relatively peaceful docking actions at previously acquired or neutral Chinese ports, as well as through aggressive naval assaults against defended cities.

The 1937 Battle of Shanghai provides a prominent example of one such naval assault. The build-up towards this conflict began on 11 August, when the Japanese amassed a fleet of twenty-one warships in the Yangtse estuary, including four cruisers, sixteen destroyers, and an aircraft carrier.[36] By 13 August this number had increased to thirty-two ships, including five troop transports and 8,000 combat personnel.[37] A Kuomintang force of at least 40,000 soldiers responded to this incursion

[35]Hsi-sheng Ch'i, "The Military Dimension, 1942–1945," in Hsiung and Levine, eds., *China's Bitter Victory*, p. 179.

[36]Julius Bigner, "Reminiscences of an Old-Timer in War-Torn Shanghai," *The China Journal* 28:3 (1937), pp. 137–40.

[37]van de Ven, *War and Nationalism in China*, pp. 197.

by launching a preemptive attack, following the order of Chiang Kai-shek to "drive the enemy into the sea, block off the coast, and resist landings."[38] One commentator present at the time later argued that the Japanese Navy was instrumental for blunting this onslaught: "The Chinese ought actually to have driven this small band [of Japanese personnel] into the river, but such was the work of the Japanese air force, operating from aircraft carriers in the river, and such the weight of the metal laid down by Japanese warships, that this small force was able to hang on until a large reinforcement of regular troops arrived from Japan."[39]

Two wartime journalists who worked for the Chungking bureau of *Time* further described the mayhem wrought by Japanese naval superiority, even as the level of Chinese involvement increased:

> On August 13, 1937, he [Chiang Kai-shek] flung the best units of his German-trained army into action against the Japanese marine garrison in Shanghai. For a few days Chinese flesh and numbers compressed the Japanese into a narrow strip by the banks of the Whang-poo River . . . The Japanese moved their fleet to offshore anchorages, marshalled their air force at Formosa, and proceeded to pump steel at the massed Chinese troops in overwhelming tonnages. Not even today is there any accurate estimate of the carnage at Shanghai; Chinese casualties mounted to the hundreds of thousands as the blood and courage of the soldiers absorbed the shock of Japan's barrages.[40]

Soon after, the Japanese landed 50,000 reinforcements directly inside Shanghai, under the protective cover of their naval guns, a figure which rose to 120,000 by mid-September.[41] They also landed troops at Woosung and several other places to the north of Shanghai, again under the protection of heavy barrages from their warships, which could reach up to thirteen kilometers inland, and close air support from nearby aircraft carriers.[42] Gradually, the Chinese defenders were overwhelmed and forced to fall back, but the fate of the city was only sealed when the Japanese carried out a further amphibious operation. This time they landed 30,000 soldiers and marines at Hangzhou Bay, south of the city, and sent troop transports yet farther up the Yangtse, threatening the defenders with

---

[38]Oliver, *Special Undeclared War,* p. 139; Chiang Kai-shek, quoted in van de Ven, *War and Nationalism in China,* p. 197.

[39]Oliver, *Special Undeclared War,* p. 138.

[40]Theodore H. White and Annalee Jacoby, *Thunder out of China* (London: Victor Gollancz, 1947), pp. 56–57.

[41]Oliver, *Special Undeclared War,* pp. 141 and 146.

[42]Bigner, "Reminiscences of an Old-Timer," pp. 137–40.

encirclement.[43] Bombarded by naval artillery, harassed by unfettered airpower, and put upon from nearly every side by land forces that were buoyed by seemingly endless reinforcements from naval transports, the Chinese defenders were slaughtered as they fled. In addition to the 187,200 military casualties cited in official Kuomintang reports, "The retreating Chinese had lost their weapons, ammunition, and rations, and their morale had collapsed."[44]

The impact of naval power was crucial for Japan's victory. Prior to the battle, the German military adviser had informed Berlin that "The Chinese infantry is good. The Chinese Air Force is about equal to the Japanese . . . The morale of the Chinese Army is high. They will put up a bitter fight."[45] Dutch military observer, Henry de Fremery, reported, however, that Chiang Kai-shek had underinvested in the Chinese navy and had made no attempt to create a seaworthy fleet to help defend against the Japanese.[46] Furthermore, while the Chinese had managed to build or obtain a number of "coastal guns of different calibres, set up in batteries along the Yangtze and at various places along the coast . . . these batteries were lacking in even the absolutely essential equipment in the form of instruments, while the munitions were out of date and the soldiers in charge of them were not up to firing these batteries in a modern way."[47] De Fremery concluded that the imbalance in naval power and poor quality of China's coastal defenses were crucial factors in blunting the initial Chinese preemptive attack, which in turn helped enable the Japanese to achieve a swift and decisive victory.[48] Chiang himself later conceded in his diary that his neglect of the defense of Hangzhou Bay against Japan's naval landings "was our biggest strategic mistake."[49] The kind of naval assault employed during the Battle of Shanghai was repeated numerous times by the Japanese along the Chinese coast, including during their capture of Tsingtao and Canton in January and October 1938 respectively.[50]

---

[43]White and Jacoby, *Thunder out of China,* p. 57; Fenby, *Penguin History of Modern China,* p. 279; van de Ven, *War and Nationalism in China,* p. 216; Oliver, *Special Undeclared War,* p. 149.

[44]Yang Tianshi, "Chiang Kai-shek and the Battles of Shanghai and Nanjing," in Peattie, Drea, and van de Ven, eds., *The Battle for China,* p. 154.

[45]Alexander von Falkenhausen, quoted in van de Ven, *War and Nationalism in China,* p. 212; Bigner, "Reminiscences of an Old-Timer," pp. 137–40; Oliver, *Special Undeclared War,* pp. 138 and 143.

[46]Henri de Fremery, "Report no. 4," in Ger Teitler and Kurt W. Radtke, eds., *A Dutch Spy in China: Reports on the First Phase of the Sino-Japanese War (1937–1939)* (Leiden: Brill, 1999), p. 66.

[47]Ibid., p. 58.

[48]de Fremery, "Report no. 6," pp. 110–11.

[49]Chiang Kai-shek, quoted in Yiang Tianshi, "Chiang Kai-shek and the Battles of Shanghai and Nanjing," in Peattie, Drea, and van de Ven, eds., *The Battle for China,* p. 154.

[50]Oliver, *Special Undeclared War,* pp. 196 and 228.

The military impact of Japanese naval power was not, however, limited purely to coastal areas. The Chinese countryside is riddled with thousands of rivers and tributaries, many of which flow for vast distances deep into the heart of the country. Japanese forces followed up their capture of the Kuomintang capital of Nanking in December 1937 with an aggressive thrust up the Yangtze River. Using their naval forces, they burst through heavily pre-prepared defenses that included forts and a large defensive boom that had been stretched across the river.[51] The Japanese then quickly seized a series of further cities, including Anking, Matang, and Juichang, culminating in the capture of the three cities that comprise Wuhan from their Kuomintang and Communist defenders in late 1938.[52] During this campaign, the navy dispatched supply boats ahead of advancing land forces, shelled Chinese defensive positions, and conducted amphibious landings so that the Japanese land troops could outmaneuver their adversaries.[53] The rapid deployment and redeployment of Japanese forces across China by both river and sea was commonplace throughout the early years of the war, providing the Japanese with impressive strategic and tactical maneuverability. One historical analyst compared the actions of the Japanese Navy at the Battle of Wuhan with those of a German Panzer force.[54]

Japanese naval power influenced the military side of the war later in the conflict as well. One example occurred during the Ichigo offensive of 1944, which caused a staggering 750,000 casualties amongst the Kuomintang's armies.[55] This campaign was underpinned by a logistics operation that involved the transportation of weapons, food, fuel, and other resources upon large cargo carriers along the Yangtze. When American planes responded by bombing these vessels, the Japanese demonstrated their belief that the waterways were essential for supplying their forces by refusing to abandon the river and switching to using smaller boats to ferry their provisions.[56]

The use of massed air power by Japan within China was also facilitated by the use of naval assets. One month after the war began, Japanese aircraft carriers launched planes to strike Kuomintang bases along the Chinese coast, as well as to bomb Nanking.[57] During the battles for Shanghai, Nanking, and other coastal cities that followed, planes operating from Japanese carriers were used to supply recon-

---

[51]Ibid., p. 150; Marvin Williamsen, "The Military Dimension 1937–1941," in Hsiung and Levine, eds., *China's Bitter Victory,* pp. 145–47.

[52]Williamsen, "The Military Dimension 1937–1941," pp. 145–47.

[53]Stephen Mackinnon, "The Defense of the Central Yangtze," in Peattie, Drea, and van de Ven, eds., *The Battle for China,* pp. 196–200.

[54]van de Ven, *War and Nationalism in China,* p. 226.

[55]Hara Takeshi, "The Ichigo Offensive," in Peattie, Drea, and van de Ven, eds., *The Battle for China,* p. 402.

[56]Ibid., p. 393.

[57]Paine, *The Wars for Asia,* p. 132.

naissance, assault enemy positions, provide close air support, and undertake counter-air operations.[58] The following year, the Japanese Navy used carrier-based planes as part of the Wuhan campaign to lure Chinese fighters away from their land-based bombers or to conduct preemptive strikes ahead of larger bombing raids.[59] The use of carrier-based planes by the Japanese Navy in China was almost always complemented by the use of land-based planes (from both the Army and Navy), and as the war progressed, the former was gradually phased out in favor of the latter. Nevertheless, Japanese planes continued to be transported across to China by ship and navy planes were periodically brought back to the carriers for training.[60] There is little doubt that both the presence and effectiveness of Japanese air power in China was increased as a result of its naval efforts in the region.

In addition to providing Japan with the ability to strike broadly and deeply into China, naval power provided Japan with the ability to entrench indefinitely. The majority of both past and present analysts are fairly united in their certainty that the strength of Japan and its naval empire meant that there was no way that either the Kuomintang or the Communists could have ever decisively defeated the Japanese and ejected them from their strongholds along the coast of the country.[61] The power of the Japanese Navy and its ability to continue importing troops, supplies, and equipment from Japan and their Pacific possessions would prevent the Chinese from ever being able to achieve more than an unhappy stalemate, even with the greatest of military fortunes. Japanese naval power had to be defeated before the Japanese could be forced out of the country and the Chinese lacked the ability to achieve this feat. The two wartime journalists from *Time* later remarked on this very fact: "The Chinese could never do more than pinch off the Japanese salients and force them back into their dug-in bases; to do more than that would have required a weight of metal and equipment that Wong Wen-hao's transplanted industry could not hope to provide. The result was the permanent exhausting stalemate known as the China war."[62] Instead, they had to rely on the naval power of the Americans, brought into the fray after the Japanese attack on Pearl Harbor, to first stall and then ultimately overpower Japan's naval power. To cite one perhaps slightly melodramatic secondary source commentator: "All the Chinese people owe the Americans a debt of gratitude, for it was due to their efforts and those of their allies . . . that the Japanese were gradually forced to retreat. . . . There is no

---

[58]Hagiwara Mitsuru, "The Japanese Air Campaigns in China, 1937–1945," in Peattie, Drea, and van de Ven, eds., *The Battle for China*, pp. 241–43; Mark Peattie, *Sunburst: The Rise of Japanese Naval Air Power, 1909–1941* (Annapolis: Naval Institute Press, 2007), pp. 103–04 and 108.

[59]Hagiwara, "The Japanese Air Campaigns," p. 245.

[60]Peattie, *Sunburst*, p. 126.

[61]van de Ven, *War and Nationalism in China*, p. 292.

[62]White and Jacoby, *Thunder out of China*, p. 66.

doubt that despite the Red Army's brave efforts, the Japanese would never have been defeated by the Chinese alone."[63] An interesting parallel can be drawn with the American Revolution, wherein it was only after the British navy was defeated by the French at the Battle of the Chesapeake that the British were finally compelled to admit defeat and abandon the New World.[64]

The military impact of Japanese naval power was therefore quite discernable. It allowed the Japanese to become embroiled within the struggle for China, transforming a local conflict into a regional affair and then—following the entry of America, Britain, and Holland into the fray—into part of a wider, global Second World War. Japanese naval power also helped to weaken the Kuomintang by contributing to the destruction of their armies and the capture of their cities, which in turn helped to readjust the balance of military power between the Kuomintang and the Communists to the detriment of the former. The latter did not escape unscathed however, as naval force was used against those Communist forces who worked initially with the Kuomintang to defend the country's coastline, rivers, and surrounding areas, including at the battle of Wuhan where—as already discussed—Japanese naval power played a crucial role.[65] Naval power was also used against those Communists who fought exclusively for their own side in Communist-infiltrated coastal provinces, such as Shangdong and Hebei, particularly during the Japanese attacks of 1939–40 and the Hundred Regiments Offensive that was unleashed by the Communists as retaliation.[66] Finally, it is worth noting that while Japanese naval power was rarely used against Communist guerrilla forces directly, it was used against Kuomintang guerrillas who—despite Communist propaganda

---

[63] Elizabeth Mauchline Roberts, *Mao Tse-tung and the Chinese Communist Revolution* (London: Methuen, 1970), p. 54.

[64] Paul Kennedy, *The Rise and Fall of the Great Powers: Economic Change and Military Conflict from 1500 to 2000* (London: Unwin Hyman, 1988), p. 118; Fletcher Pratt presses this argument even further, suggesting that Britain could have continued fighting in the American War indefinitely had their fleet not received a mauling by the French at the Battle of Trincomalee in 1782. Pratt asserts that, in the face of this French naval aggression, the British had the choice to commit the remainder of their navy to either protect India or to engage further in America. India was considered more important and so America was abandoned. Fletcher Pratt, *The Battles that Changed History* (New York: Doubleday, 1956), pp. 241–49.

[65] Cultural China, "Relationship between the Nationalists and Communists in the Second Sino-Japanese War," http://history.cultural-china.com/en/34H7545H12590.html; T'ien-wei Wu, "The Chinese Communist Movement," in Hsiung and Levine, eds., *China's Bitter Victory,* pp. 81–86 and 96–98; Nigel de Lee, "The Far Eastern Experience," in Ian Beckett, ed., *The Roots of Counter-insurgency: Armies and Guerrilla Warfare, 1900–1945* (London: Blanford Press, 1988), p. 143.

[66] Nigel de Lee, "Southeast Asia: The Impact of Mao Tse-tung," in Pimlott, ed., *Guerrilla Warfare,* p. 39.

to the contrary—were present throughout the entire conflict in sizable numbers, a fact that even Stillwell felt begrudgingly compelled to acknowledge in 1944.[67] Later historians suggested that in 1941 there were 476,200 of these Kuomintang guerrillas.[68] The use of Japanese naval power against these partisans made it possible for both the Kuomintang and the Communists to see how naval power could be used during anti-guerrilla operations.

The impact of naval power was not, however, limited purely to military actions. Prior to the war, the majority of Chinese industry and commerce had developed within the cities along the coastline and the major rivers. The Japanese seizure of these focal points deprived the Chinese of the majority of their trading and manufacturing capabilities, to the extent that they preserved just 10% of their textile capacity and 40% of their machine shops and heavy industry.[69] Furthermore, at the very outset of the war, Japan laid down a heavy blockade of the Chinese ports, which made it incredibly challenging for the Chinese to import the equipment, weapons, or other resources that they needed to fight.[70] The inability to obtain sufficient quantities of petroleum was just one of many issues that plagued the Chinese economy as a result of the blockade. Prior to the war, China had relied heavily on imported gasoline, diesel oil, and motor oil, and during the war years, domestic production simply could not meet the demand. It is estimated that local oil fields—themselves only discovered in the 1930s—produced roughly seven million gallons of gasoline, kerosene, and diesel fuel, in contrast to an estimated demand of fifty million.[71]

The seizure of the coastal ports and the accompanying naval quarantine had a crippling effect on the Chinese economy as a whole. It is impossible to quantify the exact damage brought to bear by these actions, in no small part because the Japanese savaged China's economy in many other ways, including physically destroying inland factories and severing China's trade routes via rail and road. However, they did undoubtedly contribute heavily to the dramatic and painful crumbling of Chinese industry, agriculture, and commerce during the war. Just one example of this collapse can be seen by the hyperinflation of the economy, in which the Chinese currency fell from a rate of two fapis to one U.S. dollar in 1937 to 1,200 fapis

---

[67]Joseph Stillwell, *The Stillwell Articles* (New York: Da Capo, 1991), p. 316.

[68]In fact, the Kuomintang retained guerrilla forces until the very end of the Civil War in 1949, to the extent that when Chiang Kai-Shek and his remaining forces fled to Formosa (Taiwan) in December 1949, the only remnants of Kuomintang forces remaining were guerrillas in Sikang and Yunnan. See Chassin, *Communist Conquest of China*, p. 237; van de Ven, *War and Nationalism in China*, pp. 235 and 283–86.

[69]White and Jacoby, *Thunder out of China*, p. 60.

[70]Fenby, *Penguin History of Modern China*, p. 194.

[71]William Kirby, "The Chinese War Economy," in Hsiung and Levine, eds., *China's Bitter Victory*, p. 194.

to one dollar in 1945.[72] In his 1947 Report to the American People on China, William Bullet explained how this "inflation now cripples China" and "the cost of living today expressed in Chinese dollars is approximately thirty thousand times what it was in 1936, before Japan attacked China. The middle class is ruined."[73] In addition to damage from lost trade, the Japanese made it impossible for China to replace its damaged industries and depleted resources from outside sources, entrenching the impact of Japan's devastation further.[74]

The Kuomintang suffered greatly as a result of this damage to the economy. The inability to import military weapons and ammunition prevented them from being able to arm their forces at the same level of the Japanese, which helped to ensure that the Japanese retained a qualitative superiority and were never threatened by a supply of high quality American or British weapons. As late as October 1944, Chiang Kai-Shek could justifiably complain that not a single rifle or artillery piece had been received from the Lend-Lease program (although Chinese units in India and Burma did receive substantial amounts of U.S. equipment).[75] Even during the height of Operation Ichigo—Japan's last devastating offensive into the heart of China—the Chinese received just sixty mountain guns, 320 tank guns, and 506 bazookas from the Americans.[76]

The rapidly contracting economy also forced the Kuomintang to centralize the collection of land taxes.[77] The British wartime reporter, George Hogg, gave a useful description of the impact of the tax burden imposed upon the peasantry of one Chinese region:

> The wheat and grain taxes are especially burdensome, for the peasants are shorthanded and agricultural production is actually dropping off . . . transported to Loyang largely at the expense of local taxation, this wheat represents a colossal war contribution from the people of Honan. Simultaneously with the imposition of this completely new tax burden, the real burden of the old taxes has risen by several hundred percent since the Central Government decided to protect itself from currency depreciation by collecting all taxes in grain. There have been additional levies of wood for defence works, charcoal for army fuel and of labour.[78]

---

[72]Chassin, *Communist Conquest of China,* p. 40.

[73]William C. Bullitt, "A Report to the American People on China," *Life* (13 October 1947), quoted in Schurmann and Schell, eds., *China Reader: Republican China,* p. 348.

[74]van de Ven, *War and Nationalism in China,* p. 275.

[75]Hsi-sheng Ch'i, "Military Dimension," p. 172.

[76]Ibid.

[77]Kirby, "Chinese War Economy," p. 192.

[78]George Hogg, *I See a New China* (Leicestershire: F. A. Thorpe, 1986), pp. 155–56.

These new levies brought with them the traditional problems associated with taxes during imperial times, including evasion by the rich, extortion by tax collectors, and an increase on the burdens carried by the peasantry. The tremendous weight of the taxes, the lack of a suitable political explanation for their necessity, and the growing belief that this fiscal onus was being borne primarily by the lower classes, all helped to turn the peasantry away from the Kuomintang and either drive them towards the Communists, or at the very least make them more sympathetic to alternative ideas.[79] The absence of income from trade also forced the Kuomintang to lean more heavily on conservative elements within the country for their economic base, a move which further isolated them from the peasantry.[80] Furthermore, the damage to the economy helped to undermine the viability of a successful post-war state. Finally, at the end of the war, many people within China expected the economy to enjoy a quick recovery and for the poverty that they had suffered to be quickly put behind them.[81] However, the damage done to the economy impeded the possibility of a swift recovery, with the result that the population became further disillusioned with the Kuomintang and, again, either turned towards the Communists or at the very least limited their support towards the Kuomintang.[82]

As a result of Japan's embargo, the only outlet through which external supplies could be reliably received was through Burma. The air lift over "the Hump" proved insufficient—providing in 1942, for example, a mere one hundred tons of materials monthly, a tiny fraction of the 5,000 tons that the Chinese expected.[83] In order to push for the reopening of the overland road through Burma, through which supplies could be more substantively channelled, Chiang Kai-shek was persuaded by the British and Americans to despatch the Kuomintang's elite forces into the area. This further undermined the domestic credibility of the Kuomintang as a nationalist power. The Chinese population was wont to ask why their supposedly nationalist leader was sending forces to fight in far off Burma, while their own homes and livelihoods were under attack. This reduced the loyalty of the people to the Kuomintang and gave further credence to the Communist claim to be the real representatives of the nation.

While the Japanese occupation of the coastal cities and the naval quarantine almost certainly affected the Communist economy to some degree, they were

---

[79]Hogg, *I See a New China,* p. 153; van de Ven, *War and Nationalism in China,* p. 278; Westad, *Decisive Encounters,* pp. 79–83.

[80]White and Jacoby, *Thunder out of China,* p. 71; Peter Townsend, *China Phoenix: The Revolution in China* (Oxford: Alden Press, 1955), pp. 129–30.

[81]Westad, *Decisive Encounters,* p. 86.

[82]Fenby, *Penguin History of Modern China,* pp. 327 and 334–35; van de Ven, *War and Nationalism in China,* p. 293.

[83]Hsi-sheng Ch'i, "Military Dimension," p. 172.

better positioned and more doctrinally able to endure its effects by developing and exploiting local agriculture and industry.[84] As Lionel Chassin, French Vice Chief of Staff for National Defence between 1946 and 1949, noted:

> The inflation of Nationalist currency had, of course repercussions in the Communist zone, but the Communists were cushioned against its effects by the economic compartmentalization of their zone, in which each region had its own bank issue, as well as by the effectiveness of their ever-present controls . . . As the party did not want to lose the good will of the population, it was reluctant to raise taxes. Thus, except for inflation, the only solution [for funding their regime] was to make the administration and the army not only productive, but self supporting. This was the solution adopted . . . Each official, with Mao Tse-tung himself in the lead, was required to work his small plot; each army unit when not engaged in combat tilled the land assigned to it, or lent a hand to the peasants.[85]

They had also already endured the stifling of trade and supplies by the Kuomintang prior to the outbreak of war with Japan, with the result that the naval quarantine came to represent simply more of the same, rather than a new factor associated with naval power. Furthermore, they were able to secure additional funds by growing and selling illegal opium, a trade which by 1943 involved the sale of nearly one million boxes of opium a year, and which contributed up to 40% of the party's revenue.[86] According to Chang and Halliday, "In 1943 the Russians estimated Mao's opium sales at 44,760 kg, worth an astronomical 2.4 billion fabi (roughly U.S. $60 million at then current exchange rates, or some U.S. $630 million today)."[87]

For the Japanese, possession of naval power provided a wide range of economic and logistical benefits. These included the ability to transfer the necessary oil for waging the war in China from their oil-rich conquest of the Dutch East Indies (which naval hard power enabled the Japanese to conquer in 1942).[88] In the later years of the war, when America first reduced and ultimately destroyed Japan's naval power—including its ability to transport troops or supplies—the country's military efforts in China suffered accordingly.[89] By 1945, the level of petroleum that Japan could transport from the Indies to China had plummeted to almost

[84]T'ien-wei Wu, "Chinese Communist Movement," pp. 91–96 and 103.

[85]Chassin, *Communist Conquest of China*, pp. 42–43.

[86]Fenby, *Penguin History of Modern China*, pp. 309.

[87]Jung Chang and Jon Halliday, *Mao: The Unknown Story* (New York: Anchor Books, 2005), p. 271.

[88]Thompson, *Empires on the Pacific*, pp. 141–45; and van de Ven, *War and Nationalism in China*, p. 22.

[89]van de Ven, *War and Nationalism in China*, p. 249.

nothing, with the result that Japan's armored warfare in China ground nearly to a halt.[90]

It is therefore clear that naval power had a distinctive impact on the Communists during the war with Japan. At a direct level, Communist forces came face-to-face with Japanese naval power in several major confrontations, including at the battle of Wuhan and during their operations in Shangdong and Hebei. At an indirect level, naval power played a significant role in radically reshaping the strategic landscape around them, not only by adding a third opposing faction who had to be defeated before victory could be won and by linking the conflict with the global Second World War, but also by inflicting a huge amount of military and economic damage onto their erstwhile—and soon to be again—enemy, the Kuomintang. It is interesting to note that, several years after securing final victory in the Chinese Civil War, Mao thanked the Japanese, asserting that without the damage they inflicted onto the Kuomintang, the Communists could not have triumphed over them in the years that followed Japan's departure.[91] The surrender of Japan in 1945 did not, however, end the relevance of naval power for the Communists and it continued to play a role in the Civil War that resumed immediately after.

Civil War: 1945–1949

The surrender of the Japanese in 1945 brought about a period in which neither of the remaining two protagonists, the Kuomintang and the Communists, possessed significant naval military resources. The Kuomintang did, however, possess a small navy comprised of a few warships, a collection of gunships, and a number of landing craft, all supplemented by a small number of former Japanese and British warships that were provided by the Americans and British.[92] Their naval power was further bolstered by the establishment of a U.S. Naval Base at Tsingtao, which provided training for 300 Kuomintang naval officers and 10,000 enlisted sailors.[93] Despite the presence of prolifically low morale and widespread corruption, this nonetheless provided the Kuomintang with total naval superiority over the Communists, who had no naval force at all.[94] This imbalance remained until the very end of the fighting in 1949.[95]

---

[90]Thompson, *Empires on the Pacific*, p. 343.

[91]Mao Zedong, *Mao Zedong on Diplomacy* (Beijing: Foreign Languages Publishing House, 1998), p. 353.

[92]Chassin, *Communist Conquest of China*, p. 46; Bruce W. Watson, *The Changing Face of the World's Navies: 1945 to the Present* (London: Arms and Armour Press, 1991), p. 58.

[93]Chassin, *Communist Conquest of China*, p. 47.

[94]Ibid., pp. 46–47.

[95]Ibid., p. 209.

The military impact for the Communists of possessing no naval power was demonstrated several times during the conflict. One example occurred after the Communists had beaten a Kuomintang army at the Battle of Liaoning and were advancing on Mukden. While the city quickly surrendered, 20,000 Kuomintang troops in the nearby port of Yingkow were able to escape unharmed via the sea, a maneuver against which the Communists could do nothing.[96] Another occurred during the Communist advance over the Yangtze River. Despite having secured nearly complete superiority on land, to the point where they were able to advance almost entirely unchecked on the majority of fronts, the Communists were nonetheless stalled for four long months before being able to cross the Yangtze.[97] When the Communists did finally push across this massive channel, they were forced to do so using crude river vessels, such as junks, sampans, and fishing boats. These transports traveled without any naval escorts, which allowed Kuomintang machine gun posts and artillery pieces to fire at them with impunity as they bobbed across the water. Sheer numbers and momentum, combined with Kuomintang treachery and incompetence, enabled the Communists to brush past these obstacles and resume their victorious march towards the south.[98] The battle was easily won, but notable nonetheless for having taken place and for the preparations having consumed such considerable time, particularly given that the Kuomintang forces were everywhere else in a state of retreat or total collapse. This weakness of naval power was again exposed by the inability of the Communists to effectively pursue Chiang Kai-shek and his remaining forces when they fled across the South China Sea into their final redoubt at Formosa (Taiwan).[99]

While the Americans did not intervene to help the Kuomintang with direct military force, they did assist them through other means. One method of support given was the provision of naval transportation for Chiang Kai-Shek's troops into Manchuria soon after the Japanese surrender. Chassin explained the significance of this assistance: "The Communist Eighth Route Army, by active resistance as well as destruction of railroads, effectively barred the movement of Chiang's armies into Manchuria by land. Some other way had to be found to prevent Manchuria from falling prey to the Communists; and it was the U.S. Navy that offered the solution to the Generalissimo's problem. Under the command of Vice Admiral Barbey, the transports of Admiral Kinkaid's Seventh Fleet embarked two Chinese armies at ports in Southern China and set forth for Manchuria."[100]

---

[96]Ibid., p. 191.

[97]Stanley Karnow, *Mao and China: Inside China's Cultural Revolution* (New York: Viking, 1984), pp. 52–53; Mauchline Roberts, *Mao Tse-tung*, p. 57.

[98]Townsend, *China Phoenix*, pp. 46–51.

[99]Westad, *Decisive Encounters*, p. 255.

[100]Chassin, *Communist Conquest of China*, p. 63.

By 23 March 1946, the Americans had transported seven of Chiang's armies to Manchuria, an act which had a huge impact on the Kuomintang war effort by enabling them to compete with the Communists over who would control Manchuria.[101] The impact of the American fleet was apparent even to the prominent Communist Chu Teh, who expressed to the journalist Robert Payne in 1947 his belief that without American troop transports the Kuomintang would not have been able to contest Manchuria.[102] A second method of support was the supply of masses of military equipment. Between 1946 and 1949 this amounted to a staggering six billion dollars worth of expenditure, a large proportion of which was delivered over the ocean.[103] This meant that the Kuomintang forces encountered by the Communists were armed and supported by modern American weapons and equipment, much of which massively outclassed those available to either side from domestic sources. The Communists themselves gained access to these armaments and equipment when the various Kuomintang armies they faced either surrendered or withdrew and left their supplies behind.[104] The degree of aid given by the Americans contributed strongly towards the increasing skepticism felt by the Communists towards America's efforts to mediate between the two warring sides. This disillusionment ultimately led to a complete rejection of peace efforts by the Communists and a commitment to total war.[105]

Given the importance that foreign naval power had played in China's affairs since the 1840s, it is perhaps fitting that the end of the War in China involved a final encounter between the Communists and Western naval power. At the same time that the Communists were crossing the Yangtze River in 1949, the British despatched the 1,480-ton frigate *Amethyst* up the Yangtze towards Nanking, with the intent of bringing supplies and protection to British subjects living in the capital. Catching sight of the ship, the Communists opened fire with field batteries, killing fifty sailors and grounding the boat along the sandbanks, after which it was impounded by the Communist authorities.[106] Three other ships sent to its rescue were similarly attacked and forced to withdraw.[107] In perhaps a final, lingering reminder of the significance of naval power, the incident ended three months later when *Amethyst* succeeded in making a daring escape, out from under the nose of numerous Communist artillery batteries and river forts.[108] The affair represented

[101] Ibid., p. 77.

[102] Payne, *Journey to Red China,* pp. 14–23.

[103] Mauchline Roberts, *Mao Tse-tung,* p. 56.

[104] Ibid.

[105] Thompson, *Empires on the Pacific,* p. 384; Payne, *Journey to Red China,* p. 174.

[106] Chassin, *Communist Conquest of China,* pp. 218–19.

[107] Ibid.

[108] *The Canberra Times,* "Amethyst Commander to be Decorated by the King," 2 August 1949.

the last major naval encounter between the Communists and Western naval power and, despite Communist efforts to the contrary, it ended with this last vestige of Western naval power living on to fight another day.

It is apparent from this appraisal that naval power once again played a noteworthy role in determining the outcome of a struggle that involved the Chinese Communists. While in itself it was not a decisive factor, it enabled the Kuomintang to fight against the Communists in Manchuria, helped to equip their Kuomintang adversaries (and ironically the Communists themselves), sabotaged attempts at mediation by the Americans, and impeded the final march of the Communist forces southwards. Perhaps most importantly, it prevented the capture of the final Kuomintang redoubt in Formosa (Taiwan), which to this day remains independent of the Communist regime, albeit now primarily under the protection of American rather than Kuomintang naval power.

The Significance of Naval Power

It is clear, then, that naval power was both influential and important during the struggles that accompanied the ascent of Mao. The strength of "gunboat diplomacy" had been amply demonstrated by the West and Japan during much of the nineteenth century and this set the context for the subsequent rise of both the Kuomintang and the Communist parties. Naval power helped the Communists to reach the conclusion that they needed to build up their own armed forces, rebuffed their early attempts at urban insurrection, and made their escape on the Long March considerably more difficult. It then played a huge role in the War with Japan by enabling the Japanese to secure staggering victories over both Kuomintang and Communist forces, to embed themselves firmly on the Chinese coast and along its major rivers—a position that only came undone when the Americans destroyed their naval power and compelled them to surrender—and to cause a sufficient level of damage to the Kuomintang's economic, social, and military position that the Communists could overwhelm them in the subsequent Civil War. During the Civil War itself, naval power played several distinct roles, including transporting Kuomintang armies to Manchuria, arming the Kuomintang (and indirectly the Communists), delaying the Communists in their final march south, and preventing the final capture of Kuomintang-held Formosa (Taiwan). These events should all have made a deep and enduring impression upon Mao. This was not, however, the case.

II. How Mao Overlooked the Sea

Absence of Naval Power

Contrary to what might be expected from the accounts given above, naval power featured exceedingly sparsely within Mao's writings. A simple word frequency count is illuminating: throughout the entire collection of Mao's *Selected Works* and *Collected Works*—a series of documents that span over two million words—the word "navy" appears a mere forty-eight times, compared to the word "army," which features 5,823 times.[109]

One of the only substantial references to the navy or sea power in relation to defeating the enemy during an insurgency came during Mao's 1938 treatise, *On Protracted War*. However, even here it seems that Mao was merely providing a statement regarding the general situation faced by China, rather than commenting specifically on naval power: "Japan is using dozens of army divisions of fairly high combat effectiveness (now numbering thirty) and part of her navy to encircle and blockade China from both land and sea, and is using her air force to bomb China. Her army has already established a long front stretching from Paotow to Hangchow and her navy has reached Fukien and Kwangtung; thus exterior-line operations have taken shape on a vast scale."[110] He continued the piece by laying out a generalized strategy for guerrilla warfare within the hinterlands of China, without further reference to the naval situation. Even when Mao did address naval issues directly, he tends to focus primarily on non-naval aspects. During his denunciation of the British protest against the Chinese capture of *Amethyst* in April 1949, for example, he focused far more on the legal and political aspects of the incident than the actual naval events that occurred.[111] More notably still, he provided no subsequent comment regarding the military embarrassment embodied by the failure of the Communists to prevent *Amethyst* from escaping three months later, leaving this instead to his subordinates to manage.

A small number of his naval references applied to his vision of China in the post-Civil War era. One such mention, made just weeks before the formal declaration of victory in 1949, was as follows: "Our national defence will be consolidated and no imperialists will ever again be allowed to invade our land. Our people's armed forces must be maintained and developed with the heroic and steeled People's Liberation Army as the foundation. We will have not only a powerful army but also a powerful

[109]*Selected Works of Mao Tse-tung, 1961–1990s,* 9 vols. (Peking: Foreign Languages Press, 1961–77; Secunderabad and Hyderabad: Kranti Publications and Sramikavarga, 1990–94); *Collected Works of Mao Tse-tung,* 10 vols. (Arlington: Joint Publications Research Service, 1978). Both numbers may be slightly understated here, as on occasion the electronic version has combined two words together. Similar words and plurals of words were also examined, with similarly sparse results, e.g. navies (3 mentions), naval (31 mentions), sea (53 mentions), seas (14 mentions), ocean (22 mentions), oceans (1 mention), fleet (18 mentions), fleets (0 mentions), ship (21 mentions), ships (6 mentions), boat (22 mentions), boats (3 mentions), warship (1 mention), and warships (26 mentions).

[110]Mao Tse-tung, "On Protracted War: May 1938," *Selected Works,* Vol. II, p. 158.

[111]Mao Tse-tung, "On the Outrages by British Warships—Statement by the Spokesman of the General Headquarters of the Chinese People's Liberation Army: April 30th 1949," *Selected Works,* Vol. IV, pp. 401–03.

air force and a powerful navy."[112] This statement related not to any guidance about the role of naval power during the Communists struggle to gain power—either by the Communist insurgents or by the Kuomintang counterinsurgents—but rather to the conventional defense of mainland China, now that it has been united. Naval power is not given any special consideration either, but is referred to as part of a generic triad of land, air, and naval assets.

Many of the other mentions hold little or no relevance to naval power at all. Some were used when discussing other topics that involve naval agencies acting as the representative agencies of the Kuomintang or foreign powers in areas that are at best tangentially related to issues of sea power, whereas others were made when discussing the different groups that may be affected by a particular law or decree. In his 1931 work entitled *Election Law of the Chinese Soviet Republic—Adopted at the First Plenary Session of the Central Executive Committee,* for example, Mao specified those groups of people who can elect and be elected into the Soviet, a broad cross-section of society which he stated should include peasants, urban workers, other exploited peoples, the army, the air force, and the navy.[113] This mention holds little or no relevance for naval power.

From this examination it is clear that Mao gave only very limited attention to naval matters. Even when he did make reference to naval affairs, his remarks on the topic tended to be included as part of wider descriptions of the strategic environment; be mentioned during dialogues addressing political, social, or legal matters that happened to have minor naval aspects; or be related to conventional naval warfare against the Western powers after the end of the Civil War. Given the important role that naval power played during the War in China and the numerous ways in which it influenced the Communists and their adversaries during their rise to power, this paucity of coverage demands explanation.

Why Mao Missed the Boat

Explaining the rationale behind why an author neglected to address a particular topic is always a challenging task, prone to accusations of speculation. However, while it is important to recognize that it is impossible to say with precise certainly why Mao paid such scant attention to naval power, it has nevertheless been possible to advance two informed propositions about why this neglect occurred and demonstrate why four alternative explanations are incorrect.

The first reason why Mao paid no heed to naval affairs was his lack of strategic vision. Mao often failed to pay attention to events beyond those that he experienced

---

[112] Mao Tse-tung, "The Chinese People have Stood Up: September 21, 1949," *Selected Works,* Vol. V, pp. 15–18.

[113] Mao Tse-tung, "Election Law of the Chinese Soviet Republic—Adopted at the First Plenary Session of the Central Executive Committee," *Collected Works,* Vol. III, p. 16.

firsthand, a reality aptly illustrated by the fact he never mentioned the Rape of Nanking—despite the immense scale of the disaster and its massive impact on the Chinese nation—but did provide copious commentary on Wang Ming's short-term success over Mao in the contest for leadership of the Communist Party, which occurred at the same time.[114] During the naval events mentioned above, Mao was rarely, if ever, in a position where he experienced the events in person, and disregarded them as a result. Throughout much of the late 1920s and early 1930s, Mao was located far away from the naval events that transpired, often working closely with peasants in rural provinces such as Hunan.[115] Similarly, during the War with Japan, while Communists located on the coast and the fringe of Mao's territories did experience the impact of Japanese naval power firsthand, Mao and the central core of his forces were situated deep within the mountains and hinterlands of China. Embedded in this relatively secluded terrain, thousands of miles away from the coast or major rivers, Mao only ever encountered unsupported Japanese ground forces and never had to fight against the awesome might of the Japanese Navy. The removal of Japan's naval power by the Americans meant that Mao was able to achieve his conquest of China without ever needing to worry about how to tackle this formidable threat. On an economic front, while the Kuomintang relied heavily on traditional trade and industry—both of which were hammered by the Japanese—the Communists were better able and more doctrinally more suited to attaining self-sustainability. This muffled the economic impact of Japan's naval blockade on their lives and wartime endeavors. Similarly, while the absence of Communist naval strength during the Civil War led to a delay prior to crossing the Yangtze river in the face of entrenched Kuomintang positions, the length of this delay was likely to have been viewed as inconsequential by Mao, who had by this time mostly delegated military functions to sub-commanders such as Lin Biao and was anyway preoccupied with rebuffing the final doomed peace initiatives launched by the Kuomintang and enjoying the satisfaction of imposing obviously unacceptable terms of surrender upon them.

The second reason was a decision to provide a selective account of events that concentrated on those elements of warfare in which the Communists excelled over those in which they did not. This was in no small part because Mao wanted to provide a "victor" narrative, which rejected prior interpretations of modern history as humiliating for China and instead emphasized the glorious triumph of the Chinese masses over their corrupt and imperialist foes.[116] Mao's reasons for wanting to

---

[114]Chang and Halliday, *Mao: The Unknown Story,* pp. 202–03.

[115]Ibid., pp. 38–39.

[116]Peter Gries, *China's New Nationalism: Pride, Politics and Diplomacy* (London: University of California Press, 2004); Zheng Wang, "National Humilitation, History Education, and the Politics of Historical Memory: Patriotic Education Campaign in China," *International Studies Quarterly* 52:4 (2008), pp. 783–806. Thanks go to Daniel Owen Spence for his very helpful guidance about this area of research and its relation to the issues discussed within this chapter.

perpetuate a "victor" narrative were to help harness popular support for his war efforts in the 1930s and 1940s, and then to support his nation-building goals during the 1950s, 1960s, and 1970s.[117] Naval power, which had been used repeatedly by foreign powers—and to a lesser extent the Kuomintang—against the Communists, but which had been barely used by the Communists themselves, did not fit within this "victor" narrative. Land power, however, was an area that Mao wanted to focus on because the Communists had proved so successful in converting nationalist fervor into fresh recruits for their armies. Even after being forced by Chiang Kai-shek's fifth encirclement campaign to flee six thousand miles across often hazardous terrain, Mao managed to retain an army of five thousand soldiers and used this to carve out a new realm in Yenan. By 1946 he had built up an army of nearly one million soldiers, supported by over half a million guerrillas, and by 1949 this force had grown yet further to more than two million soldiers and nearly two million guerrillas.[118] Consequently, to advance a "victor" narrative, to galvanize his own people, and to impress the outside world, Mao felt compelled to emphasize the role of land warfare and marginalize the impact of naval power.

A good example of selective reporting to support the "victor" narrative can be seen in the Communist accounts of the 1949 *Amethyst* incident, which downplayed the relevance of naval power. Specifically, the Communists claimed that *Amethyst* was only able to flee from its Communist captors by using the Chinese passenger boat *Kiang Ling Liberation* as a human shield against Communist shelling and by sinking this civilian ship once they were challenged to stop, as a way of diverting the Communists' attention onto helping the survivors rather than shooting at *Amethyst*.[119] This ran contrary to the facts, as reported by the surviving master of *Kiang Ling Liberation*. In a newspaper interview on 20 August, this experienced sailor indicated that his ship had been sunk by shore-based artillery fire and that *Amethyst* had been riding behind his vessel, not alongside it.[120] The bitter reality that their forces had been unable to stop *Amethyst* because they were powerless to do so was not a humiliation that the Communists were willing to admit because it did not fit into the new "victor" narrative.

These two reasons combine to suggest that Mao was mostly unaware of the impact of naval power because it did not affect him personally very often and much of its impact was removed by forces beyond his control. Those elements of

---

[117] Gries, *China's New Nationalism*, p. 48.

[118] Ibid.

[119] "P. L. A. Commander Speaks on Atrocities Perpetrated by *Amethyst*," New China News Agency, 3 August 1949, quoted in Phillips, *Escape of the Amethyst*, pp. 264–67; Mao Tse-tung, "On the Outrages by British Warships," pp. 401–03.

[120] *South China Morning Post*, 20 August 1949, quoted in Malcolm H. Murfett, *Hostage on the Yangtze: Britain, China and the Amethyst Crisis of 1949* (Annapolis: Naval Institute Press, 1991), pp. 217–18.

naval power that he did become aware of he chose to ignore because they fitted more neatly into a narrative of "national humiliation" rather than the "victor" narrative he was trying to build. While various alternative explanations could be offered, these all possess substantial flaws and should be rejected. They include the idea that Mao believed that:

1) Winning an insurgency relied purely on winning the hearts and minds of the population and that naval power is irrelevant for this goal. There are several problems with this explanation. The first is that we know that the correct application of naval power can influence the hearts and minds of a population: ships can be used to patrol areas under threat from insurgents, thereby providing a positive sense of protection and commitment by the government, and they can be used to deliver aid, transport local dignitaries, and carry out state-building functions.[121] Furthermore, the simple physical presence of a ship can help to demonstrate the military strength and economic credibility of the government, which may in turn deter the population from supporting the visibly weaker insurgent. The second is that Mao never explicitly states that naval power is irrelevant, which would leave any proponent of this argument to have to infer that this was his perspective. Given both the prominence of naval power on the period during which Mao rose to power and the extensive amount that Mao wrote during his life, it seems difficult to believe that if this had been his belief that he would not have provided extensive written instructions to explain his position.

2) He should only discuss purely strategic level concepts—such as the importance of politics for winning an insurgency—and avoid getting bogged down in the minutiae of operational and tactical level issues such as the relative value of naval versus land or air power. The most glaring problem with this thesis is that Mao did frequently delve into the finer details of land combat in his works, sometimes to the point of near absurdity, showing that he was not averse to outlining his thoughts in depth on both tactical and operational matters. Examples of this include Mao's in-depth analysis and instructions with regards to the establishment of base areas, the equipment of guerrillas, and the Three Rules and Eight Remarks regarding the manner with which Communist guerrillas should interact with the population.[122]

3) Insurgent forces lack either the fixed base or the material power needed to develop their own naval power. There are two difficulties with this argument. The first is that Mao did in fact enjoy a fixed base for much of his insurgent career. The second is that if Mao had genuinely believed that the construction of a navy was impossible, it seems unlikely that he would simply have ignored it. He would have

---

[121]Tim Benbow, "Maritime Forces and Counter-insurgency," *Contemporary Security Policy* 28:1 (2007), pp. 80–95.

[122]Mao Tse-tung, "On Protracted War," pp. 93–102; Mao Tse-tung, "On Guerrilla Warfare," in U.S. Marine Corps, FMFRP 12–18, *Mao Tse-tung On Guerrilla Warfare* (Washington, DC: Department of the Navy, 1989), pp. 71–93.

included suggestions as to how an insurgent force could work to combat the enemy's naval power, including possibly the sorts of methods tried in the American Revolution (described later in the chapter), a conflict which had impressed many of the senior Communist hierarchy and which Mao himself had studied in great depth.[123]

4) The nature of insurgent warfare meant that the weapons possessed by the enemy were entirely irrelevant, because China had the time, space, and population needed to outlast any occupying enemy. While this is appealing because it fits somewhat with Mao's own words, it does not stand up to closer scrutiny.[124] While it was hoped at the start of the war that the Chinese could outlast the Japanese in a war of attrition, in fact the opposite occurred.[125] As late as 1944, with their empire crumbling around them, the Japanese still managed to launch the devastating Ichigo Offensive, a campaign in which they conquered eight provinces—a population of one hundred million people—and inflicted nearly half a million military casualties.[126] Although the Japanese focused their efforts more heavily against the Kuomintang throughout the war, the harsh reality about Japan's strength and staying power can nonetheless have scarcely escaped even Mao's strategically shortsighted notice when the Japanese General Yasuji Okamura caused massively heavy losses to the Communists—possibly to the edge of extinction—during his campaign to "Kill All, Burn All and Destroy All," which killed more than 100,000 communist insurgents.[127] This comprehensive effort only ground to a halt when the Japanese High Command was forced to redeploy Okamura's best troops to fight against the Americans in other parts of the Empire.[128] Furthermore, Mao's model of insurgent warfare envisioned that victory would be secured through a campaign of mobile warfare fought by a fully developed insurgent army against the weakened forces of the opponent. It did not envisage winning by sapping the political will of the enemy until they left of their own accord; that development is usually associated with later insurgencies in Vietnam and Algeria. Consequently, Mao's own model would suggest that he would ultimately have had to come face-to-face with Japanese naval power and defeat it before final victory could be achieved.

---

[123] Payne, *Journey to Red China,* pp. 21 and 39; Esme Hawes, *The Life and Times of Mao Tse-Tung* (Bristol: Siena, 1996), p. 7.

[124] Mao Tse-tung, "On Protracted War," pp. 117 and 174–78.

[125] van de Ven, *War and Nationalism in China,* p. 292.

[126] John H. Boyle, *China and Japan at War, 1937–1945: The Politics of Collaboration* (Stanford: Stanford University Press, 1972), p. 319.

[127] Robert B. Asprey, *War in the Shadows: The Guerrilla in History* (New York: William Morrow, 1994), p. 445.

[128] Asprey, *War in the Shadows,* p. 445; de Lee, "Southeast Asia," pp. 147–49.

III. Global Amnesia about Naval Power in Insurgencies

In his recent pioneering work, "Maritime Forces and Counter-insurgency," Tim Benbow brings to light the peculiar absence of any substantial body of academic or military literature that discusses the role of naval power within insurgencies. Benbow argues that this is an unjustifiable omission because naval power both can and has played an immensely rich diversity of roles within insurgent conflicts. These include allowing a counterinsurgent to deploy forces to a conflict theater, transport supplies, redeploy forces, provide logistics support, supply command and control, provide communications, provide intelligence, enable surveillance, conduct naval bombardments, launch amphibious operations, maintain a general presence through patrolling, blockade the conflict area, deter other forces from intervening in the conflict, and offer sea defense of coastal targets.[129] Numerous examples exist to confirm Benbow's thesis. During the 1948–1958 Malayan Emergency, for example, the Royal Navy (including *Amethyst*) was used to carry out a blockade that helped to starve the insurgents of weapons, conduct patrols, transport security forces and local officials, visit areas that had suffered from insurgent attacks, and provide fire support.[130] It is also possible to find examples of insurgents engaging in naval activity. During the American Revolution, the Americans employed numerous forms of asymmetric naval warfare, including the use of individual state navies that were comprised mainly of shallow-draft barges, galleys, and gunboats and which "acted as maritime militia, fending off British naval raids to gather provisions and preventing Loyalists from supplying ships lying offshore."[131] Other asymmetric naval efforts by the Americans involved the deployment of privateers, private naval forces, new weapons such as submarines and sea mines, and even the construction of an (admittedly mostly useless) conventional Continental Navy.[132] While the Royal Navy was far too large and powerful to be directly threatened, the Americans were nonetheless able to hurt the British war effort by "attacking its lucrative seaborne commerce and disrupting its military lines of supply and communication."[133]

The curious absence of naval power within modern writings about insurgencies can be explained, at least in part, as stemming from Mao's failure to address the topic. Modern analysts have relied heavily on the theories of Mao Zedong when developing their opinions about the features that characterize an insurgency, with the result that Western theorists may have mimicked Mao's omission of naval

[129]Benbow, "Maritime Forces," pp. 81–83.

[130]Ibid., pp. 84–86.

[131]Allan Millett and Peter Maslowski, *For the Common Defense: A Military History of the United States of America* (New York: The Free Press, 1984), p. 76.

[132]Ibid., p. 75.

[133]Ibid., pp. 75–78.

power from insurgencies. A good example of this can be seen in *The U.S. Army/Marine Corps Counterinsurgency Field Manual,* which has been criticized for depending too heavily on Mao for its conceptions of insurgency.[134] Despite being co-authored by the U.S. Marine Corps, the manual fails to make a single substantial reference to the potential role of naval power in insurgencies.[135] The landlubber hand of Mao is fiercely evident.

Using this yard stick, it may be possible to tentatively identify a growth in the influence of Mao upon the thinking of insurgent and counterinsurgent analysts by tracing the gradual depletion of the presence of naval power within military academic analysis over the past sixty years. For example, unlike the current U.S. counterinsurgency manual, the 1965 U.S. Marine Corps publication FMFM 8-2 (*Operations Against Guerrilla Units*) does in fact contain several chapters on the use of landing forces, naval support, and mobile sea bases for tackling insurgencies, suggesting that at this time the influence of Mao and his neglect of naval power upon counterinsurgent thought may have been less prominent than in current times.[136] This suggests an interesting path for future research.

Finally, Mao's failure to address naval issues also draws into question his relevance for other modern insurgencies, despite the frequent claims made by analysts that his teachings have helped to evolve—or even revolutionize—the nature of insurgent warfare. It is difficult to understand, for example, how an insurgency in a country such as the Philippines, which consists of many thousands of islands, can truly have been that heavily influenced by "the Great Helmsman" who forgot about the sea.

Conclusion

By conducting a dedicated appraisal of the interplay between naval power and the rise of the Communist Party of China, this chapter has shown how naval power played an important role in shaping the armed struggles that accompanied and ultimately led to the ascent of Mao Zedong as leader of the country. These struggles included the early clashes with the Kuomintang and the foreign powers, the War with Japan, and the subsequent Civil War. Despite its continued impact on both the character and the outcomes of these struggles, however, Mao neglected to discuss the issue in any depth within his copious writings. This absence occurred

---

[134]Frank Hoffman, "Neo-classical Counterinsurgency?" *Parameters* (Summer 2007), pp. 71–74; Ralph Peters, "Progress and Peril: New Counterinsurgency Manual Cheats on the History Exam," *Armed Forces Journal* (February 2007).

[135]*The U.S. Army/Marine Corps Counterinsurgency Field Manual* (Chicago: University of Chicago Press, 2007).

[136]U.S. Marine Corps, *Operations Against Guerrilla Units* (Washington, DC: Department of the Navy, 1965).

as a result of Mao's short-sighted strategic vision and his decision to focus on a "victor" narrative, rather than as a result of any particular military or political genius. Mao's failure to discuss naval power in his works helps to explain the peculiar absence of a robust discourse about the role of the navy in counterinsurgency literature and helps to undermine the popular-held belief that Mao substantially altered the nature of modern insurgencies.

10

The Army-Navy Contest for Control of Land-Based Antisubmarine Aviation and the Military Unification Debate, 1942–1948

George H. Monahan

This chapter explores the history of the dispute between the U.S. Army and Navy during World War II over which service would control land-based antisubmarine aviation in the Battle of the Atlantic. Additionally, the chapter seeks to explore the relationship of this wartime dispute to the post-war military unification movement which culminated in the passage of the National Security Act of 1947. Many lessons learned through the management of the extemporized wartime defense bureaucracy convinced political and military leaders of the need to develop an integrated post-war military-intelligence-diplomatic apparatus for the purpose of defending against foreign threats to national security.

Most studies of the 1947 National Security Act tend to focus on national security issues which gave rise to lingering questions touching upon politics and military policy into the Cold War era. These include: the scope and limitations of authority of the Central Intelligence Agency; the control and use of nuclear weapons; the roles and missions of the Army, Navy, and independent Air Force; the institutionalization of the Joint Chiefs of Staff; and the extent of power exercised by the newly-created Secretary of Defense over the unified armed forces.

This chapter in no way seeks to detract from these issues as the overriding concerns shaping the debate leading to passage of the 1947 National Security Act. While acknowledging their primacy, it is the contention of this study that the wartime dispute over which service would have control of land-based antisubmarine aviation has largely been overlooked as a significant contemporary issue which helped shape the debate leading to the passage of the National Security Act of 1947.

A Brief Historiography of the Battle of the Atlantic and Post-War Military Unification

There are numerous secondary works devoted to the Battle of the Atlantic as well as to the National Security Act of 1947; however, very few tie the two issues together.

Of those works that do, the story of the Navy's efforts to retain control of its land-based antisubmarine aviation is considered as a far less significant factor in shaping the debate over the merits of the legislation than other, more weighty issues such as the role and authority of the Secretary of Defense, the potential Prussianization of the military, and the potential dangers to individual liberty posed by the creation of a Central Intelligence Agency. A brief survey of this literature is warranted.

Most secondary works devoted to the Battle of the Atlantic overwhelmingly concentrate on the role that surface craft played in antisubmarine operations. Most, however, give some attention to the role aviation played in combating the U-boats. Samuel Eliot Morison's *The Battle of the Atlantic, September 1939–May 1943*[1] and *The Atlantic Battle Won, May 1943–May 1945*[2] published respectively in 1947 and 1956, remain the standard American account of the battle. Morison devotes part of chapter 10 of the first volume to the role of antisubmarine aviation, yet he pays very brief attention to the interservice conflict over control of long-range, land-based aircraft. The official British history is presented in Stephen Roskill's three-volume work, *The War at Sea, 1939–1945,*[3] published between 1954 and 1961. Roskill discusses the role that the land-based RAF Coastal Command played in combating the U-boats while dealing with American land-based antisubmarine aviation in only a cursory manner. Many works on the Battle of the Atlantic have been published subsequent to Morison's and Roskill's classics; however, no significant new ground is broken regarding the role that aviation played in antisubmarine operations. The most recent of these include Andrew Williams' 2002 work, *The Battle of the Atlantic,*[4] Bernard Ireland's *The Battle of the Atlantic,*[5] published in 2003, and *Bitter Ocean: The Battle of the Atlantic, 1939–1945,*[6] published in 2006 by David Fairbank White.

The short-lived contribution of the U.S. Army Air Forces Antisubmarine Command in the Battle of the Atlantic is discussed in chapter 15 of Wesley Craven and James Cate's *The Army Air Forces in World War II,* Vol. I, *Plans and Early Operations, January 1939 to August 1942,*[7] published in 1948. This work briefly mentions

---

[1]Samuel Eliot Morison, *The Battle of the Atlantic, September 1939–May 1943* (Boston: Little, Brown and Company, 1947).

[2]Samuel Eliot Morison, *The Atlantic Battle Won, May 1943–May 1945* (Boston: Little, Brown and Company, 1956).

[3]Stephen Roskill, *The War at Sea, 1939–1945* (London: HMSO, 1954–61).

[4]Andrew Williams, *The Battle of the Atlantic* (New York: Basic Books, 2002).

[5]Bernard Ireland, *The Battle of the Atlantic* (Annapolis: Naval Institute Press, 2003).

[6]David Fairbank White, *Bitter Ocean: The Battle of the Atlantic, 1939–1945* (New York: Simon and Schuster, 2006).

[7]Wesley Craven and James Cate, *The Army Air Forces in World War II,* Vol. I, *Plans and Early Operations, January 1939 to August 1942* (Washington, DC: Government Printing Office, 1948).

the interservice conflict but focuses more on the operational history of Army air units in combating the U-boats. The role played by the RAF Coastal Command in antisubmarine warfare is examined at length in a number of secondary monographs, the most recent of which is Andrew Hendrie's 2006 work, *The Cinderella Service: RAF Coastal Command, 1939–1945*.[8]

Jeffrey Barlow deals largely with the consequences of unification shortly after the passage of the National Security Act of 1947 in *Revolt of the Admirals: The Fight for Naval Aviation, 1945–1950*,[9] published in 1994. In the first chapter of this work, Barlow examines how the competing air power doctrines of the Army Air Force and Navy affected the post-war service unification debate as well as the interservice agreement leading up to the passage of the 1947 National Security Act which allowed the Navy to retain its land-based aviation. This story, however, is presented as a terse preface to the events surrounding the cancellation of the super-carrier *United States* and the subsequent interservice recriminations over the perceived shortcomings of the B-36 bomber. Barlow takes up the theme of interservice rivalry during the nascent Cold War more broadly in his 2009 work, *From Hot War to Cold: The U.S. Navy and National Security Affairs, 1945–1955*.[10] Although this work provides an indispensable background for understanding the functioning of the naval bureaucracy in the post-war period, it only touches upon the controversy over control of land-based antisubmarine aircraft during World War II in a cursory fashion.

The interservice difficulties arising from the extemporized managerial structure of the Joint Chiefs of Staff is examined in Mark Stoler's *Allies and Adversaries: The Joint Chiefs of Staff, the Grand Alliance and U.S. Strategy in World War II*,[11] published in 2000. Although this work only briefly mentions the U-boat menace as one of the problems plaguing Stimson's War Department during the Bolero buildup, it provides an indispensable background for understanding the parochialism and personality conflicts of the American and the British service chiefs as they collaborated in the formation of Allied grand strategy.

Douglas T. Stuart's 2008 work, *Creating the National Security State: A History of the Law that Transformed America*,[12] offers perhaps the best in-depth account of the events leading to the passage of the National Security Act in 1947 and of the subse-

---

[8]Andrew Hendrie, *The Cinderella Service: RAF Coastal Command, 1939–1945* (Barnsley: Pen and Sword, 2006).

[9]Jeffrey Barlow, *Revolt of the Admirals: The Fight for Naval Aviation, 1945–1950* (Washington, DC: Naval Historical Center, 1994).

[10]Jeffrey Barlow, *From Hot War to Cold: The U.S. Navy and National Security Affairs, 1945–1955* (Stanford: Stanford University Press, 2009).

[11]Mark Stoler, *Allies and Adversaries: The Joint Chiefs of Staff, the Grand Alliance and U.S. Strategy in World War II* (Chapel Hill: University of North Carolina Press, 2000).

[12]Douglas T. Stuart, *Creating the National Security State: A History of the Law that Transformed America* (Princeton: Princeton University Press, 2008).

quent development of the national security bureaucracy up to 1960. Stuart contends that until the Japanese attack on Pearl Harbor, very few in the United States appreciated the need to develop an integrated military-intelligence-diplomatic apparatus for the purposes of monitoring, assessing, and defending against foreign threats. Stewart explains how the shock of Pearl Harbor and the lessons learned through the management of an *ad hoc* wartime defense bureaucracy shaped the debates that resulted in the passage of the National Security Act. Moreover, Stuart discusses how the post-1947 national security bureaucracy was in turn shaped by inter-agency rivalries over funding and the definition of roles and missions. Although Stuart mentions the wartime dispute over the control of land-based antisubmarine aviation, he does not develop an argument about this issue's significance in the post-war arguments for and against military unification.

The Interservice Conflict Over Antisubmarine Warfare

Between the world wars, the military services began to formulate new roles for aviation in modern warfare. The air doctrines which emerged, however, reflect very different military philosophies. Naval aviation from its inception was generally embraced by the old guard Navy leadership as a tactical support to fleet operations. Conversely, Army airmen strove for autonomy from a service they believed hidebound in antiquated cavalry and infantry traditions and sought to develop an Army Air Force based on the principles of strategic air power. The Army air power advocates also claimed that an Air Force of long-range, land-based bombers would have the capability of neutralizing an approaching enemy fleet, and thus, the Air Force would come to replace the Navy as the nation's first line of maritime defense.[13]

In the face of these ambitious claims, the services continued to compete for the lion's share of congressional funding to expand their respective aviation components. As a cost saving measure, the Army airmen argued that Congress should drastically cut spending on warship construction and concentrate on expanding strategic Army aviation.[14] To demonstrate the soundness of this proposal, in 1921 General Billy Mitchell directed his Army air units to participate in a Navy experiment to survey the effect of aerial bombardment on the obsolete German battleship, *Ostfriesland.* Army bombers did indeed sink *Ostfriesland;* however, Mitchell's claims that strategic bombers had rendered surface fleets obsolete was far from conclusive. The *Ostfriesland* incident served to heighten tensions between the services during the interwar period since the Navy believed the Army airmen were

[13]H. H. Arnold, *Global Mission* (New York: Harper and Brothers, 1949), p. 157; See also Tami Davis Biddle, *Rhetoric and Reality in Air Warfare* (Princeton: Princeton University Press, 2002), p. 136.

[14]Alfred F. Hurley, *Billy Mitchell: Crusader for Air Power* (New York: Franklin Watts, 1964), p. 60.

actively seeking to curtail the size of the Navy and reallocate all military aircraft into an independent strategic Air Force.

As the competition for congressional funding continued, the respective services attempted to establish proprietary jurisdiction over specific missions, going so far in 1935 as to issue the *Joint Action of the Army and the Navy* (JAAN), officially outlining the boundaries where one or the other service could conduct air operations.[15] The restrictions in this joint action plan were meant to serve the parochial interests of the respective services more than the national security interests of the nation. Lest the Army encroach on the Navy's domain, the JAAN restricted Army flights over the ocean beyond a one hundred mile limit. Conversely, for fear that the Navy might try to develop its own strategic bombing, the JAAN imposed restrictions on development of all naval land-based aviation, excluding that necessary for training. The net result was that by the time the U.S. entered World War II, the Navy was severely handicapped with regard to its long-range antisubmarine and reconnaissance air capabilities, while the Army, which was jealously excluded from operating over the ocean, actually possessed the most suitable aircraft for antisubmarine operations.

In January 1942, Germany launched a devastating U-boat offensive against Allied shipping in the western Atlantic. At the time, the Navy had very few aircraft suitable for long-range antisubmarine patrols.[16] As a consequence, while the Navy scrambled to deploy antisubmarine surface craft and organize shipping convoys, land-based bomber units of the Army Air Forces were dispatched on antisubmarine patrols over the western Atlantic as a stop-gap measure.[17] In spite of this immediate national security threat, a partisan interservice disagreement arose over the propriety of employing Army units for a mission within the operational jurisdiction of the Navy. Claiming that its aviators were better trained in antisubmarine tactics, the Navy sought to reallocate hundreds of Army bombers to the Navy and retrofit them for antisubmarine warfare.[18] The Army Air Forces, meanwhile, intended to deploy all its aircraft overseas for the strategic bombing campaign

---

[15]The Joint Board, *Joint Action of the Army and the Navy* (Washington, DC: Government Printing Office, 1936), p. 8.

[16]Archibald D. Turnbull and Clifford L. Lord, *History of United States Naval Aviation* (New Haven: Yale University Press, 1949), pp. 313–14. See also Naval Historical Center, Washington, DC, Operational Archives Branch, The Official Papers of Fleet Admiral Ernest J. King (hereinafter King Official Papers), D. B. Duncan, Memorandum for Admiral King, 17 January 1942 (accessed on microfilm at U.S. Naval War College, Newport, RI).

[17]War Diary, North Atlantic Naval Coastal Frontier, January 1942, Chapter 3, pp. 1–2 (accessed on microfilm at U.S. Naval War College, Newport, RI).

[18]King Official Papers, E. J. King, Memorandum for General H. H. Arnold, 20 February 1942; C. E. Weakley, Memorandum for Secretary of War, 30 January 1948, enclosure A, p. 2. War Diary, North Atlantic Naval Coastal Frontier, January 1942, Chapter 4, p. 6 (accessed on microfilm at U.S. Naval War College, Newport, RI).

against the enemy heartland, which it hoped would demonstrate the decisive nature of strategic air power in actual wartime conditions. The Army Air Forces feared, however, that by ceding land-based bombers to the Navy, the strategic bombing campaign would be weakened, thus rendering the Army Air Forces incapable of proving the efficacy of its strategic air power doctrine. In an effort to thwart the Navy's bid to commandeer its bombers, the Army proposed establishing an Army Air Forces unit tasked with conducting antisubmarine operations.[19] The Army leadership argued that its airmen could attack submarines just as well as naval aviators, making the cession of bomber aircraft to the Navy unnecessary. By retaining control of its bombers assigned to antisubmarine warfare, the Army Air Forces hoped to redeploy its aircraft for strategic bombardment once the U-boat menace had been neutralized. A bitter interservice contest ensued as the Army Air Forces vehemently opposed all efforts by the Navy to reallocate these land-based aircraft.

While the Navy haggled with the Army over the reallocation of aircraft, a secondary dispute emerged over what tactics Army air units must employ in antisubmarine warfare. According to the principle of unity of command, Army Air Forces antisubmarine units were under the operational control of the Navy, and Admiral Ernest J. King, the Chief of Naval Operations, favored defensive tactics involving air patrols in the vicinity of ship convoys.[20] Alternatively, the Army leadership, under the close supervision of Secretary of War Henry Stimson, favored offensive tactics against enemy submarines as they transited "bottleneck" areas of the ocean.[21] The disagreement over tactics, along with the Navy's insistence that Army bombers be reallocated to the Navy, drew Secretary of War Stimson and Admiral King into bitter personal conflict.

Stimson's interest in antisubmarine warfare stemmed from his overall management of the shipment of material and manpower in Great Britain for the cross-channel invasion. Frustrated by the Navy's perceived failure at curbing U-boat attacks on shipping, the Secretary of War took matters into his own hands and sponsored the creation of the Army Air Forces Antisubmarine Command (AAFAC), an American unit patterned after RAF Coastal Command, which was tasked with developing more effective offensive antisubmarine tactics than the defensive tactics favored by the Navy.[22] However, the War Department's plan to put

---

[19] King Official Papers, H. H. Arnold, Letter to Admiral King, 16 March 1942.

[20] War Diary, Eastern Sea Frontier, April 1942, Chapter 2, pp. 2–3 (accessed on microfilm at U.S. Naval War College, Newport, RI).

[21] Samuel Eliot Morison, *The Battle of the Atlantic, 1939–1943* (Edison, NJ: Castle Books, 1947), p. 243.

[22] Yale University Library, Manuscripts and Archives, The Diary of Henry L. Stimson (hereinafter Stimson Diary), 21 May 1942. See also King Official Papers, E. J. King, Letter to General McNarney, 28 May 1942; Library of Congress, Washington, DC, Manuscript Division, Henry Harley Arnold Papers, Reel 203, H. H. Arnold, Memorandum for the Joint

offensive tactics into practice was stymied by Admiral King who insisted that the AAFAC operate under Navy command and employ the Navy's defensive tactical doctrine. Believing that offensive tactics would prove superior if tried, Stimson attempted to force an obdurate Admiral King into permitting an Army-led antisubmarine offensive in the Navy's oceanic domain. Lacking the authority to force King's compliance, Stimson requested that Navy Secretary Frank Knox act on his behalf and order King to cooperate.[23] Knox refused, believing it inadvisable to interfere in purely military matters which he thought best left to the uniformed leadership.[24] In frustration, Stimson went so far as to request presidential intervention. To his disappointment, Roosevelt refused to take sides on the matter.[25] Although a limited AAFAC offensive was eventually launched in the Bay of Biscay in 1943, Stimson believed that Navy obstructionism had hampered decisive victory in the operation. Stimson's aggravation was compounded when he learned that the Navy, in an effort to frustrate War Department plans to develop the AAFAC into an American version of Coastal Command, established a competing antisubmarine unit known as the Tenth Fleet, insisting that all Army antisubmarine units be subsumed within its organizational structure. With bitter resentment towards Admiral King, Stimson and his War Department subordinates thought it advisable to withdraw from an interservice conflict they could not win and finally agreed to cede all Army land-based antisubmarine aircraft to the Navy. This, the so-called Arnold-McNarney-McCain agreement, provided a timetable for transfer and meant the ultimate divestiture of the Army Air Forces from antisubmarine operations.[26]

By the summer of 1943, the Allies gained the initiative against the U-boats largely because of the full-scale introduction of escort carriers for convoy protection. Yet, in spite of this success, Stimson harbored a lingering bitterness towards the Navy over how the antisubmarine air war was managed, and his acrimony endured for years after the war had ended.

Chiefs of Staff, 19 October 1942.

[23]U.S. National Archives and Records Administration (hereinafter NARA), Record Group (hereinafter RG) 107, Formerly Top Secret Correspondence of Secretary Stimson 1940–1946, Navy File, Henry L. Stimson, Memorandum for the Secretary of the Navy, 7 July 1942. According to Stimson's Diary entry for 7 July, this memorandum was written by Robert Lovett.

[24]RG 107, Formerly Top Secret Correspondence of Secretary Stimson 1940–1946, Navy File, Henry L. Stimson, Memorandum for the Secretary of the Navy, 7 July 1942.

[25]Stimson Diary, 26 March 1943. See also RG 107, Formerly Top Secret Correspondence of Secretary Stimson 1940–1946, ASW File, Henry L. Stimson, Letter for President Roosevelt, 26 March 1943. This letter forcefully argues for the need to establish an American Coastal Command on the model of the RAF Coastal Command.

[26]King Official Papers, E. J. King, Memorandum for the Chief of Staff, U.S. Army, 14 June 1943.

Disagreement over how best to prosecute the antisubmarine war revealed a deep divide between the way civilian authority operated *vis-à-vis* the uniformed leadership in the War and Navy Departments. During the height of the interservice conflict, Stimson repeatedly wrote in his diary of his belief that the Navy Department was operating under an antiquated and flawed organizational model desperately in need of reform.[27] He criticized what he described as a clique of secretive admirals in control of a byzantine bureaucracy where the Secretary of the Navy functioned more as a bookkeeper than as the delegated voice of the President as Commander in Chief. By contrast, Stimson considered his own position as Secretary of War a vital link in the chain of command between the President and the Army Chief of Staff. In the War Department, Stimson was used to exercising a significant, if limited, degree of direct authority over military operations and enjoyed a close working relationship with Army Chief of Staff George Marshall.[28] No such relationship existed, however, between Navy Secretary Frank Knox and Admiral King, where the latter would rarely inform the former of operational details, and only begrudgingly if pressed to do so. Stimson believed that the Navy Department's antiquated and flawed command model was desperately in need of reforms similar to those effected under Elihu Root, who, as Secretary of War in 1904, directed the Army to adopt a General Staff command model. Root's reforms had the effect of fostering strong civilian control within the War Department, where the Army Chief of Staff functioned under the clear direction of the Secretary of War. By contrast, the department structure of the Navy remained virtually unchanged since the Spanish-American War, having no general staff comparable to that of the Army. Moreover, the Secretary of the Navy exercised virtually no command authority over the Chief of Naval Operations, and the uniformed bureau chiefs controlled their own administrative areas like private fiefdoms.

The Movement Toward Military Unification

During the course of World War II, civilian policy-makers and military leaders alike grew concerned over wasteful spending on duplicative efforts in the prosecution of the war. Likewise, many expressed concern with the perceived imperfections of the extemporized command structure of the Joint Chiefs of Staff. In an effort to address these concerns, a movement began as early as 1943 to streamline military administration through the establishment of a single Department of Defense, under a single Secretary of Defense, having overarching control over a single Chief of Staff for both the Army and the Navy. Pursuant to this end, in April of 1944, the House of Representatives Select Committee on Post-War Military

[27] For example, see Stimson Diary, 7 June 1942; 21 July 1942; 23 July 1942; 27 July 1942.

[28] Elting E. Morison, *Turmoil and Tradition: A Study of the Life and Times of Henry L. Stimson* (Boston: Houghton Mifflin, 1960), p. 505.

Organization (the so-called Woodrum Committee) convened to investigate the benefits of enacting changes to the national defense bureaucracy. Over the course of nine months, the Woodrum Committee explored the merits of Air Force independence, as well as the possible advantages that might be gained by merging the Army, Navy, and Air Force into a single department. As Secretary of War, Henry Stimson was privileged to testify before the committee as the first witness.

On the whole, Stimson's prepared testimony was unremarkable. As General Marshall had advised, the Secretary spoke in broad terms of the need to consolidate the armed forces into a single Department of Defense for the sake of efficiency. Stimson spoke in laudatory terms of the Navy's cooperation with the Army in the war effort and of the need for a post-war reorganization so as to guarantee that future military leaders would continue their cooperative efforts in managing the national defense.[29] Nothing in his remarks could be considered insulting to the Navy or indicative of the War Department's desire to strip the Navy of its air component. A far more telling indication of Stimson's lingering antipathy towards the Navy is found in the text of an undated *aide memoire* titled "Brief for Woodrum Committee in Executive Session." In it, Stimson outlines the deleterious conditions which prevailed in the War Department before the reforms enacted by Elihu Root. The document covers in detail how Root's reforms strengthened civilian control over the War Department and notes the beneficial aspects of having a well-organized General Staff. The *aide memoire* also touched upon the merits of an independent Air Force. Yet, of all the issues Stimson's *aide memoire* covered, the most telling referred to the necessity of subordinating the Army and the Navy under a unified, post-war "War Department," with a single Chief of Staff having authority over all land, air, and sea operations. The advantage, Stimson noted, was that an overarching Chief of Staff "can in the name of the President order cooperation of the three elements instead of merely persuading."[30] Clearly, Stimson is hinting at instances where the War Department's efforts were blocked by Admiral King when he refused to cooperate with the War Department's plans for an antisubmarine offensive in the Atlantic. Although Stimson's notes do not mention the antisubmarine war explicitly, it is important to consider that the document in question was meant to serve as a memory aid during closed discussions and it is likely that Stimson discussed these matters in great detail with the members of the Woodrum Committee in executive session. Moreover, it is essential to note that Stimson's frequent references to Elihu Root in the *aide memoire* echo entries from his diary

[29]Princeton University Library, Department of Rare Books and Special Collections, Public Policy Papers, James V. Forrestal Papers (hereinafter Forrestal Papers), Box 144, Folder 9, Statement by the Secretary of War before the House Select Committee on Post War Military Policy, 25 April 1944.

[30]RG 107, Formerly Top Secret Correspondence of Secretary Stimson 1940–1946, Post-War Military Policy File, Brief for Woodrum Committee in Executive Session, Post-War Departmental Reorganization, undated.

written during the height of his imbroglio with Admiral King in 1943. In these diary entries, Stimson clearly ponders the desirability of forcing Root-like administrative reforms on the Navy. As we shall see, Stimson will return to this theme in 1947 when making the case for military unification in his memoir, *On Active Service in Peace and War.*

Most of the War Department witnesses that followed Stimson in testifying before the Woodrum Committee spoke of the merits of military reorganization without specifically raising issues involving the future of naval aviation. If there was a single bombshell dropped during the hearings, however, it happened when Assistant Secretary of War for Air, Robert A. Lovett, testified. In his opening remarks, Lovett stated that he favored Air Force independence within a unified Department of Defense. He discussed how the urgency of war allowed for the rise of two air arms resulting in "a considerable amount of overlapping activities, certain uneconomical duplication and competition and, in some instances, to blurred lines of responsibilities between the Services."[31] Lovett made clear his arguments were not *necessarily* War Department policy, but were his own opinions formed during his management of the Army Air Forces during the war. Lovett stated that he believed the post-war Navy should retain a highly specialized "Fleet Air Force," consisting mostly of "carrier and ship based aircraft." However, according to Lovett, all land-based aircraft would become the responsibility of an independent Air Force.[32] Moreover, Lovett testified that all aircraft, whether deployed against sea or land targets, would be allocated in the interest of "national defense" by a unified military General Staff, "and not on the tortured interpretation of antiquated documents dealing with vague theories and doctrines which have to be thrown away the moment war breaks out."[33] Clearly, at the highest level, Lovett is declaring the War Department's belief that the 1935 *Joint Action of the Army and the Navy* must be considered null and void and that the Navy should be stripped of its land-based airplanes. In spite of his assertion that his testimony was only personal opinion and not War Department policy, one must understand that War Department witnesses coordinated their testimony prior to the commencement of the Woodrum Committee hearings. Secretary Stimson knew exactly what Lovett planned to say and it is almost certain he fully approved of it. By Lovett couching his statement as "personal opinion," the War Department had some leeway in dealing with the Navy leadership which was sure to react hostilely to the Assistant Secretary's statement.

Although the Woodrum Committee intended to investigate a variety of issues pertinent to post-war military organization, the future status of naval aviation

---

[31] Forrestal Papers, Box 144, Folder 9, Statement by Robert A. Lovett, Assistant Secretary of War for Air, before the House Select Committee on Post War Military Policy, House of Representatives, 26 April 1944.

[32] Ibid.

[33] Ibid.

emerged as the most important and divisive issue during the hearings. Following Assistant Secretary Lovett, almost every Navy witness testified in opposition to Air Force independence and affirmed the Navy's need to retain its land-based aviation component.[34] Likewise, the Navy men opposed an overarching Department of Defense, fearing that military unification might lead to the empowerment of an anti-Navy, pro-Air Force Defense Secretary intent on denuding the Navy of its aviation. In sum, the Navy men considered the military unification movement as just another ploy by the strategic air power advocates to strip the Navy of its traditional role as the nation's first line of defense. Yet for all the testimony offered for and against military unification, the Woodrum Committee adjourned without reporting a bill to Congress and agreed to defer all matters pertaining to military reorganization until after the war had ended.

The Post-War Unification Debates

Congress took the matter of military unification up again after the war and a series of unification bills were introduced in Congress between 1945 and 1947. As before, witnesses from the War and Navy Departments testified regarding the merits of an overarching Defense Department and an independent Air Force, and likewise, as before, interservice rivalry engendered recriminations on both sides. As prior to the Woodrum Committee, one of the most contentious issues to emerge during the post-war unification hearings was the future status of naval aviation. During these hearings, it became evident that War Department animosity over the Navy's obstruction of an Army-led antisubmarine offensive and the subsequent reallocation of Army antisubmarine aircraft to the Navy in 1943 had not abated. As a result, those favoring Air Force independence adamantly opposed the Navy's retention of all land-based aviation, claiming even antisubmarine warfare and sea reconnaissance should properly be assigned to an independent Air Force. The Navy stood firm on its need to retain certain land-based aviation, claiming that antisubmarine warfare and sea reconnaissance were intrinsic to the Navy's mission and required specialized naval training. Thus, between 1945 and early 1947, efforts to pass military unification bills were frustrated largely due to disagreement in Congress over the future of land-based naval aviation.

During the interwar period, Army airmen objected to the Navy's development of land-based airplanes for fear that the Navy planned to develop its own strategic air arm capable of attacking land-based targets, thereby encroaching on the Air Force's operational jurisdiction. During the post-war unification debates, the Navy's assurance that such fears were unfounded did not wholly satisfy the Army

[34] *Hearings Before the Select Committee on Post-War Military Policy, House of Representatives, Seventy-Eighth Congress, Second Session* (Washington, DC: Government Printing Office, 1944), p. 224. See also Forrestal Papers, Box 131, Folder 10, General Board of the Navy, Memorandum to SecNav, et al., 26 May 1944.

airmen, since the Navy had in fact used a small number of land-based airplanes originally appropriated for antisubmarine purposes to attack Japanese land targets in the Pacific. However, in 1947, with the long-cherished goal of Air Force independence in sight, the Army air power advocates were willing to compromise and dropped their opposition to the development of certain naval land-based aviation. This turnabout was contingent on the Navy's agreeing to formally limit its land-based aviation capability specifically to antisubmarine warfare, reconnaissance, and protection of shipping.

Thus, in 1947, a compromise between the War and Navy Departments was hammered out which allowed military unification to go forward. Representatives from both services drafted a unification bill which provided for an overarching Secretary of Defense and an independent Air Force, with an attending Executive order that explicitly provided for the Navy's retention of both carrier and certain land-based aviation necessary for antisubmarine warfare and reconnaissance.[35] Committees in both the House and Senate conducted investigations on this compromise bill and its attending Executive order, and as in earlier investigations, heard testimony from civilian and uniformed witnesses regarding the proposed legislation.

In spite of the War Department's sincere efforts to guarantee the Navy's retention of its land-based aviation, voices of opposition were still raised from within the upper echelons of the Navy. Opponents of unification repeated the claim that the compromise bill was really just a "Trojan Horse" meant to empower a pro-Air Force Defense Secretary intent on slashing funding for naval aviation, thereby eliminating it through neglect. By this late date, a migration of issues becomes noticeable in the debate. This shift of focus had largely to do with emerging questions pertaining to what roles the Air Force and the Navy might play in any future wartime deployment of atomic weapons.[36]

Conventional strategic bombardment during World War II failed to live up to the pre-war claims made by the air power advocates of its ability to hasten victory through the destruction of the enemy's war-making capability.[37] The reasons for this were hotly debated during the war. Although beyond the scope of this present study, it is worthwhile to note that one argument claimed that World War II-era precision bombing technology was incapable of delivering conventional bombs on target more than a small percentage of the time. The air power advocates believed such questions had been rendered moot at Hiroshima and Nagasaki, when, in

---

[35]*Hearings Before the Committee on Armed Services on S. 758, United States Senate, Eightieth Congress, First Session* (Washington, DC: Government Printing Office, 1947), pp. 2–7.

[36]*Hearings Before the Committee on Expenditures in the Executive Departments, House of Representatives, Eightieth Congress, First Session* (Washington, DC: Government Printing Office, 1947), pp. 510, 524–25.

[37]For example, see Chapter 10 of Michael S. Sherry, *The Rise of American Air Power* (New Haven: Yale University Press, 1987); and Davis Biddle, *Rhetoric and Reality in Air Warfare.*

August 1945, pinpoint accuracy was no longer necessary given the enormous destructive capability of the atomic bomb. The swift surrender of Japan which followed seemed to prove that effective strategic bombardment had indeed broken the enemy's will to resist. Here, so the air power advocates claimed, was the unimpeachable justification for the development of an independent strategic Air Force tasked with atomic bombardment of strategic targets in the enemy's heartland. Given the size of atomic bombs in the late 1940s, the only delivery systems capable of carrying these weapons were large, land-based airplanes such as the B-29 and the B-36, the latter of which was in development at the end of the war. As long as the Air Force monopolized these types of airplanes, it faced no competition from the Navy in the delivery of strategic atomic bombs. This was all but guaranteed by the terms of the latest unification bill which strictly limited the Navy's land-based aviation to antisubmarine and reconnaissance missions. Furthermore, in a manner redolent of the controversial *Ostfriesland* tests in 1921, the air power advocates' optimism that an independent Air Force would emerge as the nation's preeminent military branch was heightened by the results of the atomic bomb tests on obsolete warships anchored at Bikini Atoll in 1947. Thus, the guarantee of the Air Force's continued monopoly of atomic bomb delivery made concessions to the Navy regarding its land-based antisubmarine and reconnaissance aviation an insignificant sacrifice. However, technological developments in carrier-based aircraft and ship construction by 1947 indicated that the Air Force's atomic monopoly might be short-lived.

The Navy was moving forward with a program to build the next generation of carriers capable of supporting aircraft large enough to carry an atomic payload. Navy officers were also quick to point out that the Air Force's new long-range bomber, the much-touted B-36, was not only extremely expensive, but was based on a technology nearing obsolescence. Moreover, the Navy men raised speculation about the possible role unmanned guided missile technology might play in future wars. Thus, by the final stage of the unification debate, the issue had migrated far from whether or not the Navy should retain control of land-based antisubmarine aviation towards whether the nation's defense interests were best served by creating an independent Air Force, and thereby institutionalize World War II-style strategic bombing at a time when this mode of warfare might be passing into obsolescence.[38] In many ways, the final stage of the unification debate presaged the interservice conflict that would emerge in 1949, when Secretary of Defense Louis Johnson canceled construction of the super-carrier *United States,* resulting in the imbroglio known as the "revolt of the admirals." Nevertheless, passage of the National Security Act in July 1947 generally signaled the end of the interservice debate over the future of naval land-based antisubmarine aviation.

[38] Davis Biddle, *Rhetoric and Reality in Air Warfare,* pp. 275–77.

Henry Stimson and the Final Phase of the Unification Debate: The Battle of the Atlantic Revisited

In April 1946, when an earlier version of the unification bill was stalled in Congress, Secretary of War Robert Patterson approached his predecessor, Henry Stimson, about publishing an article in a popular magazine such as *Reader's Digest* or *The Saturday Evening Post* on the merits of military unification in an effort to win popular support for the flagging bill.[39] Although Stimson strongly supported military unification, he declined to act, citing his poor health resulting from a recent heart attack though he did not rule out writing a supportive article once his health improved. By the following July, Stimson began collaborating with McGeorge Bundy in the writing of his memoir, eventually published under the title *On Active Service In Peace and War.* At the same time, the latest version of a unification bill slowly began making its way through Congress, and Stimson considered it an opportune time to publicly lend his support to the legislation. He directed Bundy to prepare Chapter 20 of his memoir as a stand-alone article for publication in the popular magazine *Ladies' Home Journal,* with the intended purpose of rallying public support just as Robert Patterson suggested in April 1946 during the height of the earlier unification debate in the Senate.

In spite of Stimson and Bundy's efforts, the National Security Act became law in July 1947, half a year before a condensed version of Chapter 20 was published in the February 1948 issue of *Ladies' Home Journal.* It must be understood that this article was conceived at a time when the bill's passage was uncertain. To make his case for the necessity of military unification, Stimson could have chosen any number of subjects around which to craft his argument, but the story he chose to relate was that of the missed opportunity to launch a decisive Army Air Forces antisubmarine offensive in the Atlantic. Stimson's larger intent was to argue that had there been an overarching Secretary of Defense having authority over both the War and Navy Departments, Admiral King could have been ordered to cooperate with the Army Air Forces and launch an antisubmarine offensive as the War Department had hoped. Since the National Security Act had already become law, Bundy included an addendum to the article stating that the new overarching Secretary of Defense would streamline the nation's military command structure and prevent recalcitrant admirals like Ernest King from defiantly protecting the parochial interests of the Navy at the expense of the national security.

Leaving aside the merits of Stimson's argument, it is important to note that with the publication of this article, Stimson was unmistakably tying the wartime controversy over control of land-based antisubmarine aviation to the need for military unification. In his mind, the two issues had become inextricably linked, and

---

[39]Library of Congress, Washington, DC, The Robert P. Patterson Papers (hereinafter Patterson Papers), Box 22, General Correspondence, Robert P. Patterson, Letter to Henry L. Stimson, 22 April 1946.

as stated above, references in his diary to the military reforms of Elihu Root indicates that this impression was formed as early as 1943.

By the time the National Security Act passed both houses of Congress on 26 July 1947, work on the first draft of Chapter 20 of Stimson's memoirs was nearly, if not completely, finished. While reviewing this earlier draft, Stimson criticized Bundy's mild tone, arguing that he desired to publish an account "dramatically much more adverse to the Navy."[40] He directed Bundy to make a sharper contrast between what he considered the lackluster Navy tactic of aerial antisubmarine defense compared to the War Department's preference for offensive antisubmarine tactics. Although Bundy mentioned RAF Coastal Command and its preference for offensive tactics in the earlier draft, to Stimson's chagrin, he apparently omitted mention of the Army Air Forces Antisubmarine Command and of the War Department's plan to launch an American antisubmarine offensive in the Bay of Biscay in 1943. Stimson wanted to accentuate the desirability of offensive tactics, but feared Bundy's omission of the AAFAC initiative would leave the reader with the false impression that the difference in tactics was simply an American vs. British contrast. By including the story of Navy opposition to the AAFAC's Biscay offensive, the public would learn that the contrast in antisubmarine tactics was not British vs. American, but rather Army vs. Navy.[41]

In an ironic turnabout, as the final version of the National Security Act was being debated in Congress, for political reasons Stimson thought it prudent for Bundy to tone down certain inflammatory and *ad hominem* comments in Chapter 20. One very incendiary passage found in a draft written some time before May 1947 compared the Joint Chiefs of Staff to the Security Council of the United Nations and likened the Navy Department to "Soviet Russia," which employed "a rigorous insistence on unanimity as a means of defending its own interests."[42] This passage was changed in the published version as a less confrontational reference to a vague military "officer" who, as a minority of one, "could employ a rigorous insistence on unanimity as a means of defending the interests of his own service."[43] An unflattering personal reference to Admiral King as a rigidly parochial guardian of the Navy's interests was also removed before publication.[44]

Sometime after the unification bill passed the Senate, but before James Forrestal was confirmed as the first Secretary of Defense, Bundy added a conclusion to the manuscript which was drawn mostly from Stimson's 21 April 1947 letter to

---

[40] Ibid.

[41] Ibid.

[42] John F. Kennedy Presidential Library, Boston, MA, The McGeorge Bundy Papers (hereinafter Bundy Papers), Series 2, Henry L. Stimson, *On Active Service,* Notes and Drafts Chapter XX (1 of 2), "A Case History and its Lessons," undated.

[43] Henry L. Stimson and McGeorge Bundy, *On Active Service in Peace and War* (New York: Harper and Brothers, 1947), p. 515.

[44] Bundy Papers, "A Case History and its Lessons," undated.

Senate Armed Services Committee chairman, Chan Gurney.[45] Bundy's initial draft of the conclusion was pared down for the sake of brevity, but in essence it presented the argument Stimson put forth in his letter to Senator Gurney that military unification was a continuation of the reforms begun by Elihu Root when he instituted the Army General Staff in 1904.[46] The chapter was reworked by Bundy at least two more times but only minor grammatical changes were made. The final working draft was nearly identical to what would be published in *On Active Service* in early 1948.

An Argument for the Next Wave of Military Reform: The Liability of Divided Leadership Made Manifest in the Antisubmarine War

Stimson's argument of the need for military unification laid out in the *Ladies' Home Journal* article was published under the title "A Time of Peril," and a nearly identical but somewhat longer version was published in *On Active Service* under the chapter title "The Army and the Navy." The work begins with token acknowledgment of the cooperative efforts of the two services in amphibious operations, further noting how Stimson admired the Navy for its courage and skill in battles such as Midway and Leyte Gulf. "This much said," Bundy wrote, "we must proceed in this chapter to a discussion of Army-Navy relationships in which the less pleasant side of the story will be emphasized."[47] He noted that differences during the war between the services were frequent, conceding that many were the inevitable clashes of "two agencies of strong will." However, Bundy wrote that Stimson believed some clashes "Grew mainly from the peculiar psychology of the Navy Department, which frequently seemed to retire from the realm of logic into a dim religious world in which Neptune was God, Mahan his prophet, and the United States Navy the only true Church. The high priests of this Church were a group of men to whom Stimson always referred as 'the Admirals.' These gentlemen were to him both anonymous and continuous. . . ."[48]

In spite of the alienation he felt from career Navy officers, Stimson was able to forge an "intimate and friendly contact" with the civilian leadership of the Navy Department. He found in Navy Secretary Frank Knox a man "without any trace of pettiness," and was gratified that their friendship "was not shaken by their occasional disagreement."[49] Nevertheless, Stimson was frustrated by what he perceived

---

[45] *Hearings Before the Committee on Armed Services on S. 758, United States Senate, Eightieth Congress, First Session, Part Two* (Washington, DC: Government Printing Office, 1947), pp. 459–60.

[46] Stimson and Bundy, *On Active Service in Peace and War*, pp. 519–23.

[47] Ibid., p. 503.

[48] Ibid., p. 506.

[49] Ibid.

as Knox's inability to bring the uniformed naval leadership in line. Stimson did not fault Knox for this, for he recognized that even the President could not control the admirals. In his estimation, the chief problem was that the entire Navy Department had never benefited from the kind of reforms carried out in the War Department by the progressive Secretary of War Elihu Root. As a consequence, "'the Admirals' had never been given their comeuppance."[50] This summarized what Stimson saw as the fundamental shortcoming of the Navy. "The Navy in World War II had in Knox, Forrestal and King three strong men at its head; they accomplished much in moving their department forward," Bundy wrote. "But in Stimson's mind it was no discourtesy to remark of them that not one was another Elihu Root."[51]

Reference to Elihu Root's War Department reforms served as a brief preface to the chapter's major theme, which was the disagreement of the Army and the Navy on matters critical to the war effort. It is the transition from these prefatory remarks to the balance of the chapter that reveals Stimson's propagandistic intentions. Bundy wrote, "What seemed important to Stimson, in retrospect, was to look behind the disagreements towards their causes, in an effort to prevent or minimize their future occurrence. The best way to do it is to study one particular disagreement in some detail. And the one with which Stimson was most deeply concerned was the prolonged struggle over antisubmarine warfare."[52] Bundy subsequently recounts at length Stimson's views regarding the War Department's disagreement with the Navy over the proper conduct of the antisubmarine war and of the Army's intention to develop the AAFAC into an offensive antisubmarine unit patterned on RAF Coastal Command. In Stimson's mind, the primary reason the American antisubmarine effort languished was because the Navy was led by bullheaded and intractable admirals like Ernest King. Moreover, the reason the Navy fell prey to such recalcitrant men was because, as stated earlier, it had never benefited from the reforming efforts of a powerful Navy Secretary as had the Army under the leadership of Elihu Root. The remedy, in Stimson's opinion, was two-fold. First, the Navy should be compelled by legislative act to accept administrative reforms similar to those which Elihu Root forced on the Army. This would go a long way towards forcing "the Admirals" to submit to civilian authority. However, this reform alone would not ensure Navy cooperation with Army initiatives since the Navy would remain under its own department secretary, who might, as in the case of Frank Knox, refuse to assert control over military affairs. Therefore, a second and equally important reform must entail the unification of the armed forces into a single Department of Defense under an overarching Defense Secretary, who, in the spirit of Elihu Root, would bring "the Admirals" firmly under civilian control and force interservice cooperation.

[50] Ibid.

[51] Ibid., p. 507.

[52] Ibid., p. 508.

Bundy outlined five major lessons Stimson believed the public should learn from a retrospective consideration of the antisubmarine war. To begin, Stimson believed that the record of Allied antisubmarine activity proved that the Army was right and the Navy was wrong on the tactical issues involved in combating the U-boat menace. Bundy pointed out that after 1942, when airborne radar came into extensive use, "five-sixths of the submarines destroyed from the air were killed by shore-based aircraft."[53] Moreover, according to Bundy, "The vast majority of these shore-based kills were accomplished by aircraft flying under the control of Slessor's Coastal Command in accordance with the principles of air autonomy and aggressive search so long and vainly argued by Stimson on the American Navy. The early Navy notion that convoy escort was the only way of fighting the submarine was in Stimson's view completely exploded by the brilliant operations of the Navy's own hunter-killer groups in 1943 and afterwards, not to mention the shore-based campaigns of Coastal Command first in the Bay of Biscay and later in Norwegian waters."[54]

A second, related lesson was of the need for military men to listen closely to the scientists, since their contribution to antisubmarine warfare was enormous and extended beyond mere technical advice. "Scientists like [Edward] Bowles and [Vannevar] Bush proved themselves to be capable of sound strategic comment and of the tactical control and use of antisubmarine weapons," Bundy wrote. Stimson found these men "far wiser than either naval or air officers who had become wedded to a limited strategic concept."[55]

The third lesson was of the importance for clear lines of command responsibility. Stimson faulted the Navy for not adopting the British policy of having "a vigorous and independent group of senior officers conducting antisubmarine warfare in a continuous campaign"[56] until 1943. He charged that the Navy was more concerned with fighting the war in the Pacific than against the U-boats, and that antisubmarine operations in the Atlantic "were left to commanders not always chosen from the top drawer."[57]

A fourth lesson of even greater import was that the Joint Chiefs of Staff was an "imperfect instrument" for top-level decision making because it was incapable of "enforcing a decision against the will of any one of its members." As stated earlier, Stimson compared the Joint Chiefs of Staff to the post-war Security Council of the United Nations: "Any officer, in a minority of one, could employ a rigorous insistence on unanimity as a means of defending the interests of his own service. Quite aside from the question of which service was right as to antisubmarine tactics, there was no justification for a situation in which the Army and the Navy worked

[53] Ibid., p. 514.

[54] Ibid.

[55] Ibid.

[56] Ibid., p. 515.

[57] Ibid.

at cross-purposes for more than a year, each appearing to the other as an ignorant, presumptuous, interfering bungler."[58]

Stimson credited the success of the wartime Joint Chiefs of Staff with General Marshall who, he believed, exercised a level-headed determination to stand firm on the most important issues, while compromising on those of lesser importance for the sake of maintaining an amicable working relationship with the Navy leadership. In other words, the success of the Joint Chiefs of Staff during the war was mostly attributable to Marshall's character. Had another, lesser man been Army Chief of Staff, there was nothing in the Joint Chiefs of Staff structure that prevented disagreements among its members from bringing the war effort to a grinding halt. By raising this point, Stimson argues that reforms were necessary which would make all the service chiefs accountable to a single Secretary of Defense, since there was no guarantee that future Army Chiefs of Staff would be as successful as Marshall at effecting compromise and cooperation.

The fifth, and according to Stimson, the "most important" lesson of the antisubmarine war was that it served as a perfect example "of the destructive effect of the mutual mistrust of the two services."[59] Unfortunately for the Allied cause, questions concerning tactics were "surrounded by all sorts of interservice recriminations," and opened up wounds that had been festering since the early 1920s:

> It was unfortunate that the Army side of the question should have been mainly an Air Forces operation, for the Navy and the Air Forces had a mutual grudge of over twenty years' standing—the Navy feared that the Air Forces wished to gain control of all naval aviation, while the Air Forces saw in the Navy's rising interest in land-based planes a clear invasion of their prescriptive rights. The Air Forces considered the Navy a backwards service with no proper understanding of air power; the Navy considered the Air Forces a loud-mouthed and ignorant branch which had not even mastered its own element.[60]

As a result, when military leaders should have been considering what antisubmarine methods were the most tactically sound, "at all echelons" emotionally charged accusations of incompetence and encroachment clouded rational discussion on this most vital issue.[61]

Bundy concluded with a prescription of how the longstanding interservice rivalries which came to the forefront during the antisubmarine war could be remedied. "To Stimson and others thinking of the future," he wrote, "it seemed evident

---

[58] Ibid.

[59] Ibid., p. 516.

[60] Ibid.

[61] Ibid.

that the primary objective of the postwar period in military affairs must be to end this division of feeling."[62] Bundy briefly discussed the Congressional efforts to bring about service unification beginning with the Woodrum Committee hearings in the House of Representatives in 1944: "Stimson, like most of his War Department colleagues, believed that the consolidation of the armed forces into a single department would be enormously helpful in reducing friction and duplication of effort. He saw it as a means of eliminating the waste of time and money involved in the necessarily cumbersome method of 'co-operation' and as a way of insuring action when and if 'co-operation' ceased to exist."[63] However, Stimson knew that any move towards unification would be strongly opposed by "the well-known Admirals." Nevertheless, in the spirit of his mentor, Elihu Root, he became an ardent supporter of Congressional legislation to bring about military unification in a single Department of Defense under a single civilian Secretary of Defense.

Fact-Checking and Publication

Shortly before the publication of "A Time of Peril" in the *Ladies' Home Journal,* Bundy solicited feedback from Edward Bowles, Stimson's wartime science advisor on matters pertaining to airborne radar.

Bowles understood that Stimson's memoir was meant to serve as propaganda in support of the unification legislation that was then wending its way through Congress, and specifically suggested that Bundy more forcefully tie Navy intransigence regarding the War Department's planned antisubmarine offensive in the Bay of Biscay to the "subsequent discussion on unification."[64] Bowles outlined for Bundy the problem faced by General Marshall, Stimson, and the War Department staff with getting President Roosevelt to intervene and stipulate which service should have command of the antisubmarine air war. According to Bowles, General Marshall eventually realized that this was a lost cause. "The Commander in Chief," he wrote, "did not elect to resolve the difficulties of command, which were the real issues."[65] Bowles also remarked that he believed an important point needing greater clarification was the shortcomings of the Joint Chiefs of Staff in their exercise of effective command. With regard to the impasse over command of the antisubmarine air war, he wrote,

> To me, here you hit on one of the most striking demonstrations of the weakness of the Joint Chiefs of Staff to come out during the war. I understand there were about fifteen similar illustrations where this weakness

---

[62]Ibid., p. 518.

[63]Ibid.

[64]Ibid.

[65]Ibid.

> made it impossible for anything more than a compromise action by the Joint Chiefs of Staff. In the instance of the antisubmarine impasse a whole year was spent in disputation at a time when our nation was exposed to what was probably its greatest peril of the war, only to determine the Army Air Forces and the Navy could not work together on this problem. To think that we would suffer so much unnecessary waste of life and shipping by such oppositionist tactics is a credit neither to the Joint Chiefs of Staff nor to the Commander-in-Chief.[66]

Bowles tied the whole issue to the contemporary unification debate. However, Bowles was unconvinced that military unification would solve the problems intrinsic to the Joint Chiefs of Staff command model. He continued, "Even with unification this situation remains unchanged. I believe Stimson would hold to the view that unification will not be completely assured until there is such a sublimation of interests that none of the three component forces will look with suspicion on the idea of an over-all Chief of Staff."[67] Bundy expressed appreciation for Bowles' critique; however, he made little change to the text as it then stood.[68]

By prior arrangement, as stated, Chapter 20 of *On Active Service* was slated for publication in the February 1948 issue of the popular magazine, *Ladies' Home Journal,* two months before the book's release in April. By mid-December 1947, the copy editors at the magazine were making the final changes to Bundy's manuscript. Although the magazine was committed to publishing the article, Bundy expressed concern that the copy editors planned to cut the section arguing for service unification.[69] Allowing "A Time of Peril" to go to press without tying the unification issue to the wartime Army-Navy dispute over antisubmarine tactics denuded it of its value as a work of pro-unification propaganda. For this reason, Bundy proposed inserting the following truncated conclusion: "His wartime experience with the difficulties of Army-Navy relations made Stimson an early and ardent advocate of the unification of the services which was finally achieved in 1947. He hoped that unification would reduce waste and conflict to a minimum; he hoped still more that it might lead to a gradual ending of the interservice rivalry which was so deeply rooted in the tradition of the armed forces. The Army, the Navy, and the Air Force, he believed, must now learn, or be taught, to live together in genuine friendship."[70]

---

[66] Ibid.

[67] Ibid.

[68] Bundy Papers, Henry L. Stimson, Book Criticisms and Critiques (file 2 of 3), McGeorge Bundy, Letter to Edward Bowles, 8 December 1947.

[69] Bundy Papers, Henry L. Stimson, Book Criticisms and Critiques (file 2 of 3), McGeorge Bundy, Letter to Miss Conling, 12 December 1947.

[70] Ibid.

A final, minor change involved the moving of a humorous anecdote about the Army-Navy football game to the last paragraph of the magazine article.[71] When the article was published in the February 1948 issue of *Ladies' Home Journal*, it carried Bundy's suggested changes and was otherwise identical to the published chapter in *On Active Service.*

On 4 February 1948, Robert Patterson wrote to Henry Stimson after reviewing the proof of *On Active Service.* "I regard it," he wrote, "as the best piece of writing that has come out of the war. I am sure that it will be most useful from the viewpoint of history."[72] It is true that the unification legislation had already been signed into law, and that the article was thus no longer needed as a way of bringing public pressure to bear on Congress. Nevertheless, the unification legislation was far from universally popular, and Stimson's article, though it missed the opportunity to influence the outcome of the unification debate in Congress, might still serve to shape the opinions of those who remained undecided as to whether the new law was necessary and beneficial to the national defense.

Conclusion

With the 1948 publication of "A Time of Peril" in *Ladies' Home Journal* and the release of the full text of *On Active Service in Peace and War,* Henry Stimson had the last word in the unification debate regarding which military service should control land-based antisubmarine aviation. In Stimson's mind, the issues of military unification and the control of land-based antisubmarine aviation were inextricably linked. Clearly, Stimson believed that military unification would prevent stubborn Navy leaders from defying civilian leadership in the future as Admiral King had done *vis-à-vis* the Secretary of War. In spite of the merits of Stimson's argument, the mismanagement of the antisubmarine war did not become the central issue cited by the other proponents of military unification when testifying before Congress. One is forced to consider the question as to why. During World War II, the interservice contest for control of land-based antisubmarine aviation engendered powerful feelings of antipathy and distrust between the military and civilian leaders of the War and Navy Departments, and for a time threatened to cripple the American war effort in Europe. Stimson's anger towards Admiral King over this specific issue had not abated even though the war had ended, and the former Secretary of War believed that the story of Navy obstructionism was strong and compelling evidence of the need for military unification. Why, then, did the

---

71 Henry L. Stimson and McGeorge Bundy, "Time of Peril," *Ladies' Home Journal* (February 1948), p. 211.

72 Patterson Papers, Box 46, General Correspondence, 1947–1952, Robert P. Patterson, Letter to Henry Stimson, 4 February 1948.

other proponents of military unification not trumpet this wartime interservice dispute as loudly as Stimson?

There were many and varied arguments put forward by proponents of military unification in support of their position. For men such as Harry Truman, whose ideas were shaped while chairing the wartime Senate Special Committee to Investigate the National Defense Program, unification was considered as a means towards eliminating wasteful spending on procurement duplication. For Generals Marshall and Eisenhower, unification was looked upon as closely associated with the creation of a single military Chief of Staff, which they believed would make the military command structure more streamlined and efficient. For Generals Arnold and Spaatz, military unification, although seemingly paradoxical, was considered a way of obtaining independence from the Army and thereby achieve the longtime dream of the air power advocates of the creation of a truly strategic Air Force. To be sure, there were many other reasons for supporting military unification which could be added to those mentioned here. It is important to note that the varied arguments advanced by the supporters of unification were not mutually exclusive. It was possible for one to sympathize with any or all of these issues while championing one specific argument. Thus, it was not necessary for every proponent of military unification to be speaking with a single voice on the subject.

The movement towards unification took shape over a long period of time and was not born of an isolated event. Like any political movement, its success relied on gathering pressure brought to bear on policymakers by concerned individuals who, in support of a common effort, hoped to effect change in an area of specific personal concern. For the many proponents of military unification during the late 1940s, Henry Stimson's arguments were seen as valid, but not necessarily the most timely ones supporting of unification. Proponents of unification such as Robert Patterson thought it proper for his voice to be added to the chorus calling for military reorganization; however, others thought it neither necessary nor desirable that Stimson's argument be advanced as the prime justification for military unification.

Henry Stimson seems to stand out most prominently among those advocating military unification owing to his rare position and influence as a two-term Secretary of War. Stimson's admiration for his mentor, Elihu Root, bordered on hero-worship and caused him to romanticize the military reform movement of the early twentieth century which strengthened the office of Secretary of War *vis-à-vis* the uniformed Army leadership. Stimson believed that Root's mantle had fallen to him and that it was incumbent upon him to continue the cause of military reform. Stimson, as Secretary of War, had the authority to reprimand and remove recalcitrant Army officers. However, for all his power over the Army, he was virtually powerless when opposed by a uniformed officer of the Navy. Stimson believed that Admiral King's obstinate refusal to bend to his will regarding the Army-led antisubmarine offensive was the most egregious example of bureaucratic backwardness and military malfeasance of World War II, convincing him that the entire Navy command organization needed reform. As we have seen, Stimson's papers indicate that he intended to raise the issue before the Woodrum Committee in 1944. It was in this forum that Stimson for the first time publicly tied Navy recalcitrance during the antisubmarine war to the need for Root-like reforms that would strengthen civilian control over a unified military command structure. Nevertheless, the Woodrum Committee failed to bring about military unification, and as a consequence, the Navy's unpunished defiance of the War Department would haunt Stimson into his retirement. As the post-war movement toward military unification unfolded, it is likely that Stimson would have continued to make his case for unification with reference to King and the antisubmarine war. However, during most of this period, Stimson was out of the fray due to ill health. It was only in 1947 after a period of convalescence that Stimson jumped back into the fight for military unification.

Robert Patterson, Stimson's successor as Secretary of War, was a strong supporter of military unification and was personally responsible for much of the interdepartmental haggling and bureaucratic compromise which resulted in the ultimate passage of unification legislation. However, for all of his support of unification, Patterson's papers betray no deep passion about the Navy's control of land-based antisubmarine aviation. No doubt, as Under Secretary of War, Patterson shared Stimson's disappointment and anger over the Navy's obstruction of the War Department's plans for an antisubmarine offensive. However, Patterson's support for military unification was inspired by different priorities. Unlike Stimson, Patterson and the next generation of War Department leaders sensed no personal connection to the 1904 reform movement of Elihu Root, nor did they feel the need to measure up to Root's image as a reformer. Finally, unlike Stimson, Patterson and the next generation were looking to the future and did not consider military unification a form of retribution against the Navy for its wartime defiance. Thus, Patterson saw no reason to present Navy recalcitrance in the antisubmarine war as "Exhibit A" in the court of public opinion regarding unification.

Of the variety of concerns raised during the post-war unification debates, the Navy's control of land-based antisubmarine aviation remained a paramount and contentious issue. However, land-based naval aviation was raised with specific reference to future budgetary concerns and as an issue affecting the argument for Air Force independence. The next generation of War Department leaders remembered Stimson's fight with the Navy for control of land-based antisubmarine aviation, but for them it was a fading historical event, serving to highlight the hazards of what might happen when two military branches worked at cross purposes. There is no indication that anyone among the next generation of War Department or Army leaders connected the question of which military service should control land-based antisubmarine aviation with the bygone reform movement of Elihu Root. This connection was only made by Stimson, and for this reason, he looked to unification as a way of vindicating the War Department's failed antisubmarine offensive. Although Admiral King got his way in 1943, Stimson, through his support for unification, was hoping to have the final say on the antisubmarine war. The next generation, however, was focused on questions concerning the disposition of land-based antisubmarine aviation as an issue affecting the future. To them, the question had more to do with proposed Air Force independence and budgetary concerns. Furthermore, when any reference to the Battle of the Atlantic was raised during the post-war unification debates, it was usually brought up by the Navy to warn against the consequences of stripping it of its air arm as the British had done when establishing the RAF. However, as the record shows, such references were not only infrequent, but they were raised by the Navy to oppose unification rather than by the War Department to support it.

It is impossible to say if the unification debate would have been framed any differently had Stimson stayed actively involved immediately following his retirement as Secretary of War. It is likely that he would have produced a memoir such as *On Active Service* much earlier than 1948 which argued for military unification so as to force Root-like reforms on the Navy. However, Stimson did not lend his voice to the post-war unification debate until fairly late, after a lengthy convalescence. It is clear that Stimson's memoir, written with Patterson's encouragement, was intended to serve as pro-unification propaganda specifically at a time when unification was a hotly contested issue. Stimson believed his fight with Admiral King over control of land-based antisubmarine aviation was a compelling argument of the need to force Root-like reforms on the Navy and intended to make this case in the court of public opinion. However, as circumstances would have it, "A Time of Peril" and *On Active Service* were published too late to affect the outcome of the unification debate. The National Security Act of 1947 passed into law half a year earlier and Stimson's memoir played no role in swaying public opinion at the time of the congressional vote.

By the time *On Active Service* was published in 1948, the contentious issues surrounding land-based antisubmarine aviation had become *passé*. The compromise unification legislation and its attendant Executive order defining the role of naval aviation had settled these issues. In spite of this, evidence from the debate before the House Committee on Expenditures in Executive Departments in 1947 indicates that the emerging hot-button issue affecting interservice relations was which branch of the military would control delivery of strategic nuclear weapons. In this next wave of interservice rivalry, carrier aviation came under assault by a covetous Air Force which feared that carrier-based planes would soon be capable of delivering a nuclear payload. Thus, as early as 1947, the first rumblings of the B-36 controversy can be heard in the House unification debate. As is well known, this issue would erupt as a major controversy in 1949 after Secretary of Defense Louis Johnson cancelled construction of the super-carrier *United States*. The issue further migrated during the Korean War, when the proponents of strategic air power realized that nuclear weapons were no panacea for national defense. Stimson died shortly after the outbreak of the Korean War and by that time, the control of antisubmarine aviation during the Battle of the Atlantic had long since faded from view as an issue pressing upon the nation's proper military organization.

For these reasons, most historical studies of the unification debates were written from a Cold War perspective—during a time when national security issues were seen as the prime motivating factor underlying the need for military unification. Many historians of the post-war unification movement were active at a time when fears of American nuclear vulnerability were genuine. It is perhaps for this reason that studies of the unification movement tend to emphasize the Japanese attack on Pearl Harbor as a major motivating factor underlying calls for unification. Such studies tend to emphasize the perceived need to create a national security apparatus having powerful intelligence organizations and a centrally organized Department of Defense so as to detect and prevent future Pearl Harbor-like nuclear attacks. For many historians, the consequence of such nascent Cold War thinking was the passage of the 1947 National Security Act, and most studies tend to promote this Cold War-centric perspective of military unification.

This chapter has attempted to restore a pre-Cold War perspective of the military unification debates, when issues of service parochialism mattered as much as issues of overall national security. By consistently returning to Stimson as a central figure in the story of military unification, one may gain a clearer understanding of issues often overlooked by historians which were at play during the unification debates—when the disposition of land-based antisubmarine aviation mattered as much as which service would control nuclear weapons in the post-war period.

Contributors

G. H. Bennett is Associate Professor in History at the University of Plymouth. The majority of his research concentrates on the Second World War and he is particularly interested in maritime history (both civilian and military) and the on-going relevance of the past to an understanding of today's maritime/naval issues. He has contributed numerous articles to journals and magazines and is the author of more than twenty books, including *Destination Normandy: Three American Regiments on D-Day* (Westport, CT: Praeger, 2007); *The RAF's French Foreign Legion: De Gaulle, the British and the Re-emergence of French Airpower, 1940–45* (London: Bloomsbury, 2011); and *Admiralty Despatches: The Story of the War from the Battlefront, 1939–45* (Stroud: Amberley, 2017). Professor Bennett has appeared frequently on television programs on the BBC and other channels.

Peter J. Dean is Professor and Pro Vice-Chancellor (Education) at the University of Western Australia. He is also a Senior Fellow at the Perth USAsia Centre. Professor Dean is the author of numerous works, including *The Architect of Victory: The Military Career of Lieutenant-General Sir Frank Horton Berryman, 1894–1981* (Melbourne: Cambridge University Press, 2011); *MacArthur's Coalition: US and Australian Operations in the Southwest Pacific Area, 1942–1945* (Lawrence: University Press of Kansas, 2018); and editor of *Australia 1942: In the Shadow of War* (Melbourne: Cambridge University Press, 2012); and *Australia 1943: The Liberation of New Guinea* (Melbourne: Cambridge University Press, 2013).

Marcus Faulkner is an international historian with interests in the development of navies and seapower in the twentieth century, intelligence, and strategic history. His current projects include a history of German naval intelligence between the world wars and a volume on the German aircraft carrier *Graf Zeppelin*. He holds a BA in War Studies (2000) from King's College London and an MA in Politics, Security and Integration from the School of Slavonic and East European Studies, UCL (2003). He completed a PhD entitled "Intelligence, Policy and the *Kriegsmarine* in the Interwar Period" with the Department of War Studies, King's College London in 2008. He is the author of *The Great War at Sea: A Naval Atlas, 1914–1919* (Barnsley: Seaforth, 2014); *War at Sea: A Naval Atlas, 1939–1945* (Barnsley: Seaforth, 2012); and, with Michael H. Clemmesen, editor of *Northern European Overture to War, 1939–1941* (Leiden: Brill, 2013). He currently teaches in the Department of War Studies, King's College London.

Rear Admiral James Goldrick had sea service around the world in the RAN and on exchange service with the British Royal Navy. An anti-submarine specialist, he commanded HMA Ships *Cessnock* and *Sydney* (twice), the Australian Surface Task Group and the multinational maritime interception force in the Persian Gulf in 2002, Australia's inter-agency Border Protection Command, the Australian Defence Force Academy (ADFA), and the Australian Defence College. He is an Adjunct Professor in the Strategic & Defence Studies Centre at The Australian National University and at UNSW Canberra (ADFA). He has published in many academic and professional journals and contributed chapters to more than forty books. He is the author of *Before Jutland: The Naval War in Northern European Waters, August 1914–February 1915* (Annapolis: Naval Institute Press, 2015), which is a much revised and extended version of his earlier work, *The King's Ships Were at Sea: The War in the North Sea, August 1914–February 1915* (Annapolis: Naval Institute Press, 1984); *No Easy Answers: The Development of the Navies of India, Pakistan, Bangladesh and Sri Lanka, 1945–1996* (Hartford, WI: Spantech & Lancer, 1997); and, with Jack McCaffrie, *Navies of South-East Asia: A Comparative Study* (London: Routledge, 2012).

Francis Grice is an Assistant Professor of Political Science and International Studies at McDaniel College in Maryland and has a Ph.D. in Defence Studies from King's College London. He researches and teaches in the fields of International Relations and Comparative Politics, with a concentration on Security Studies in Asia and the Pacific. He is the author of *The Myth of Mao Zedong and Modern Insurgency* (Cham: Palgrave Macmillan, 2019).

C. I. Hamilton is a naval historian and has been successively a member of staff in the Universities of Cambridge, Durham, and the Witwatersrand. He retired from the latter recently and moved to London to be close to The National Archives, Kew. His most recent book on naval history is *The Making of the Modern Admiralty: British Naval Policy-Making, 1805–1927* (Cambridge: Cambridge University Press, 2011). He is currently engaged on a study of the Admiralty in the Second World War.

Iain E. Johnston-White is Lecturer in British History at the University of Roehampton. His research interests lie at the intersection of the British Empire and war in the twentieth century. He is the author of *The British Commonwealth and Victory in the Second World War* (Basingstoke: Palgrave Macmillan, 2017).

Evan Mawdsley is an Honorary Professorial Research Fellow at the University of Glasgow. He is the author of a number of works, including *Thunder in the East: The Nazi-Soviet War, 1941–1945* (London: Hodder Arnold, 2005); *World War II: A New History* (Cambridge: Cambridge University Press, 2009); and *December 1941: Twelve Days That Began a World War* (New Haven: Yale University Press, 2011).

Donald K. Mitchener, Ph.D., is a Lecturer in the Department of History at the University of North Texas.

George H. Monahan, Ph.D., is Associate Professor of History on the Eastern Campus of Suffolk County Community College in Riverhead, New York.

Alessio Patalano is Lecturer in War Studies in the Department of War Studies, King's College London, and specializes in East Asian security and Japanese naval history and strategy. He is the author of *Post-war Japan as a Sea Power: Imperial Legacy, Wartime Experience and the Making of a Navy* (London: Bloomsbury, 2015) and editor of *Maritime Strategy and National Security in Japan and Britain: From the First Alliance to Post-9/11* (Leiden: Global Oriental, 2012).

Alan D. Zimm was a member of the Johns Hopkins University Applied Physics Laboratory's Principal Professional Staff, serving as Section Leader in the Aviation Systems and Advanced Concepts Group. A former naval officer, he received a D. P. A. from the School of Policy, Planning, and Development at the University of Southern California, with a concentration in Policy Analysis and Strategic Planning. He is the author of *Attack on Pearl Harbor: Strategy, Combat, Myths, Deceptions* (Philadelphia: Casemate, 2011).

Index

New Perspectives on the Second World War

Series Editor: Robert von Maier

andarta books
an imprint of the University Press of Kentucky

Brécourt Academic